Acclaim for Leonard Susskind's

THE COSMIC LANDSCAPE

STRING THEORY AND THE ILLUSION OF INTELLIGENT DESIGN

"In this extraordinary work, Leonard Susskind ushers us to the mind-bending edge of a possible paradigm shift." — *Booklist*

"Susskind walks us through the strange world of physics and cosmology, starting with the big bang and ending with his own groundbreaking notion of looking at the cosmos as a landscape of possible worlds rather than a single universe that is bounded in space and time."

— Jonathan Kirsch, *Los Angeles Times Book Review*

"Once in a great while scientists enter into a debate not so much over which theory is right, but over what they want from a theory at all. String theorists are in such a conflict now: Should they be looking for a unique theory picked out by the ferociously stringent constraints of mathematical-physics? Or should they learn to accept the irreducible fact that many theories are allowed, and we happen to live in one universe (among many) where one of those allowed theories happens to be realized? Leonard Susskind has contributed tremendously to the development of string theories – here, in *The Cosmic Landscape*, he offers a crystal-clear, frankly committed vision to this fundamental debate over one of the fundamental theories of modern science."

— Peter Galison, Mallinckrodt Professor of History of Science and of Physics, Harvard University

"Susskind is without any doubt a leading figure in theoretical physics, with a track record of prescient ideas spanning more than thirty years."

— Sir Martin Rees, author of *Our Final Hour, Just Six Numbers,* and *Before the Beginning*

"For years, the students and colleagues of Leonard Susskind have wished that the wider world of laypersons could experience what they have: an extraordinary and expansive scientific imagination combined with the gift of lucid and witty exposition. Here is the fulfillment of that wish, a book in which, against the background of modern physical and cosmological theory, Susskind addresses the fundamental debate in both contemporary science and the wider culture: Is the delicate fine tuning which makes human life possible a result of chance or design?"

— Van A. Harvey, professor (emeritus) of religious studies, Stanford University

"*The Cosmic Landscape* gives a great overview of this important terrain, as seen from an enthusiast's viewpoint." — *Nature*

"Susskind is among the most influential and admired theoretical particle physicists of his generation. In a field dominated by fashion and trends, he has always gone his own way, and he is one of the very few who have managed to take the field with him."

— Lee Smolin, author of *The Life of the Cosmos*

"*The Cosmic Landscape* may end up irking both sides of the current debate, becoming the most talked-about popular science title in recent times." — *Telegraph* (Calcutta, India)

THE COSMIC LANDSCAPE

STRING THEORY AND THE ILLUSION OF INTELLIGENT DESIGN

Leonard Susskind

BACK BAY BOOKS
Little, Brown and Company
New York Boston London

Back Bay Books / Little, Brown and Company
Hachette Book Group USA
237 Park Avenue, New York, NY 10017
Visit our Web site at www.HachetteBookGroupUSA.com

Originally published in hardcover by Little, Brown and Company, December 2005
First Back Bay paperback edition, December 2006

Illustration credits: Einstein (p. 70), California Institute of Technology Archives; Rube Goldberg machine (p. 112), Rube Goldberg is the ® and © of Rube Goldberg Inc.; Kepler's model (p. 120), from J. Kepler, *Mysterium Cosmographicum* (1596); Crab Nebula (p. 181), ESO — European Southern Observatory; eclipse (p. 188), Fred Esperak; snowflake (p. 242), Kenneth G. Libbrecht, Professor of Physics, Cal Tech; Calabi Yau (p. 290), Jean-François Colonna, CMAP (Centre de Mathématiques Appliquées); Mercator projection (p. 311), *An Album of Map Projections*, USGS Professional Paper 1453, by John P. Snyder and Phillip M. Voxland (USGPO, 1989); Escher (p. 312), M. C. Escher's *Limit Circle IV*, © 2005 The M. C. Escher Company — Holland. All rights reserved. www.mcescher.com.

Library of Congress Cataloging-in-Publication Data

Susskind, Leonard.
Cosmic landscape : string theory and the illusion of intelligent design / Leonard Susskind.
p. cm.
Includes bibliographical references and index.
HC ISBN 978-0-316-15579-3
PB ISBN 978-0-316-01333-8
1. Cosmogony. 2. Astrophysics. 3. Intelligent design (Teleology).
4. String models. I. Title.
QB981.S886 2005
523.1'2 — dc22 2005018796

10 9 8 7 6 5 4 3 2

Q-MART

Printed in the United States of America

"Your Highness, I have no need of this hypothesis."

— Pierre-Simon de Laplace (1749–1827),
reply made to Napoléon when asked why his
celestial mechanics had no mention of God

Contents

	Preface	ix
	Introduction	3
chapter one	*The World According to Feynman*	17
chapter two	*The Mother of All Physics Problems*	63
chapter three	*The Lay of the Land*	89
chapter four	*The Myth of Uniqueness and Elegance*	111
chapter five	*Thunderbolt from Heaven*	131
chapter six	*On Frozen Fish and Boiled Fish*	169
chapter seven	*A Rubber Band–Powered World*	199
chapter eight	*Reincarnation*	229
chapter nine	*On Our Own?*	261
chapter ten	*The Branes behind Rube Goldberg's Greatest Machine*	273
chapter eleven	*A Bubble Bath Universe*	293

chapter twelve *The Black Hole War* 325

chapter thirteen *Summing Up* 343

Epilogue 377

A Word on the Distinction between Landscape and Megaverse 381

Glossary 383

Note on Terminology 389

Index 390

Preface

I have always enjoyed explaining physics. In fact it's more than just enjoyment: I need to explain physics. A lot of my research time is spent daydreaming – telling an imaginary admiring audience of laymen how to understand some difficult scientific idea. I suppose I am a bit of a ham, but it's more than that. It's part of the way I think: a mental tool for organizing my ideas and even creating new ways of thinking about problems. So it was natural that at some point I would decide to try my hand at writing a book for a general audience. A couple of years ago I decided to take the plunge and write a book about the twenty-year debate between myself and Stephen Hawking concerning the fate of information that fell into a black hole.

But just about that time, I found myself in the eye of a huge scientific hurricane. The issues involve not only the origin of the universe but also the origin of the laws that govern it. In my scientific article "The Anthropic Landscape of String Theory," I called attention to the new emerging concept that I had christened the Landscape. The paper stirred up an enormous fuss in the physics and cosmology communities that by now has spread to philosophers and even theologians. The Landscape represents an idea that cuts across boundaries and touches not only on current paradigm shifts in physics and cosmology, but also on the profound cultural questions that are rocking our social and political landscape: can science explain the extraordinary fact that the universe appears to be uncannily, nay, spectacularly, well designed for our own existence? I decided to put the black hole book on the back burner for the moment and write a popular book about this extraordinary story. Thus was born *The Cosmic Landscape*.

Some readers of this book will be aware that in the past few years the science sections of newspapers have been reporting that cosmologists are mystified by two astonishing "dark" discoveries. The first is that 90 percent of the matter in the universe is made of some shadowy, mysterious substance called dark matter. The other is that 70 percent of the energy in the universe is composed of an even more ghostly mysterious stuff called dark energy. The words *mystery*, *mysterious*, and *mystified* get a very thorough workout in these articles.

I have to admit I find neither discovery all that mysterious. To me, the word *mystery* conveys something that completely eludes rational explanation. The discoveries of dark matter and energy were surprises but not mysteries. Elementary-particle physicists (I am one of them) have always known that their theories were incomplete and that many particles remain to be discovered. The tradition of postulating new, hard-to-detect particles began when Wolfgang Pauli correctly guessed that one form of radioactivity involved an almost invisible particle called the neutrino. Dark matter is not made of neutrinos, but by now physicists have postulated plenty of particles that could easily form the invisible stuff. There is no mystery there — only the difficulties of identifying and detecting those particles.

Dark energy has more of a claim to being called mysterious, but the mystery has much more to do with its absence than its presence. Physicists have known for seventy-five years or more that there is every reason for space to be filled with dark energy. The mystery is not why dark energy exists but why so little of it exists. But one thing is clear: even a little more dark energy would have been fatal to our own existence.

The real mystery raised by modern cosmology concerns a silent "elephant in the room," an elephant, I might add, that has been a huge embarrassment to physicists: why is it that the universe has all of the appearances of having been specially designed just so that life forms like us can exist? This has puzzled scientists and at the same time encouraged those who prefer the false comfort of a creationist myth. The situation in many ways resembles biology before Darwin, when thoughtful people were unable to understand how, without the guiding hand of a deity, natural processes of physics and chemistry could possibly create anything as complex as the human eye. Like the eye, the special properties of the physical universe are so surprisingly fine-tuned that they demand explanation.

Let me be up front and state my own prejudices right here. I thoroughly believe that real science requires explanations that do not involve supernatural agents. I believe that the eye evolved by Darwinian mechanisms. Furthermore, I believe that physicists and cosmologists must also find a natural explanation of our world, including the amazing lucky accidents that conspired to make our own existence possible. I believe that when people substitute magic for rational explanation, they are not doing science no matter how loudly they claim otherwise.

In the past most physicists (including me) have chosen to ignore the elephant — even to deny its existence. They preferred to believe that nature's laws follow from some elegant mathematical principle and that the apparent design of the universe is merely a lucky accident. But recent discoveries in astronomy, cosmology, and above all, String Theory have left theoretical physicists little choice but to think about these things. Surprisingly, we may be starting to see the reasons for this pattern of coincidences. Evidence has been accumulating for an explanation of the "illusion of intelligent design" that depends only on the principles of physics, mathematics, and the laws of large numbers. This is what *The Cosmic Landscape* is about: the scientific explanation of the apparent miracles of physics and cosmology and its philosophical implications.

Who are the intended readers of this book? The answer is anyone with a lively interest in science and a curiosity about how the world came to be the way it is. But although the book is aimed at a lay audience, it is not aimed at the "lightweight" who is afraid to stretch his or her mind. I have kept the book free of equations and jargon but not of challenging concepts. I have avoided mathematical formulas, but on the other hand, I have striven to give accurate and clear explanations of the principles and mechanisms that underlie the new emerging paradigm. Understanding this new paradigm will be critical for anyone hoping to follow the further developments in answering the "big questions."

I am indebted to many people, some who didn't even know that they were helping me write this book. Among them are all the physicists and cosmologists whose ideas I have drawn on — Steven Weinberg, Gerard 't Hooft, Martin Rees, Joseph Polchinski, Raphael Bousso, Alan Guth, Alex Vilenkin, Shamit Kachru, Renata Kallosh, and above all, Andrei Linde, who has been generously sharing his ideas with me for many years.

The actual writing of the book would not have been possible without the support of my agent, John Brockman, and my friend Malcolm Griffith, who read and criticized the early mess of a manuscript that I sent him and taught me how to "juggle more than three balls" (Malcolm's way of describing the difficulties of writing a coherent book). To all the people at Little, Brown — Steve Lamont, Carolyn O'Keefe, and especially my editor, and now friend, Liz Nagle — I owe a huge debt of gratitude for her extraordinary contribution to the writing of this book. Liz's patient guidance was way beyond the call of duty.

And finally to my wife, Anne, I am grateful, beyond measure, for her continuous loving support and help.

The Cosmic Landscape

Introduction

The air is very cold and still: except for the sound of my own breathing, the silence is absolute. The dry, powdery snow crackles whenever my boot touches down. Its perfect whiteness, lit by starlight, gives the terrain a luminous, eerie brilliance, while the stars fade into a continuous glow across the black celestial dome. The night is brighter on this desolate planet than on my own home world. Beauty, but of a cold and lifeless kind: a place for metaphysical contemplation if ever there was one.

Alone, I'd left the safety of the base, to think about the day's events and to watch the sky for meteors. But it was impossible to think of anything other than the sheer enormousness and impersonal nature of the universe. The pinwheeling of galaxies, the endless expansion of the universe, the infinite coldness of space, the heat of stars being born, and their final death throes as red giants: surely this must be the point of existence. Man — life in general — seems irrelevant to the workings of the universe: a mere smudge of water, grease, and carbon on a pinpoint planet circling a star of no special consequence.

Earlier, during the short stingy sunlight hours, Curt, Kip, and I had hiked about a mile to the Russian compound to see if we could find some Ivans to talk to. Stephen had wanted to come with us but his wheelchair could not navigate the snowdrifts. The derelict compound, just a few low rusted corrugated-metal buildings, looked deserted. We banged on the doors, but no life appeared. I cracked open the door and peered into the spooky interior darkness, then decided to brave entry and have a look around. As cold inside as it was outside, the compound was completely abandoned. The hundred or so dormitory rooms were unlocked but deserted. How did a hundred men disappear so completely? In silence we hiked back to our own base.

At the bar, we found our Russian, drinking and laughing — Victor. Victor, it seems, was one of the last three Russians left on the planet. Supplies from Russia had ceased more than a year ago. They would have starved but for the fact that our own people adopted them. We never saw the other two Russians, but Victor assured us they were alive.

Victor insisted on buying me a drink, "for the cold," and asked, "How do you like this %#&*^ place?" I told him in all my travels only once had I seen the night sky even remotely as beautiful as here. Ironically, that other alien planet was so hot that the rocks would fry anything that touched them.

Of course we were not really on another planet. It only seemed that way. Antarctica is truly alien. Stephen Hawking, Curt Callan, Kip Thorne, Stan Deser, Claudio Teitelboim, myself, our wives, and a few other theoretical physicists were there for fun — as a lark — a reward for coming to Chile for a conference on black holes. Claudio, an eminent Chilean physicist, had arranged for the Chilean Air Force to fly us in one of its giant Hercules cargo planes to their Antarctic base for a couple of days.

It was August 1997 — winter in the southern hemisphere — and we were expecting the worst. The coldest I had ever experienced was 20 below zero Fahrenheit, and I was worried about how well I would handle the 60 below that can grip the base in midwinter. When the plane landed, we anxiously zipped up the heavy Arctic gear that the military had provided and prepared for the fearful cold.

Then the cargo hold opened, and Curt's wife, Chantal, bounded out of the plane, threw up her arms, and joyously yelled back, "It's about as cold as a winter day in New Jersey." And so it was. It stayed that way for the whole day while we frolicked in the snow.

Sometime during that night the beast awoke. By morning Antarctica had unleashed its fury. I went outside for a couple of minutes to get a taste of what Shackleton and his shipwrecked men had endured. Why hadn't they all perished? Not a single member of the expedition was lost. Freezing cold and soaking wet for more than a year, why didn't they all die of pneumonia? Out there in the blast of the storm, I knew the answer: nothing survives — not even the microbes that give men colds.

The other alien "planet" I'd mentioned to Victor was Death Valley — another lifeless place. No, not quite lifeless. But I wondered how much hotter it would have to get to fry all protoplasm. What Antarctica has in common with Death Valley is extreme dryness. It's too cold for much water vapor to be suspended in the air — that and the complete lack of light pollution make it possible, in both extremes, to see the stars in a way that modern man rarely can. Standing there in the Antarctic starlight, it occurred to me how lucky we humans are. Life is fragile: it thrives only in a narrow range of temperatures between freezing and boiling. How lucky that our planet is just the right distance from the sun: a little farther, and the death of the perpetual Antarctic winter — or worse — would prevail; a little closer, and the surface would truly fry anything that touched it. Victor, being Russian, took a spiritual view of the question. "Was it not," he asked, "God's infinite kindness and love that permitted our existence?" My own "mindless" explanation will become clear in good time.

In fact we have much more to be thankful for than just the earth's temperature. Without the right amount of carbon, oxygen, nitrogen, and other elements, a temperate climate would be wasted. If the sun at the center of our solar system were replaced by the more common binary star system,* planetary orbits would be too chaotic and unstable for life to have evolved. There are endless dangers of this kind. But on top of all these are the laws of nature themselves. All it takes is a small change in Newton's laws, or the rules of atomic physics, and poof — life would either be instantly extinguished or would never have formed. It seems that our guardian angel not only provided us with a very benign planet to live on but also made the rules of existence — the laws of physics and cosmology — just right for us. This is one of the greatest mysteries of nature. Is it luck? Is it intelligent and benevolent design? Is it at all a topic for science — for metaphysics — for religion?

This book is about a debate that is stirring the passions of physicists and cosmologists but is also part of a broader controversy, especially in the United States, where it has entered the partisan political discourse. On one side are the people who are convinced that the world must have

*A binary star system just means two stars orbiting around their center of mass.

been created or designed by an intelligent agent with a benevolent purpose. On the other side are the hard-nosed, scientific types who feel certain that the universe is the product of impersonal, disinterested laws of physics, mathematics, and probability — a world without a purpose, so to speak. By the first group, I don't mean the biblical literalists who believe the world was created six thousand years ago and are ready to fight about it. I am talking about thoughtful, intelligent people who look around at the world and have a hard time believing that it was just dumb luck that made the world so accommodating to human beings. I don't think these people are being stupid; they have a real point.

The advocates of intelligent design generally argue that it is incredible that anything as complex as the human visual system could have evolved by purely random processes. It is incredible! But biologists are armed with a very powerful tool — the Principle of Natural Selection — whose explanatory power is so great that almost all biologists believe the weight of evidence is strongly in favor of Darwin. The miracle of the eye is only an apparent miracle.

I think the design enthusiasts are on better ground when it comes to physics and cosmology. Biology is only part of the story of creation. The Laws of Physics and the origin of the universe are the other part, and here again, incredible miracles appear to abound. It seems hopelessly improbable that any particular rules accidentally led to the miracle of intelligent life. Nevertheless, this is exactly what most physicists have believed: intelligent life is a purely serendipitous consequence of physical principles that have nothing to do with our own existence. Here I share the skepticism of the intelligent-design crowd: I think that the dumb luck needs an explanation. But the explanation that is emerging from modern physics is every bit as different from intelligent design as Darwin's was from "Soapy" Sam Wilberforce's.*

The debate that this book is concerned with is not the bitter political controversy between science and creationism. Unlike the debate be-

*Samuel Wilberforce, an Anglican bishop, was called Soapy because of his slipperiness in ecclesiastical debate. Thomas Huxley, Darwin's chief disciple, was called Darwin's Bulldog for obvious reasons. The two squared off in 1860 to debate Darwin's *On the Origin of Species by Means of Natural Selection.* Soapy Sam gleefully asked Huxley if it was his grandmother or grandfather who had been the ape? Huxley turned on him and said, "I'd rather be descended from a monkey than from someone who would so prostitute the truth."

tween "Darwin's Bulldog" Thomas Huxley and Wilberforce, the present argument is not between religion and science but between two warring factions of science — those who believe, on the one hand, that the laws of nature are determined by mathematical relations, which by mere chance happen to allow life, and those who believe that the Laws of Physics have, in some way, been determined by the requirement that intelligent life be possible. The bitterness and rancor of the controversy have crystallized around a single phrase — the *Anthropic Principle* — a hypothetical principle that says that the world is fine-tuned so that we can be here to observe it! By itself I would have to say that this is a silly, half-baked notion. It makes no more sense than saying that the reason the eye evolved is so that someone can exist to read this book. But it is really shorthand for a much richer set of concepts that I will make clear in the chapters that follow.

But the controversy among scientists does have repercussions for the broader public debate. Not surprisingly, it does overflow the seminar rooms and scientific journals into the political debates about design and creationism. Christian Internet sites have leapt into the fray:

> The Bible says:
>
> "From the time the world was created, people have seen the earth and the sky and all that God made. They can clearly see His invisible qualities — His eternal power and divine nature. So they have no excuse whatsoever for not knowing God."
>
> This is as true today as it ever has been — in some ways, with the discovery of the Anthropic Principle, it is more true now than ever before. So the first kind of evidence that we have is the creation itself — a universe that carries God's signature — a universe "just right" for us to live in.

And from another religious site:

> In his book "The Cosmic Blueprint," the astronomer professor Paul Davies concludes that the evidence for design is overwhelming:
>
> Professor Sir Fred Hoyle — no sympathizer with Christianity — says that it looks as if a super-intellect has monkeyed with physics as well as with chemistry and biology.

And the astronomer George Greenstein says:

> As we survey all the evidence, the thought insistently arises that some supernatural agency, or rather Agency, must be involved. Is it possible that suddenly, without intending to, we have stumbled upon scientific proof of the existence of a supreme being? Was it God who stepped in and so providentially created the cosmos for our benefit?*

Is it any wonder that the Anthropic Principle makes many physicists very uncomfortable?

Davies and Greenstein are serious scholars, and Hoyle was one of the great scientists of the twentieth century. As they point out, the *appearance* of intelligent design is undeniable.† Extraordinary coincidences *are* required for life to be possible. It will take us a few chapters to fully understand this "elephant in the room," but let's begin with a sneak preview.

The world as we know it is very precarious, in a sense that is of special interest to physicists. There are many ways it could go bad — so bad that life as we know it would be totally impossible. The requirements that the world be similar enough to our own to support conventional life fall into three broad classes. The first class involves the raw materials of life: chemicals. Life is, of course, a chemical process. Something about the way atoms are constructed makes them stick together in the most bizarre combinations: the giant crazy Tinkertoy molecules of life — DNA, RNA, hundreds of proteins, and all the rest. Chemistry is really a branch of physics: the physics of the valence electrons, i.e., those that orbit the nucleus at the outer edges of the atom. It's the valence electrons hopping back and forth or being shared between atoms that gives the atoms their amazing abilities.

*I don't know the religious beliefs of Davies or Greenstein, but I would be wary of too literal an interpretation. Physicists often use terms like *design*, *agency*, and even *God* as metaphors for that which is not known — period. I have used the term *agent* in print and have been sorry ever since. Einstein often spoke of God: "God is cunning but He is not malicious." "God does not play dice." "I want to know how God created the world." Most commentators believe Einstein was using the term *God* as a metaphor for an orderly set of laws of nature.

†Will this sentence appear out of context on a religious Internet site? I hope not.

The Laws of Physics begin with a list of elementary particles like electrons, quarks, and photons, each with special properties such as mass and electric charge. These are the objects that everything else is built out of. No one knows why the list is what it is or why the properties of these particles are exactly what they are. An infinite number of other lists is equally possible. But a universe filled with life is by no means a generic expectation. Eliminating any of these particles (electrons, quarks, or photons), or even changing their properties a modest amount, would cause conventional chemistry to collapse. This is obviously so for the electrons and for the quarks that make up protons and neutrons. Without these there could be no atoms at all. But the importance of the photon may be less obvious. In later chapters we will learn about the origin of forces like electric and gravitational forces, but for now it's enough to know that the electric forces that hold the atom together are consequences of the photon and its special properties.

If the laws of nature seem well chosen for chemistry, they are also well chosen for the second set of requirements, namely, that the evolution of the universe provided us with a comfortable home to live in. The large-scale properties of the universe — its size; how fast it grows; the existence of galaxies, stars, and planets — are mainly governed by the force of gravity. It's Einstein's theory of gravity — the General Theory of Relativity — that explains how the universe expanded from the initial hot Big Bang to its present large size. The properties of gravity, especially its strength, could easily have been different. In fact it is an unexplained miracle that gravity is as weak as it is.* The gravitational force between electrons and the atomic nucleus is ten thousand billion billion billion billion times weaker than the electrical attraction. Were the gravitational forces even a little stronger, the universe would have evolved so quickly that there would have been no time for intelligent life to arise.

But gravity plays a very dramatic role in the unfolding of the universe. Its pull causes the material in the universe — hydrogen, helium, and the so-called dark matter — to clump, into galaxies, stars, and finally planets. However, for this to happen, the very early universe must

*For the experts, the weakness of gravity is equivalent to the lightness of the usual elementary particles. The smallness of particle masses is called the gauge hierarchy problem. Although interesting ideas have been posited, there is no consensus on its solution.

have been a bit lumpy. If the original material of the universe had been smoothly distributed, it would have stayed that way for all time. In fact, fourteen billion years ago, the universe was just lumpy enough — a bit lumpier or a bit less lumpy, and there would have been no galaxies, stars, or planets for life to evolve on.

Finally, there is the actual chemical composition of the universe. In the beginning there were only hydrogen and helium: certainly not sufficient for the formation of life. Carbon, oxygen, and all the others came later. They were formed in the nuclear reactors in the interiors of stars. But the ability of stars to transmute hydrogen and helium into the all-important carbon nuclei was a very delicate affair. Small changes in the laws of electricity and nuclear physics could have prevented the formation of carbon.

Even if the carbon, oxygen, and other biologically important elements were formed inside stars, they had to get out in order to provide the material for planets and life. Obviously we cannot live in the intensely hot cores of stars. How did the material escape the stellar interior? The answer is that it was violently ejected in cataclysmic supernova explosions.

Supernova explosions themselves are remarkable phenomena. In addition to protons, neutrons, electrons, photons, and gravity, supernovae require yet another particle — the ghostly neutrino previously mentioned. The neutrinos, as they escape from the collapsing star, create a pressure that pushes the elements in front of them. And, fortunately, the list of elementary particles happens to include neutrinos with the right properties.

As I said, a world full of biological phenomena is by no means a generic expectation. From the point of view of picking elementary-particle lists and strengths of forces, it is the very rare exception. But how exceptional — exceptional enough to warrant a radically new paradigm including the Anthropic Principle? If we were to base our opinions on only the things that I've explained so far, opinions would be mixed, even among those who are open to anthropic ideas. Most of the individual fine-tunings needed for life are not so very precise that they couldn't just be lucky accidents. Perhaps, as physicists have always believed, a mathematical principle will be discovered that explains the list of particles and constants of nature and a lot of lucky accidents will prove to be just that: a lot of lucky accidents. But there is one fine-tuning of nature that I will explain in chapter 2 that is incredibly unlikely. Its occurrence has been a stupendous puzzle to physicists for more than half a century. The only explanation, if it can be called that, is the Anthropic Principle.

A paradox then: how can we ever hope to explain the extraordinarily benevolent properties of the Laws of Physics, and our own world, without appeal to a supernatural intelligence? The Anthropic Principle, with its placement of intelligent life at the center of the explanation for our universe, would seem to suggest that someone, some Agent, is looking out for humankind. This book is about the emerging physical paradigm that does make use of the Anthropic Principle but in a way that offers a wholly scientific explanation of the apparent benevolence of the universe. I think of it as the physicist's Darwinism.

What are these Laws of Physics of which I've spoken? How are they formulated? Until Richard Feynman came along, the only tools that physicists had for expressing Laws of Physics were the arcane, impenetrable equations of quantum field theory — a subject so difficult that even mathematicians have trouble understanding it. But Feynman's uncanny ability to visualize physical phenomena changed all that. He made it possible to summarize the laws of elementary particles by drawing a few simple pictures. Feynman's pictures and the laws of elementary-particle physics, known to physicists as the *Standard Model*, are the subjects of chapter 1.

Is it really true that the universe and its laws are very delicately balanced? The second chapter, "The Mother of All Physics Problems," could also be called "The Mother of All Balancing Acts." When the laws of elementary particles meet the laws of gravity, the result is a potential catastrophe: a world of such violence that astronomical bodies, as well as elementary particles, would be torn asunder by the most destructive force imaginable. The only way out is for one particular constant of nature — Einstein's *cosmological constant* — to be so incredibly finely tuned that no one could possibly think it accidental. First introduced by Einstein soon after the completion of his theory of gravity, the cosmological constant has been the greatest enigma of theoretical physics for almost ninety years. It represents a universal repulsive force — a kind of antigravity — that would instantly destroy the universe if it were not astonishingly small. The problem is that all our modern theories imply that the cosmological constant should not be small. The modern principles of physics are based on two foundations: the Theory of Relativity and quantum mechanics. The generic result of a world based on these principles is a universe that would very quickly self-destruct. But for reasons that have been completely incomprehensible, the cosmological constant is fine-tuned to an astonishing degree. This, more than any other lucky "accident," leads some people to conclude that the universe must be the result of a design.

Is the Standard Model of particle physics "written in stone"? Are other laws possible? In the third chapter of this book, I explain why our particular laws are not at all unique — how they could change from place to place or from time to time. The Laws of Physics are much like the weather: they are controlled by invisible influences in space almost the same way that temperature, humidity, air pressure, and wind velocity control how rain and snow and hail form. The invisible influences are called fields. Some of them, like the magnetic field, are fairly familiar. Many others are unfamiliar, even to physicists. But they are there, filling space and controlling the behavior of elementary particles. The *Landscape* is the term that I coined to describe the entire extent of these theoretical environments. The Landscape is the space of possibilities — a schematic representation of all the possible environments permitted by theory. Over the last couple of years, the existence of a rich Landscape of possibilities has become the central question of String Theory.

The controversy is not only scientific. In chapter 4 we will talk about the aesthetic side of the debate. Physicists, particularly theoretical physicists, have a very strong sense of beauty, elegance, and uniqueness. They have always believed that the laws of nature are the unique inevitable consequence of some elegant mathematical principle. The belief is so deeply ingrained that most of my colleagues would feel an immense sense of loss and disappointment if this uniqueness and elegance turned out to be absent — if the Laws of Physics are "ugly." But are the Laws of Physics elegant in the physicist's sense? If the only criterion for how the universe works is that it should support life, it may well be that the whole structure is a clumsy, ungainly "Rube Goldberg machine."* Despite the protestations of physicists that the laws of elementary particles

*See chapter 3 for the definition of a Rube Goldberg machine.

are elegant, the empirical evidence points much more convincingly to the opposite conclusion. The universe has more in common with a Rube Goldberg machine than with a unique consequence of mathematical symmetry. We cannot fully understand the controversy and the shifting paradigms without also understanding the notions of beauty and elegance in physics, how they originated, and how they compare with the real world.

This book is about a conceptual "earthquake," but it is not the work of only theorists. Much of what we know comes from experimental cosmology and modern astronomy. Two key discoveries are driving the paradigm shift — the success of inflationary cosmology and the existence of a small cosmological constant. *Inflation* refers to the brief period of rapid exponential expansion that initially set the stage for the Big Bang. Without it the universe would probably have been a tiny Little Pop, no bigger than an elementary particle. With it, the universe grew to proportions vastly bigger than anything we can detect with the most powerful telescopes. When Alan Guth first suggested inflation, in 1980, there seemed to be very little chance that astronomical observations would ever be able to test it. But astronomy has advanced by orders of magnitude since 1980: so much so that what seemed inconceivable then is accomplished fact today.

The enormous advances in astronomy led to a second discovery that came as a thunderbolt to physicists, something so shocking that we are still reeling from the impact. The infamous cosmological constant,* which almost everyone was sure was exactly zero, isn't. It seems that the laws of nature were fine-tuned just enough to keep the cosmological constant from being a deadly danger to the formation of life, but no more than that. Chapter 5 is devoted to these discoveries. This chapter also explains all the basic astronomical and cosmological background that the reader will need.

The cosmological constant may be the "mother of all balancing acts," but there are many additional delicate conditions that seem like fantastically lucky coincidences. Chapter 6, "On Frozen Fish and Boiled Fish," is all about these lesser balancing acts. They range from the cosmological to the microscopic, from the way the universe expands to the

*Also known as dark energy.

masses of elementary particles like the proton and neutron. Once again the lesson is not that the universe is simple but that it is full of surprising, unexplained, lucky coincidences.

Until very recently, the Anthropic Principle was considered by almost all physicists to be unscientific, religious, and generally a goofy misguided idea. According to physicists it was a creation of inebriated cosmologists, drunk on their own mystical ideas. Real theories like String Theory would explain all the properties of nature in a unique way that has nothing to do with our own existence. But a stunning reversal of fortune has put string theorists in an embarrassing position: their own cherished theory is pushing them right into the waiting arms of the enemy. String Theory is turning out to be the enemy's strongest weapon. Instead of producing a single unique elegant construct, it gives rise to a colossal landscape of Rube Goldberg machines. The result of the reversal is that many string theorists have switched sides. Chapters 7, 8, 9, and 10 are about String Theory and how it is changing the paradigm.

Chapters 11 and 12 are about the startling new view of the universe that is emerging out of the combined work of astronomers, cosmologists, and theoretical physicists: the world — according to cosmologists such as Andrei Linde, Alexander Vilenkin, and Alan Guth — consists of a virtually infinite collection of "pocket universes" of enormous diversity. Each pocket has its own "weather": its own list of elementary particles, forces, and constants of physics. The consequences of such a rich view of the universe are profound for physics and cosmology. The question, "Why is the universe the way it is?" may be replaced by, "Is there a pocket in this vast diversity in which conditions match our own?" How the mechanism called *Eternal Inflation* caused this diversity to evolve from primordial chaos and how it revolutionizes the debates over the Anthropic Principle and the design of the universe are the subjects of chapter 11.

This cosmological paradigm shift is not the only one taking place in the foundations of physics. Chapter 12 concerns another titanic battle, a conflict I call the Black Hole War. The Black Hole War has played out over the last thirty years and has radically changed the way theoretical physicists think about gravity and black holes. The fierce battle was over the fate of information that falls behind the horizon of a black hole: is it permanently lost, totally beyond the knowledge of observers on the outside, or is there some subtle way in which the details are conveyed back out as the black hole evaporates? Hawking's view was that all information behind the horizon is irretrievably lost. Not even the slightest shred of information about the objects that are on the other side can ever be reconstructed. But that has turned out to be wrong. The laws of quantum mechanics prevent even a single bit from being lost. In order to understand how information escapes the prison of a black hole, it was necessary to completely rebuild our most basic concepts of space.

What does the Black Hole War have to do with the concerns of this book? Because the universe is expanding under the influence of the cosmological constant, cosmology also has its horizons. Our cosmic horizon is about fifteen billion light-years away, where things are moving so rapidly away from us that light from there can never reach us, nor can any other signal. It is exactly the same as a black hole horizon — a point of no return. The only difference is that the cosmic horizon surrounds us, whereas we surround a black hole horizon. In either case nothing from beyond the horizon can influence us, or so it was thought. Furthermore, the other pocket universes — the gigantic sea of diversity — are all beyond our reach behind the horizon! According to classical physics, those other worlds are forever completely sealed off from our world. But the very same arguments that won the Black Hole War can be adapted to cosmological horizons. The existence and details of all the other pocket universes are contained in the subtle features of the cosmic radiation that constantly bathes all parts of our observable universe. Chapter 12 is an introduction to the Black Hole War, how it was won, and its implications for cosmology.

The controversy detailed in *The Cosmic Landscape* is a real one: physicists and cosmologists feel passionately about their own views, whatever they happen to be. Chapter 13 takes a look at the current opinions of many of the world's leading theoretical physicists and cosmologists and how they individually view the controversy. I also discuss the various ways that experiment and observation can guide us toward consensus.

To Victor's question, "Was it not God's infinite kindness and love that permitted our existence?" I would have to answer with Laplace's reply to Napoléon: "I have no need of this hypothesis." *The Cosmic Landscape* is my answer, as well as the answer of a growing number of physicists and cosmologists, to the paradox of a benevolent universe.

CHAPTER ONE

The World According to Feynman

No doubt we'll never know the name of the first cosmologist to look to the sky and ask, "What is all this? How did it get here? What am *I* doing here?" What we do know is that it occurred deep in the prehistoric past, probably in Africa. The first cosmologies, creation myths, were nothing like today's scientific cosmology, but they were born of the same human curiosity. Not surprisingly these myths were about earth, water, sky, and living creatures. And of course they featured the supernatural creator: how else to explain the existence of such complex and intricate creatures as humans, not to mention rain, sun, edible animals, and plants that seemed to be placed on earth just for our benefit?

The idea that precise laws of nature govern both the celestial and terrestrial world dates back to Isaac Newton. Before Newton, there was no concept of universal laws that applied both to astronomical objects like planets and to ordinary earthly objects like falling rain and flying arrows. Newton's laws of motion were the first example of such universal laws. But even for the mighty Sir Isaac, it was far too much of a stretch to suppose that the same laws led to the creation of human beings: he spent more time on theology than physics.

I'm not a historian, but I'll venture an opinion: modern cosmology really began with Darwin and Wallace.[1] Unlike anyone before them,

1. Alfred Russel Wallace (1823–1913), a contemporary of Darwin's, was the codiscoverer of natural selection as the mechanism driving the evolution of species. It was reading a short note of Wallace's that finally induced Darwin to publish his own work.

they provided explanations of our existence that completely rejected supernatural agents. Two natural laws underlie Darwinian evolution. The first is that copying information is never perfect. Even the best reproduction mechanisms from time to time make small errors. DNA replication is no exception. Although it would take a century for Crick and Watson to uncover the double helix, Darwin intuitively understood that accumulated random mutations constitute the engine that drives evolution. Most mutations are bad, but Darwin understood enough about probability to know that every now and then, by pure chance, a beneficial mutation occurs.

The second pillar of Darwin's intuitive theory was a principle of competition: the winner gets to reproduce. Better genes prosper; inferior genes come to a dead end. These two simple ideas explained how complex and even intelligent life could form without any supernatural intervention. In today's world of computer viruses and Internet worms, it's easy to imagine similar principles applying to completely inanimate objects. Once the magic was removed from the origin of living creatures, the way lay open to a purely scientific explanation of creation.

Darwin and Wallace set a standard not only for the life sciences but for cosmology as well. The laws that govern the birth and evolution of the universe must be the same laws that govern the falling of stones, the chemistry and nuclear physics of the elements and the physics of elementary particles. They freed us from the supernatural by showing that complex and even intelligent life could arise from chance, competition, and natural causes. Cosmologists would have to do as well: the basis for cosmology would have to be impersonal rules that are the same throughout the universe and whose origin has nothing to do with our own existence. The only god permitted to cosmologists would be Richard Dawkins's "blind watchmaker."[2]

The modern cosmological paradigm is not very old. When I was a young graduate student at Cornell University, in the early 1960s, the Big Bang theory of the universe was still in hot competition with an-

2. Richard Dawkins, *The Blind Watchmaker: Why the Evidence of Evolution Reveals a Universe without Design* (New York: Norton, 1996). Dawkins invokes the metaphor of a blind watchmaker to describe how evolution blindly created the universe of biology. The metaphor could easily be extended to the creation of the cosmos.

other serious contender. The Steady State theory was, in a sense, the logical opposite of the Big Bang. If the Big Bang said that the universe began at some time, the Steady State said that it had always existed. The Steady State theory was the brainchild of three of the world's most famous cosmologists — Fred Hoyle, Herman Bondi, and Thomas Gold — who thought that the explosive creation of the universe a mere ten billion years ago was too unlikely a possibility. Gold was a professor at Cornell and had his office a few doors down from mine. At the time he was tirelessly preaching the virtue of an infinitely old (and also infinitely big) universe. I barely knew him well enough to say good morning to him, but one day, very uncharacteristically, he sat down to coffee with a few graduate students, and I was able to ask him something that had been bothering me: "If the universe is eternally unchanging, how is it that the galaxies are all receding away from one another? Doesn't it mean that in the past they were more closely packed?" Gold's explanation was simple: "The galaxies are indeed moving apart, but as they separate, new matter is created to fill the space between them." It was a clever answer, but it made no mathematical sense. Within a year or two, the Steady State universe had given way to the Big Bang and was soon forgotten. The victorious Big Bang paradigm asserted that the expanding universe was only about ten billion years old and about ten billion light-years big.[3] But one thing that both theories shared was a belief that the universe is homogeneous, which means that it is everywhere the same: governed by the same universal Laws of Physics throughout. Moreover, those Laws of Physics are the same ones that we discover in terrestrial laboratories.

It's been very exciting, over the last forty years, to watch experimental cosmology mature from a crude, qualitative art to a very precise, quantitative science. But it is only recently that the basic framework of George Gamow's Big Bang theory has started to yield to a more powerful idea. As the new century dawns, we are finding ourselves at a watershed that is likely to permanently change our understanding of the universe. Something is happening that is much more than the discovery of new facts or new equations. Our entire outlook and framework

3. One light-year is, of course, the distance light travels in a year. It comes to about ten trillion kilometers.

for thinking, the whole epistemology of physics and cosmology, are undergoing upheaval. The narrow twentieth-century paradigm of a single universe about ten billion years old and ten billion light-years across with a unique set of physical laws is giving way to something much bigger and pregnant with new possibilities. Gradually cosmologists and physicists like myself are coming to see our ten billion light-years as an infinitesimal pocket of a stupendous *megaverse.*[4] At the same time, theoretical physicists are proposing theories that demote our ordinary laws of nature to a tiny corner of a gigantic Landscape of mathematical possibilities.

The word *Landscape*, in the present context, is fewer than three years old, but since I introduced it in 2003, it has become part of the cosmologist's vocabulary. It denotes a mathematical space representing all of the possible environments that theory allows. Each possible environment has its own Laws of Physics, its own elementary particles, and its own constants of nature. Some environments are similar to our own but slightly different. For example, they may have electrons, quarks, and all the usual particles but with gravity a billion times stronger than ours. Others have gravity like ours but contain electrons that are heavier than atomic nuclei.[5] Still others may resemble our world except for a violent repulsive force (called the cosmological constant) that rips apart galaxies, molecules, and even atoms. Not even the three dimensions of space are sacred; regions of the Landscape describe worlds of four, five, six, and even more dimensions.

According to modern cosmological theories, the diversity of the Landscape is paralleled by a corresponding diversity in ordinary space. Inflationary cosmology, which is our best theory of the universe, is leading us, sometimes unwillingly, to a concept of a megaverse, filled with a prodigious number of what Alan Guth calls "pocket universes." Some pockets are microscopically small and never get big. Others are big like ours but totally empty. And each lies in its own little valley of the Landscape. The old twentieth-century question, "What can you find in the universe?" is giving way to, "What can you not find?"

4. The term *multiverse* has been widely used instead of *megaverse*. Personally, I prefer the sound of *megaverse*. My apologies to the multiverse enthusiasts.

5. In our world, atomic nuclei are thousands of times heavier than electrons.

Man's place in the universe is also being reexamined and challenged. A megaverse of such diversity is unlikely to support intelligent life anywhere but in a tiny fraction of its expanse. According to this view, many questions such as, "Why is a certain constant of nature one number, instead of another?" will have answers that are entirely different from what physicists had hoped. No unique value will be picked out by mathematical consistency, since the Landscape permits an enormous variety of possible values. Instead, the answer will be, "Somewhere in the megaverse, the constant equals *this* number; somewhere else it is *that* number. We live in one tiny pocket where the value of the constant is consistent with our kind of life. That's it! That's all! There is no other answer to the question."

Many coincidences occur in the laws and constants of nature that have no explanation other than, "If it were otherwise, intelligent life could not exist." To some it seems as though the Laws of Physics were chosen, at least in part, to permit our existence. Called the Anthropic Principle, this idea is hated by most physicists, as I noted in my introduction. To some it smells of supernatural creation myths, religion, or intelligent design. Others feel that it represents surrender, a giving up of the noble quest for rational answers. But because of unprecedented new developments in physics, astronomy, and cosmology, these same physicists are being forced to reevaluate their prejudices. There are four principal developments driving this sea change: two from theoretical physics and two from observational astronomy. On the theoretical side an outgrowth of inflationary theory called Eternal Inflation is demanding that the world be a megaverse, full of pocket universes that have bubbled up out of inflating space, like bubbles in an uncorked bottle of champagne. At the same time, String Theory is producing a Landscape of enormous diversity. The best estimates are that 10^{500} distinct kinds of environments are possible. This number (one followed by five hundred zeros) is far beyond being "unimaginably large," but even it may not be big enough to count the possibilities.

Very recent astronomical discoveries exactly parallel the theoretical advances. The newest astronomical data about the size and shape of the universe provide confirmation that the universe exponentially "inflated" to a stupendous size much bigger than the standard ten or fifteen billion light-years. There is very little doubt that we are embedded in a vastly bigger megaverse. But the biggest news is that in our pocket of space, the notorious cosmological constant (a mathematical term that Einstein originally introduced into his equations and later rejected in disgust) is not quite zero as it was thought to be. This discovery has rocked the boat more than any other. The cosmological constant represents an extra gravitational repulsion, a kind of antigravity that was believed to be absolutely absent from the real world. The fact that it is not absent is a cataclysm for physicists, and the only way that we know how to make any sense of it is through the reviled and despised Anthropic Principle.

I don't know what strange and unimaginable twists our view of the universe will undergo while exploring the vastness of this Landscape. But I would bet that at the turn of the twenty-second century, philosophers and physicists will look back to the present as a time when the twentieth-century concept of the universe gave way to a megaverse, populating a Landscape of mind-boggling proportions.

Nature Has the Jitters

> "Anyone who is not shocked by quantum theory has not understood it."
>
> — NIELS BOHR

The idea that the Laws of Physics can vary throughout the universe is as meaningless as the idea that there can be more than one universe. The universe is all there is; it may be the one noun in the English language that logically should have no plural. The laws governing the universe as a whole cannot change. What laws would govern those changes? Are they not also part of the Laws of Physics?

But I mean something much more modest by the Laws of Physics than the grand, overarching laws that regulate all aspects of the megaverse. I mean the things that an ordinary twentieth-century physicist, a physicist more interested in the laboratory than the universe, would have meant: the laws governing the building blocks of ordinary matter.

This book is about these Laws of Physics — not *what* they are but *why* they are. But before we can discuss the why, we need to know the what. Exactly what are these laws? What do they say, and how are they expressed? The task of this first chapter is to bring you up to speed on the Laws of Physics as they were understood circa the year 2000.

To Isaac Newton and those who came after him, the physical world was a precise deterministic machine whose past determined its future "as sure as night follows day." The laws of nature were rules (equations) that expressed this determinism in precise mathematical language. For example, one could determine how objects move along precise trajectories given their initial starting points (including their velocities). The great French eighteenth-century physicist and mathematician Pierre-Simon de Laplace expressed it this way:

> We may regard the present state of the universe as the effect of its past and the cause of its future. An intellect which at a certain moment would know all forces that set nature in motion, and all positions of all items of which nature is composed, if this intellect were also vast enough to submit these data to analysis, it would embrace in a single formula the movements of the greatest bodies of the universe and those of the tiniest atom; for such an intellect nothing would be uncertain and the future just like the past would be present before its eyes.

Just in case the translation from the French is unclear, Laplace was saying that if, at some instant, you (some superintellect) knew the position and velocity of every particle in the universe, you could forever after predict the exact future of the world. This ultra-deterministic view of nature was the prevailing paradigm until, at the beginning of the twentieth century, that subversive thinker Albert Einstein came along and changed everything. Although Einstein is most famous for the Theory of Relativity, his boldest and most radical move — his most subversive move — had to do with the strange world of quantum mechanics, not the Theory of Relativity. Since that time physicists have understood that the Laws of Physics are quantum laws. For that reason, I have chosen to begin this first chapter with a short course on "thinking quantum-mechanically."

You are about to enter the bizarre Alice in Wonderland world of modern physics, where nothing is what it seems, everything fluctuates and shimmers, and uncertainty reigns supreme. Forget the predictable clockwork universe of Newtonian physics. The world of quantum mechanics is anything but predictable. The early-twentieth-century revolutions in physics were not "velvet revolutions." They not only changed the equations and Laws of Physics, but they destroyed the epistemological foundations of much of classical science and philosophy. Many physicists were unable to cope with the new ways of relating to phenomena and were left behind. But a younger, more flexible generation reveled in the bizarre modern ideas and developed new intuitions and powers of visualization. So complete was the change that many theoretical physicists of my generation find it easier to think quantum-mechanically or relativistically than in the old classical ways.

Quantum mechanics was the biggest shock. At the quantum level the world is a jittery, fluctuating place of probabilities and uncertainty. But the electron does not just stagger around like a drunken sailor. There is a subtler pattern to the randomness that is best described in the arcane symbolism of abstract mathematics. However, with a little effort on my part and some patience on your part, the most important things can be translated into common language.

Since the nineteenth century, physicists have used the metaphor of a billiard table to represent the physical world of interacting, colliding particles. James Clerk Maxwell used the analogy; so did Ludwig Boltzmann. By now it's been used by scores of physicists to explain the quantum world. The first time I heard it used was by Richard Feynman, who explained things this way:

> Imagine a billiard table that is so perfectly constructed that it has no friction at all. The balls and cushions are so elastic (bouncy) that whenever a collision occurs, the balls bounce with no loss of kinetic energy. Let's also remove the pockets so that once the balls are set into motion, they will continue moving forever, colliding, bouncing off the cushions, and moving on. The game starts with fifteen balls arranged in a triangle like a two-dimensional version of a stack of cannon balls. The cue ball is sent rocketing toward the pack.
>
> What happens next is too complicated and unpredictable to follow. But why is it so unpredictable? It's because each collision

magnifies minute differences in the starting positions and velocities of the balls, so that even the tiniest deviation eventually leads to an entirely different outcome. [This kind of ultra-sensitivity to initial conditions is called chaos, and it is a ubiquitous feature of nature.] Trying to reproduce a pool game is not like reproducing a chess game. It would take almost infinite precision. Nevertheless, in classical physics the balls move along perfectly precise trajectories, and the motion is completely predictable, if only we know the initial positions and velocities of the balls with infinite precision. Of course, the longer we want to predict the motion, the more accurately we need to know the initial data. But there is no limit to the precision of this data and no limit to our ability to predict the future from the past.

By contrast, the quantum billiard game is unpredictable no matter how hard the players work to make it precise. No amount of precision would allow anything more than statistical predictions of outcomes. The classical billiard player might resort to statistics just because the initial data were imperfectly known or because solving the equations of motion might be too hard. But the quantum player has no choice. The laws of quantum mechanics have an intrinsically random element that can never be eliminated. Why not — why can't we predict the future from the knowledge of the initial positions and velocities? The answer is the famous Heisenberg Uncertainty Principle.

The Uncertainty Principle describes a fundamental limitation on how well we can determine the positions and velocities simultaneously. It's the ultimate catch-22. By improving our knowledge of the position of a ball in an effort to improve our predictions, we inevitably lose precision about where the ball will be in the next instant. The Uncertainty Principle is not just a qualitative fact about the behavior of objects. It has a very precise quantitative formulation: the product of the uncertainty of an object's position and the uncertainty of its momentum[6] is always larger than a certain (very small) number called Planck's constant.[7]

6. The momentum of an object is defined to be its velocity multiplied by its mass.

7. The symbol for Planck's constant is the letter h and its numerical value is 6.626068×10^{-34} m^2 kg/s. Here, *m*, *kg*, and *s* stand for *meter*, *kilogram*, and *second*.

Heisenberg and others after him tried to dream up ways to beat the Uncertainty Principle. Heisenberg's examples involved electrons, but he might just as well have used billiard balls. Shine a beam of light on a quantum billiard ball. The light that reflects off the ball can be focused on a photographic film, and from the image, the location of the ball can be deduced. But what about the velocity of the ball: how can it be measured? The simplest and most direct way would be to make a second measurement of the position a short time later. Knowing the position at two successive instants, it's an easy matter to determine the velocity.

Why is this type of experiment not possible? The answer goes back to one of Einstein's greatest discoveries. Newton had believed that light consisted of particles, but by the beginning of the twentieth century, the particle theory of light had been completely discredited. Many optical effects like interference could be explained only by assuming light to be a wave phenomenon similar to ripples on the surface of water. In the mid-nineteenth century James Clerk Maxwell had produced a highly successful theory that envisioned light as electromagnetic waves that propagate through space in much the same way that sound propagates through air. Thus, it was a radical shock when, in 1905, Albert Einstein proposed that light (and all other electromagnetic radiation) is made out of little bullets called quanta, or photons.[8] In some strange new way, Einstein was proposing that light had all the old wave properties — wavelength, frequency, etc. — but also a graininess, as if it were composed of discrete bits. These quanta are packets of energy that cannot be subdivided, and that fact creates certain limitations when one attempts to form accurate images of small objects.

Let's begin with the determination of the position. In order to get a good sharp image of the ball, the wavelength of the light must not be too long. The rule is simple: if you want to locate an object to a given accuracy, you must use waves whose wavelengths are no larger than the allowable error. All images are fuzzy to some degree, and limiting the fuzziness means using short wavelengths. This is no problem in classi-

8. The term *quantum* (used as a noun) is somewhat more general than *photon*. *Quantum* refers to any discrete packet of energy while *photon* is the more specific term referring to electromagnetic energy. Thus one could say that the photon is the quantum of electromagnetic radiation.

cal physics, where the energy of a beam of light can be arbitrarily small. But as Einstein claimed, light is made of indivisible photons. Moreover, as we will see below, the shorter the wavelength of a light ray, the larger is the energy of those photons.

What all of this means is that getting a sharp image that accurately locates the ball requires you to hit it with high-energy photons. But this places severe limitations on the accuracy of subsequent velocity measurements. The trouble is that a high-energy photon will collide with the billiard ball and give it a sharp kick, thus changing the very velocity that we intended to measure. This is an example of the frustration in trying to determine both position and velocity with infinite precision.

The connection between the wavelength of electromagnetic radiation and the energy of photons — the smaller the wavelength, the larger the energy — was one of Einstein's most important 1905 discoveries. In order of increasing wavelength, the electromagnetic spectrum consists of gamma rays, X-rays, ultraviolet light, visible light, infrared light, microwaves, and radio waves. Radio waves have the longest wavelength, from meters to cosmic dimensions. They are a very poor choice for making precise images of ordinary objects because the images will be no sharper than the wavelength. In a radio image a human being would be indistinguishable from a sack of laundry. In fact it would be impossible to tell one person from two, unless the separation between them was greater than the wavelength of the radio wave. All the images would be blurred fuzz balls. This doesn't mean that radio waves are never useful for imaging: they are just not good for imaging small objects. Radio astronomy is a very powerful method for studying large astronomical objects. By contrast, gamma rays are best for getting information about really small things such as nuclei. They have the smallest wavelengths, a good deal smaller than the size of a single atom.

On the other hand, the energy of a single photon increases as the wavelength decreases. Individual radio photons are far too feeble to detect. Photons of visible light are more energetic: one visible photon is enough to break up a molecule. To an eye that has been accustomed to the dark, a single photon of visible-wavelength light is just barely enough to activate a retinal rod. Ultraviolet and X-ray photons have enough energy to easily kick electrons out of atoms, and gamma rays can break up not only nuclei, but even protons and neutrons.

This inverse relation between wavelength and energy explains one of the all-pervasive trends in twentieth-century physics: the quest for bigger and bigger accelerators. Physicists, trying to uncover the smallest constituents of matter (molecules, atoms, nuclei, quarks, etc.) were naturally led to ever-smaller wavelengths to get clear images of these objects. But smaller wavelengths inevitably meant higher-energy quanta. In order to create such high-energy quanta, particles had to be accelerated to enormous kinetic energies. For example, electrons can be accelerated to huge energies, but only by machines of increasing size and power. The Stanford Linear Accelerator Center (SLAC) near where I live can accelerate electrons to energies 200,000 times their mass. But this requires a machine about two miles long. SLAC is essentially a two-mile microscope that can resolve objects a thousand times smaller than a proton.

Throughout the twentieth century many unsuspected things were discovered as physicists probed to smaller and smaller distances. One of the most dramatic was that protons and neutrons are not at all elementary particles. By hitting them with high-energy particles, it became possible to discern the tiny components — quarks — that make up the proton and neutron. But even with the highest-energy (shortest-wavelength) probes, the electron, the photon, and the quark remain, as far as we can tell, pointlike objects. This means that we are unable to detect any structure, size, or internal parts to them. They may as well be infinitely small points of space.

Let's return to Heisenberg's Uncertainty Principle and its implications. Picture a single ball on the billiard table. Because the ball is confined to the table by the cushions, we automatically know something about its position in space: the uncertainty of the position is no bigger than the dimensions of the table. The smaller the table, the more accurately we know the position and, therefore, the less certain we can be of the momentum. Thus, if we were to measure the velocity of the ball confined to the table, it would be somewhat random and fluctuating. Even if we removed as much kinetic energy as possible, this residual fluctuation motion could not be eliminated. Brian Greene has used the term *quantum jitters* to describe this motion, and I will follow his lead.[9]

9. Brian Greene, *The Elegant Universe: Superstrings, Hidden Dimensions, and the Quest for the Ultimate Theory* (New York: Norton, 2003).

The kinetic energy associated with the quantum jitters is called *zero-point energy*, and it cannot be eliminated.

The quantum jitters implied by the Uncertainty Principle have an interesting consequence for ordinary matter as we try to cool it to zero temperature. Heat is, of course, the energy of random molecular motion. In classical physics, as a system is cooled, the molecules eventually come to rest at absolute zero temperature. The result: at absolute zero all the kinetic energy of the molecules is eliminated.

But each molecule in a solid has a fairly well-defined location. It is held in place, not by billiard table cushions, but by the other molecules. The result is that the molecules necessarily have a fluctuating velocity. In a real material subject to the laws of quantum mechanics, the molecular kinetic energy can never be totally removed, even at absolute zero!

Position and velocity are by no means unique in having an Uncertainty Principle. There are many pairs of so-called conjugate quantities that cannot be determined simultaneously: the better one is fixed, the more the other fluctuates. A very important example is *energy-time uncertainty principle:* it is impossible to determine both the exact time that an event takes place and the exact energy of the objects that are involved. Suppose an experimental physicist wished to collide two particles at a particular instant of time. The energy-time uncertainty principle limits the precision with which she can control the energy of the particles and also the time at which they hit each other. Controlling the energy with increasing precision inevitably leads to increasing randomness in the time of collision — and vice versa.

Another important example that will come up in chapter 2 involves the electric and magnetic fields at a point of space. These fields, which will play a key role in subsequent chapters, are invisible influences that fill space and control the forces on electrically charged particles. Electric and magnetic fields, like position and velocity, cannot be simultaneously determined. If one is known, the other is necessarily uncertain. For this reason the fields are in a constant state of jittering fluctuation that cannot be eliminated. And, as you might expect, this leads to a certain amount of energy, even in absolutely empty space. This *vacuum energy* has led to one of the greatest paradoxes of modern physics and cosmology. We will come back to it many times, beginning with the next chapter.

Uncertainty and jitters are not the whole story. Quantum mechanics has another side to it: the quantum side. The word *quantum* implies a certain degree of discreteness or graininess in nature. Photons, the units of energy that comprise light waves, are only one example of quanta. Electromagnetic radiation is an oscillatory phenomenon; in other words, it is a vibration. A child on a swing, a vibrating spring, a plucked violin string, a sound wave: all are also oscillatory phenomena, and they all share the property of discreteness. In each case the energy comes in discrete quantum units that can't be subdivided. In the macroscopic world of springs and swings, the quantum unit of energy is so small that it seems to us that the energy can be anything. But, in fact, the energy of an oscillation comes in indivisible units equal to the frequency of the oscillation (number of oscillations per second) times Planck's very small constant.

The electrons in an atom, as they sweep around the nucleus, also oscillate. In this case the quantization of energy is described by imagining discrete orbits. Niels Bohr, the father of the quantized atom, imagined the electrons orbiting as if they were constrained to move in separate lanes on a running track. The energy of an electron is determined by which lane it occupies.

Jittery behavior and discreteness are weird enough, but the real weirdness of the quantum world involves "interference." The famous "two-slit experiment" illustrates this remarkable phenomenon. Imagine a tiny source of light — a very intense miniature lightbulb — in an otherwise dark room. A laser beam will also do. At some distance a photographic film has been placed. When light from the source falls on the film, it blackens it in the same way that an ordinary photographic "negative" is produced. Obviously if an opaque obstacle such as a sheet of metal is placed between the source and the film, the film will be protected and will be unblackened. But now cut two parallel vertical slits in the sheet metal so that light can pass through them and affect the film. Our first experiment is very simple: block one slit — say, the left one — and turn on the source.

After a suitable time a horizontal broad band of blackened film will appear: a fuzzy image of the right slit. Next, let's close the right slit and open the left one. A second broad band, partially overlapping the first, will appear.

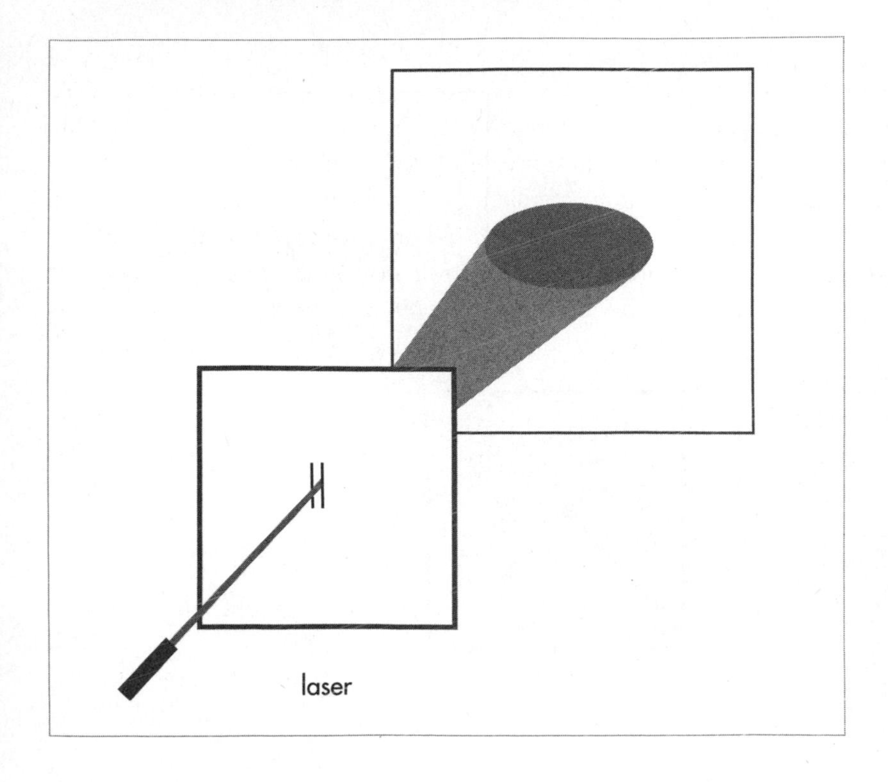

Now start over with fresh, unexposed film, but this time open both slits. If you don't know in advance what to expect, the result may amaze you. The pattern is not just the sum of the previous two blackened regions. Instead, we find a series of narrow dark and light bands, like the stripes of a zebra, replacing the two fuzzier bands. There are unblackened bands where previously the original dark bands overlapped. Somehow the light going through the left and right slits cancels in these places. The technical term is *destructive interference*, and it is a well-known property of waves. Another example of it is the "beats" that you hear when two almost identical notes are played.

If you actually try this experiment at home, you may find that it's not so easy as I made it sound. Two things make it difficult. The interference pattern can be seen only if the slits are very narrow and very close together. Don't expect to succeed by cutting holes with a can opener. Secondly, the source has to be very small. The old, low-tech way of making a small source is to pass the light through a very small pinhole

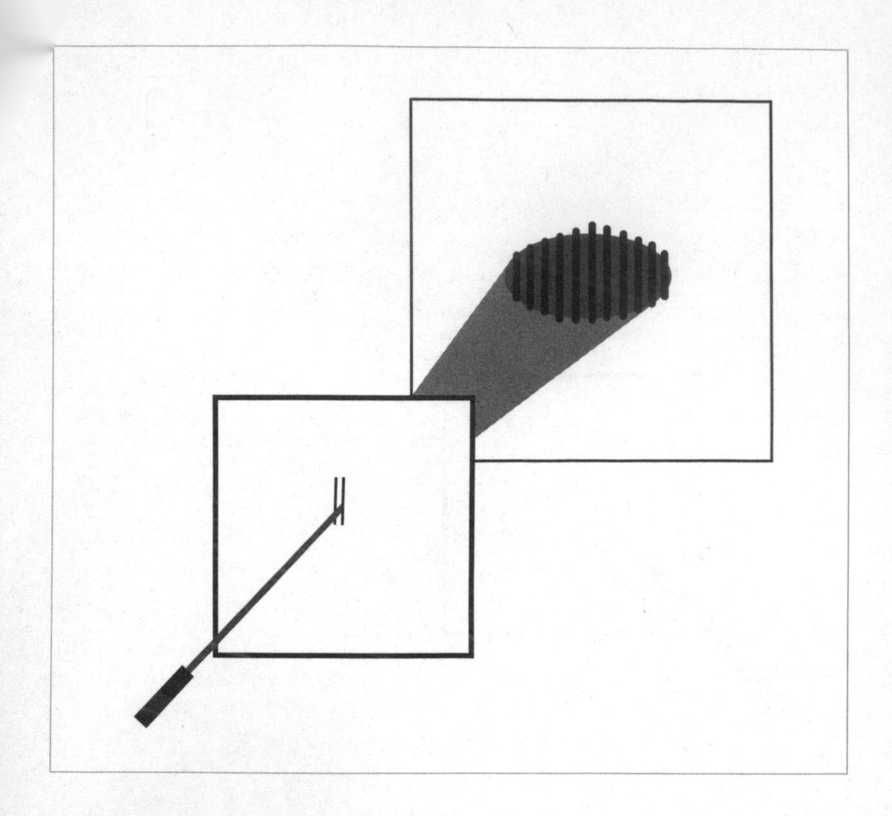

before allowing it to fall on the sheet with the slits. A much better way is to use a high-tech laser. A laser pointer is ideal. Laser light passing through very carefully made slits produces excellent zebralike interference patterns. The main problem in doing the experiment will be holding everything steady.

Now we are going do the entire optical exercise over again, but this time we will turn down the intensity of the source to such a low level that individual photons come through, one at a time. If we expose the film for a short time, a few darkened dots appear where individual photons landed on the film. Expose it again, the same way, and the dots will get denser. Eventually we will see the patterns of the first experiment reproduced. Among other things the experiment confirms Einstein's idea that light is composed of discrete photons. Moreover, the particles arrive randomly, and only when enough have accumulated do we see that a pattern is reproduced.

But these particle-like photons behave in a most unexpected way. When both slits are opened, not a single particle arrives at the locations where the destructive interference took place. This is despite the fact that photons do arrive at these places when only one slit is open. It seems that opening the left slit prevents the photons from going through the right slit and vice versa.

To express it differently, suppose the point X is a point on the film where destructive interference takes place. The photon can get to X if the left slit is open. It can get to X if the right slit is open. A sensible person would expect that if both were opened, a photon would be even more likely to get to X. But no — not a single photon appears at X no matter how long we wait. How does the photon, about to go through the left slit, know that the right slit is open? Physicists sometimes describe this peculiar effect by saying that the photon does not go through either of the slits, but instead "feels out" both paths, and at certain points, the contribution of the two paths cancel. Whether or not this helps your understanding, interference is a very weird phenomenon. You get used to the weirdness of quantum mechanics if you work with it for forty or more years. But stopping to reflect on it, it's weird!

Elementary Particles

Nature seems to be organized in a hierarchical way: big things made out of smaller things made out of yet smaller things until we reach the smallest things that we are able to uncover. The ordinary world is full of such hierarchies. An automobile is nothing but its parts: wheels, engine, carburetor, and so on. The carburetor, in turn, is built from smaller parts such as idle screws, choke levers, jets, and springs. As far as one can tell, the properties of the smaller things determine the behavior of the larger. This view, that the whole is the sum of its parts and that nature can be understood by reducing it to the simplest and smallest components, is called *reductionism*.

Reductionism has the status of a dirty word in many academic quarters. It stirs up passions almost as powerful as those that evolution excites in certain religious circles. The notion that all of existence is no more than inanimate particles touches the same insecurities as the similar idea that we humans are mere vehicles for our selfish genes. But, like it or not, reductionism does work. Every auto mechanic is a reductionist, at least during working hours. In science the power of reductionism is phenomenal.[10] The basic laws of biology are determined by the chemistry of organic molecules like DNA, RNA, and proteins. Chemists reduce the complex properties of molecules to those of atoms, and then physicists take over. Atoms are nothing but collections of electrons orbiting atomic nuclei. As we learn in elementary science courses, nuclei are composites of protons and neutrons. They, in turn, are made of quarks. How far does this "Russian nesting doll" view of nature go? Who knows? But twentieth-century physics has succeeded in pushing reductionism to the level of the so-called elementary particles. By the Laws of Physics, I mean the laws of these so-far smallest building blocks. It will be important to have a clear idea of what these laws are before we can begin to question why they are the laws.

The language of theoretical physics is mathematical equations. It's hard for physicists to conceive of any form for a theory other than an equation or a small set of equations. Newton's equations, Maxwell's equations, Einstein's equations, Erwin Schrödinger's equation — these are some of the most important examples. The mathematical framework for elementary-particle physics is called *quantum field theory*. It is a difficult mathematical subject loaded with very abstract equations. In fact the equations of quantum field theory are so complicated that one may get the feeling that equations are not really the right way to express the theory. Luckily for us the great Richard Feynman had exactly that feeling. So he invented a pictorial way to visualize the equations. Feynman's way of thinking is so intuitive that the main ideas can be summarized without a single equation.

Dick Feynman was a genius of visualization (he was also no slouch with equations): he made a mental picture of anything he was working on. While others were writing blackboard-filling formulas to express the laws of elementary particles, he would just draw a picture and figure out

10. Whether it works for the study of the mind is a contentious issue. My own view is that the behavior of animate matter is subject to the same Laws of Physics as inanimate matter. I know of no evidence to the contrary. On the other hand, the phenomenon of consciousness has yet to be fully explained by reductionist science.

the answer. He was a magician, a showman, and a show-off, but his magic provided the simplest, most intuitive way to formulate the Laws of Physics. *Feynman diagrams* (see page 38) are literally pictures of the events that take place as elementary particles move through space, collide, and interact. A Feynman diagram can be nothing more than a few lines describing a couple of colliding electrons, or it can be a vast network of interconnected branching, looping trajectories describing all of the particles making up anything from a diamond crystal to a living being or an astronomical body. These diagrams can be reduced to a few basic elements that summarize everything that is known about elementary particles. Of course there is more than just the pictures — there are all the technical details of how they are used to make precise calculations, but that's less important. For our purposes a picture is worth a thousand equations.

Quantum Electrodynamics

A quantum field theory begins with a cast of characters, namely, a list of elementary particles. Ideally the list would include all elementary particles, but that is not practical: we are fairly certain that we don't even know the complete list. But not too much is lost by making a partial list. It is like a theater performance: in reality every story involves everyone on earth, past and present, but no sane author would try to write a play with several billion characters. For any particular story some characters are more important than others, and the same is true in elementary-particle physics.

The original story that Feynman set out to tell is called Quantum Electrodynamics, or QED for short, and it involves only two characters: the electron and the photon. Let me introduce them.

The Electron

In 1897 the British physicist J. J. Thomson made the first discovery of an elementary particle. Electricity was well known before that point, but Thomson's experiments were the first to confirm that electric currents are reducible to the motion of individual charged particles. The moving particles that power toasters, lightbulbs, and computers are, of course, electrons.

For dramatic effects it's hard to beat electrons. When a giant lightning bolt rips across the sky, electrons flow from one electrified cloud to another. The roar of thunder is due to a shock wave caused by the collision of rapidly accelerated electrons with air molecules blocking their path. The visible lightning bolt consists of electromagnetic radiation that was emitted by agitated electrons. The tiny sparks and crackling noises due to static electricity, on a very dry day, are manifestations of the same physics on a smaller scale. Even ordinary household electricity is the same flow of electrons, tamed by electrically conducting copper wires.

Every electron has exactly the same electric charge as every other electron. The charge of the electron is an incredibly small number. It takes an enormous number of electrons — about 10^{19} per second — to create a common electric current of one amp. There is an oddity about the charge of the electron that has puzzled and troubled generations of undergraduates studying physics: the electron's charge is *negative*. Why is that? Is there something intrinsically negative about the electron? In fact the negativity of the electron charge is not a property of the electron but rather a definition. The trouble dates back to Benjamin Franklin, who was the first physicist to realize that electricity was a flow of charge.[11] Franklin, who knew nothing of electrons, had no way of knowing that what he called *positive current* was actually a flow of electrons in the opposite direction. For this reason we have inherited the confusing convention of a negative electron charge. As a consequence, we physics professors constantly have to remind students that when electric current flows to the left, electrons move to the right. If this boggles your mind, blame it on Ben Franklin and then ignore it.

If all electrons were suddenly to disappear, a great deal more than toasters, lightbulbs, and computers would fail. Electrons play another very profound role in nature. All ordinary matter is made of atoms, which in turn are made of electrons — each electron whirling around the atomic nucleus like a ball on a rope. Atomic electrons determine the

11. In addition to writing *Poor Richard's Almanack* and signing the Declaration of Independence, Benjamin Franklin was one of the outstanding scientists of the eighteenth century.

chemical properties of all the elements listed in the periodic table. Quantum Electrodynamics is more than the theory of electrons: it is the basis for the theory of all matter.

The Photon

If the electron is the hero of QED, the photon is the sidekick that makes the hero's deeds possible. The light emitted by a lightning bolt can be traced to microscopic events in which individual electrons shake off photons when they are accelerated. The entire plot of QED revolves around one fundamental process: the emission of a single photon by a single electron.

Photons also play an indispensable role in the atom. In a sense that will become clear, photons are the ropes that tether the electrons to the nucleus. If photons were to be suddenly eliminated from the list of elementary particles, every atom would instantly disintegrate.

The Nucleus

One of the main goals of QED was to understand the detailed properties of simple atoms, especially hydrogen. Why hydrogen? Hydrogen, having only a single electron, is so simple that the equations of quantum mechanics can be solved. More complex atoms with many electrons, all exerting forces on one another, could be studied only with the aid of powerful computers, which didn't exist when QED was being formulated. But to study any atom, one more ingredient must be added — the nucleus. Nuclei are made of positively charged protons and electrically neutral neutrons. These two particles are very similar to each other, apart from the fact that the neutron has no electric charge. Physicists group these two particles together and give them a common name: the *nucleon*. A nucleus is essentially a blob of sticky nucleons. The structure of any nucleus, even of hydrogen, is so complicated that physicists like Feynman decided to ignore it. They concentrated instead on the much simpler physics of the electron and photon. But they couldn't do away with the nucleus altogether. So they introduced it not as an actor, but as a stage prop. Two things made this possible.

First, the nucleus is much heavier than an electron. It is so heavy that it is almost immobile. No big mistake is made if the nucleus is replaced by an immovable point of positive electric charge.

Second, nuclei are very small by comparison with atoms. The electron orbits the nucleus at about 100,000 nuclear diameters and never gets close enough to be affected by the complicated internal nuclear structure.

According to the reductionist view of particle physics, all the phenomena of nature — solids, liquids, gases, living as well as inanimate matter — are reduced to the constant interaction and collision of electrons, photons, and nuclei. That's the action and the whole plot — actors crashing into one another, bouncing off one another, and here and there, giving birth to new actors out of the collision. It is this banging away of particles by other particles that Feynman diagrams depict.

Feynman Diagrams

"If you come to a fork in the road, take it."

— YOGI BERRA

We have the actors, we have the script, and now we need a stage. Shakespeare said, "All the world's a stage," and as usual, the Bard got it right. The set for our farce is the whole world: for a physicist that means all of ordinary three-dimensional space. Up-down, east-west, and north-south are the three directions near the surface of the earth. But a stage direction involves not only *where* an action takes place, but also *when* it takes place. Thus, there is a fourth direction to *space-time:* past-future. Ever since Einstein's discovery of the Special Theory of Relativity, physicists have been in the habit of picturing the world as a four-dimensional space-time that encompasses not only *the now,* but also all of the future and the past. A point in space-time — a where and a when — is called an *event.*

A sheet of paper or a blackboard can be used to represent space-time. Because the paper or the blackboard has only two dimensions, we'll have to cheat a bit. The horizontal direction on the paper will be a stand-in for all three directions of space. We will have to stretch our imagination and pretend that the horizontal axis is really three perpendicular axes. That leaves us with the vertical direction to represent time. The future is usually taken to be up and the past, down (that of course is just as arbitrary as the fact that maps place the northern hemisphere above the southern). A point on the sheet of paper is an event — a *where* and a *when:* a space-time point. This is where Feynman began: particles, events, and space-time.

Our first Feynman diagram depicts the simplest of all stage directions: "Electron, go from point a to point b." To represent this graphically, draw a line on a piece of paper from event a to event b. Feynman also put a little arrow on the line whose purpose will become clear shortly. The line connecting a with b is called a *propagator.*

The photon also can move from one space-time point to another. To depict the photon's motion, Feynman drew another line, or propagator. Sometimes the photon propagator is drawn as a wavy line, sometimes as a dashed line. I will use the dashed line.

Propagators are more than just pictures. They are quantum-mechanical instructions for calculating the probability that a particle

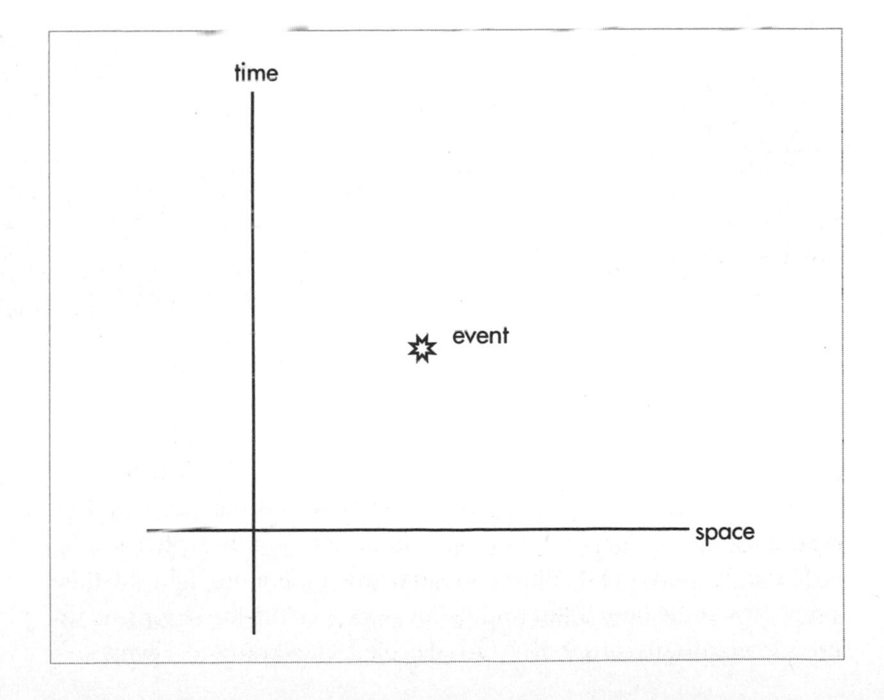

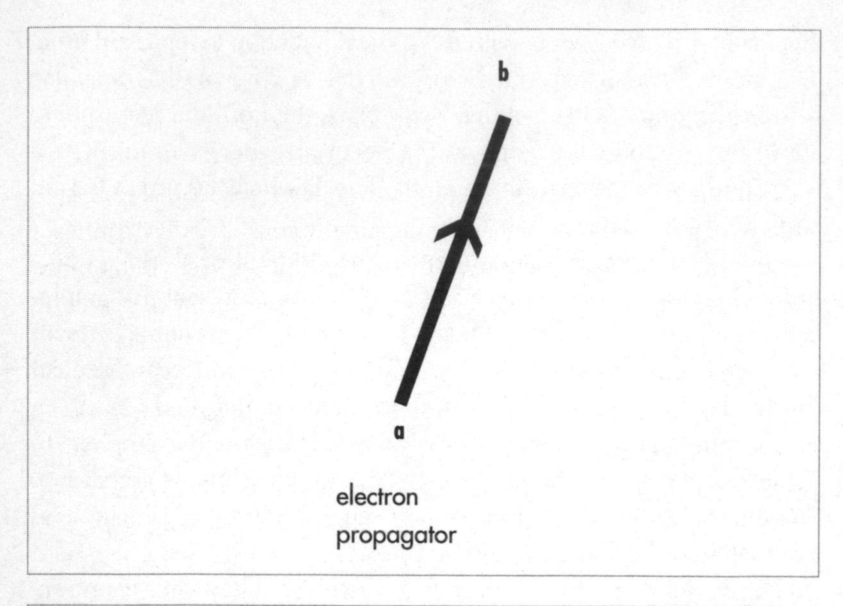

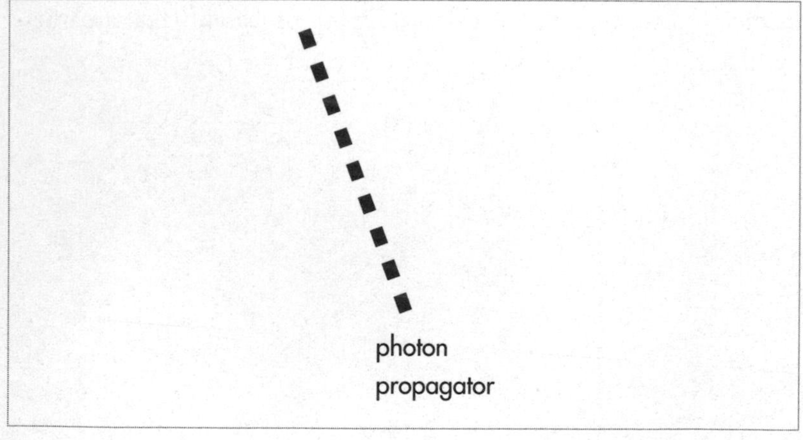

starting at point a will show up later at point b. Feynman had the radical idea that a particle doesn't merely move along a particular path: in an odd way it feels out all paths — random zigzag paths as well as straight paths. We saw a bit of this quantum weirdness in the two-slit experiment. Photons don't go through just the left slit or the right slit: they somehow sample both paths and in the process create the surprising interference patterns where they are detected. According to Feynman's theory all possible paths contribute to the probability for the particle to go from a to b. In the end a particular mathematical expression representing all possible paths between the two points gives the probability to go from a to b. All of this is implicit in the notion of a propagator.

Nothing of very great interest would ever happen if all that ever took place were the free motion of electrons and photons. But they both take part in one coordinated action that is responsible for everything interesting in nature. Recall what happens when electrons move from one cloud to another during a lightning storm. Night suddenly turns into day. Light emitted by the sudden violent electric current dramatically illuminates the sky for an instant. Where does that light come from? The answer traces back to individual electrons. When the motion of an electron is suddenly disturbed, it may respond by shaking off a photon. The process, called *photon emission,* is the basic event of Quantum Electrodynamics. Just as all matter is built of particles, all processes are built from the elementary events of emission and absorption. Thus, the electron — while moving through space-time — can suddenly shoot out a single quantum (or photon) of light. All the visible light that we see, as well as radio waves, infrared radiation, and X rays, is composed of photons that have been emitted by electrons, either in the sun, the filament of a lightbulb, a radio antenna, or an X-ray machine. Thus, Feynman added to the list of particles a second list: a list of elementary events. This brings us to a second kind of Feynman diagram.

The Feynman diagram representing the event of photon emission is called a *vertex diagram.* A vertex diagram looks like the letter Y, or better yet, a forked road: the original electron comes to the fork and shoots off a photon. Subsequently, the electron takes one path, and the photon, the other. The point where the three lines join — the event that emits the photon — is the vertex.

Here is a way to view a Feynman diagram as a short "movie." Get a square of cardboard a few inches on a side and make a long thin slit about one sixteenth of an inch wide. Now place the square over the Feynman diagram (first fill in the dashed lines) with the slit oriented in the horizontal direction. The short line segments showing through the slit represent particles. Start the slit at the bottom of the diagram. If you now move the slit up, you will see the particles move, emit, and absorb other particles and do all the things that real particles do.

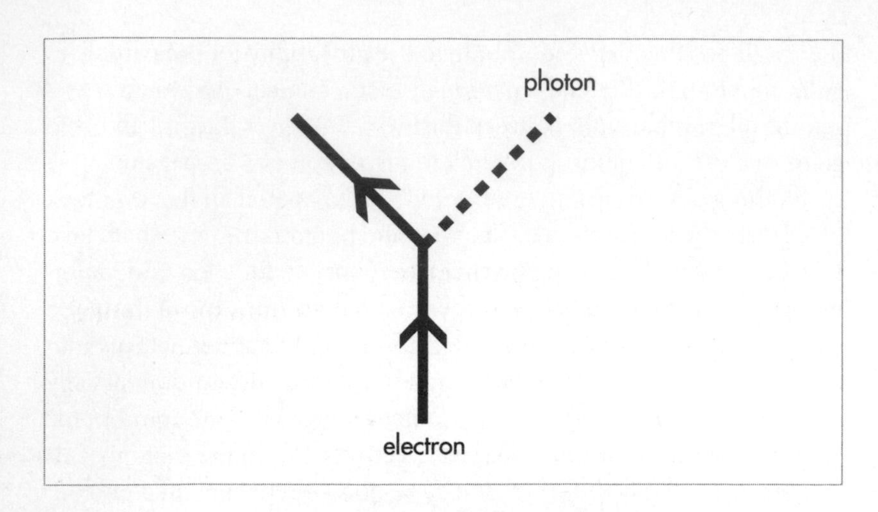
photon
electron

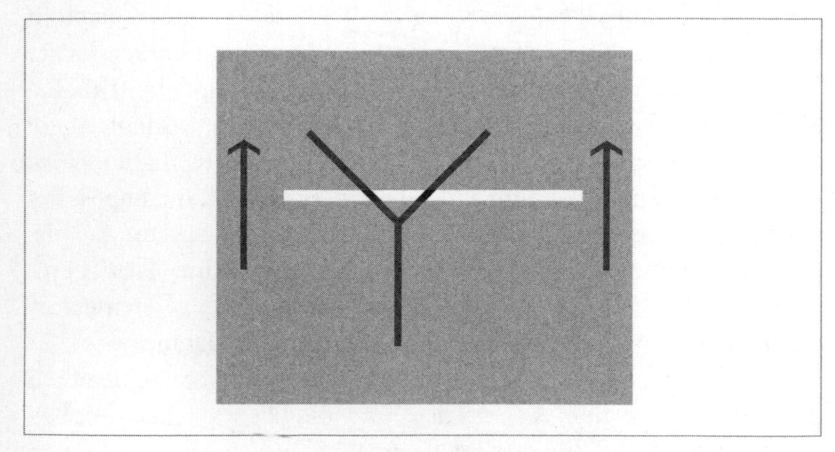

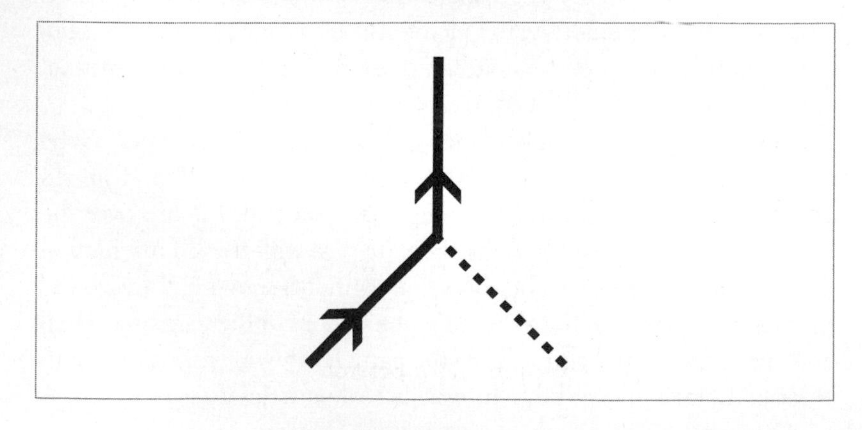

The vertex diagram can be turned upside down (remember, past is down, and future is up) so that it describes an electron and a photon approaching each other. The photon gets absorbed, leaving only the lone electron.

Antimatter

Feynman had a purpose in mind when he put little arrows on the electron lines. Each type of electrically charged particle, such as the electron and proton, has a twin, namely, its antiparticle. The antiparticle is identical to its twin, with one exception: it has the opposite electric charge. When matter meets antimatter, look out! The particles and antiparticles will combine and disappear (annihilate), but not without leaving over their energy in the form of photons.

The antiparticle twin of the electron is called the positron. It appears to be a new addition to the list of particles, but according to Feynman, the positron is not really a new object: he thought of it as an electron *going backward in time!* A positron propagator looks exactly like an electron propagator except that the little arrow points downward toward the past instead of upward toward the future.

Whether you think of a positron as an electron going backward in time or an electron as a positron going backward in time is up to you. It's an arbitrary convention. But with this way of thinking, you can flip

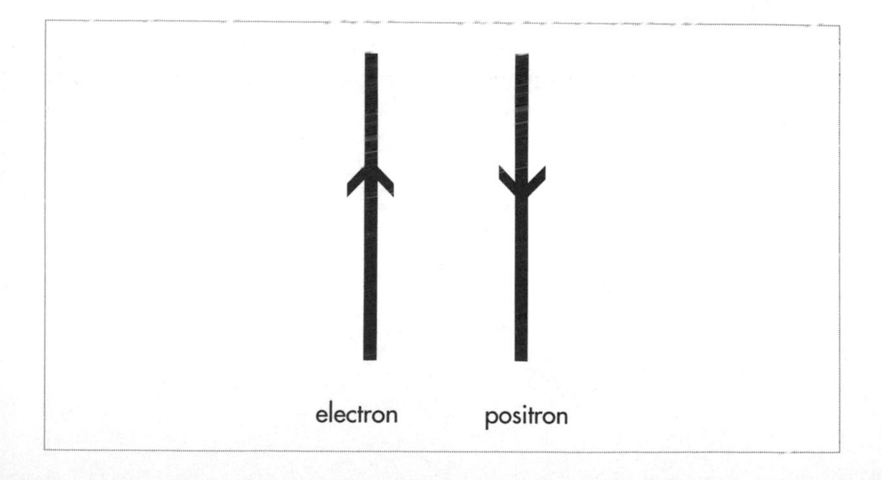

the vertex in new ways. For example, you can flip it so that it describes a positron emitting a photon.

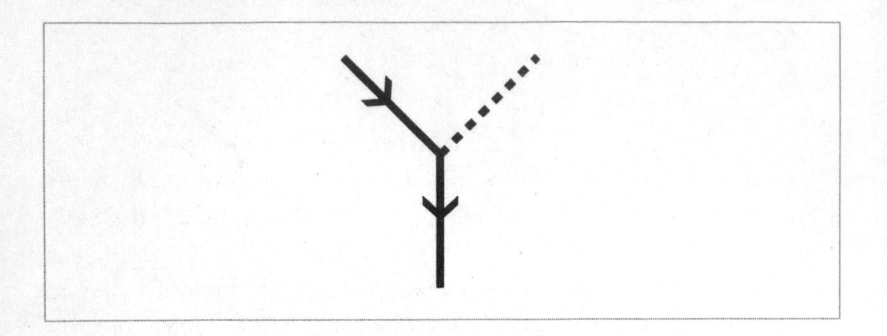

You can even turn it on its side so that it shows an electron and a positron annihilating and leaving only a single photon

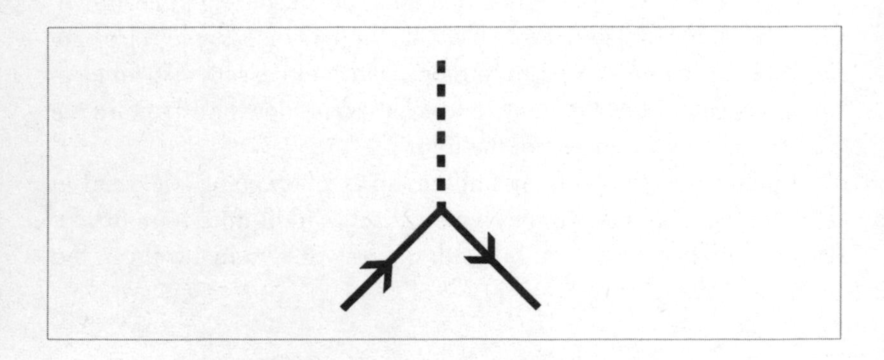

or a photon disappearing and becoming an electron and a positron.

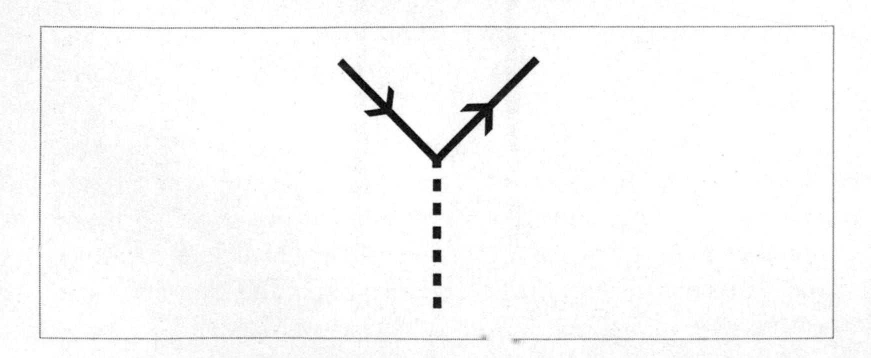

Feynman combined these basic ingredients, propagators and vertices, to make more complex processes. Here is an interesting one.

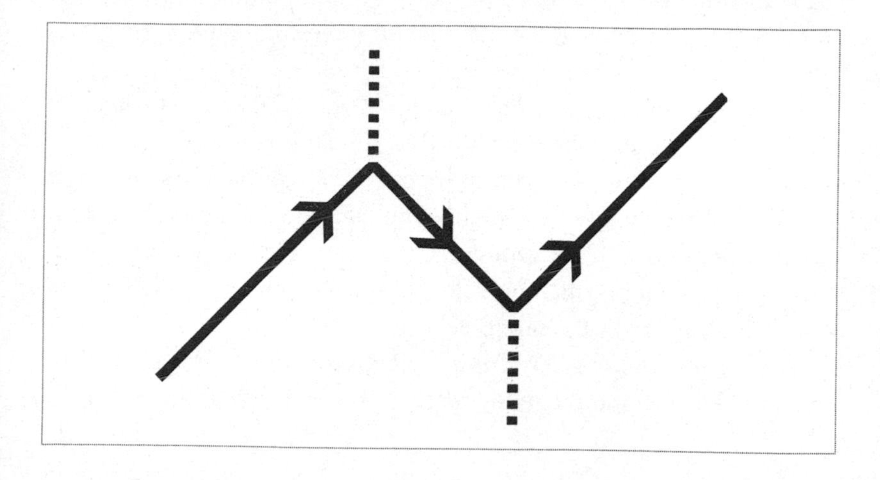

Can you see what it describes? If you use the cardboard-with-slit to view the diagram, here is what you will see: initially, at the lower part of the diagram, there are only an electron and a photon. Without warning, the photon spontaneously becomes an electron-positron pair. Then the positron moves toward the electron, where it meets its twin, and together they annihilate, leaving a photon. In the end there are a single photon and a single electron.

Feynman had another way to think about such diagrams. He pictured the incoming electron as "turning around in time" and temporarily moving toward the past, then turning around again toward the future. The two ways of thinking — either in terms of positrons and electrons or in terms of electrons moving backward in time — are completely equivalent. Propagators and vertices: that's all there is to the world. But these basic elements can be combined in an infinite variety of ways to describe all of nature.

But aren't we missing something important? Objects in nature exert forces on one another. The idea of force is deeply intuitive. It is one of the few concepts in physics that nature has equipped us to understand without consulting a textbook. A man pushing a boulder is exerting a force. The boulder is resisting by pushing back. The gravitational attraction of the earth keeps us from floating away. Magnets exert forces on pieces of iron. Static electricity exerts forces on bits of paper. Bullies shove wimps. The idea of force is so basic to our lives that evolution made sure that we had a concept of force built into our neural circuitry. But much less intuitive is the fact that all forces originate from attraction and repulsion between elementary particles.

Did Feynman have to add a separate set of ingredients to the recipe: specific rules of force between particles? He did not.

All forces in nature derive from special *exchange diagrams,* in which a particle like a photon is emitted by one particle and absorbed by another. For example, the electric force between electrons comes from a Feynman diagram in which one electron emits a photon, which is subsequently absorbed by the other electron.

The photon jumping across the gap between the electrons is the origin of the electric and magnetic forces between them. If the electrons

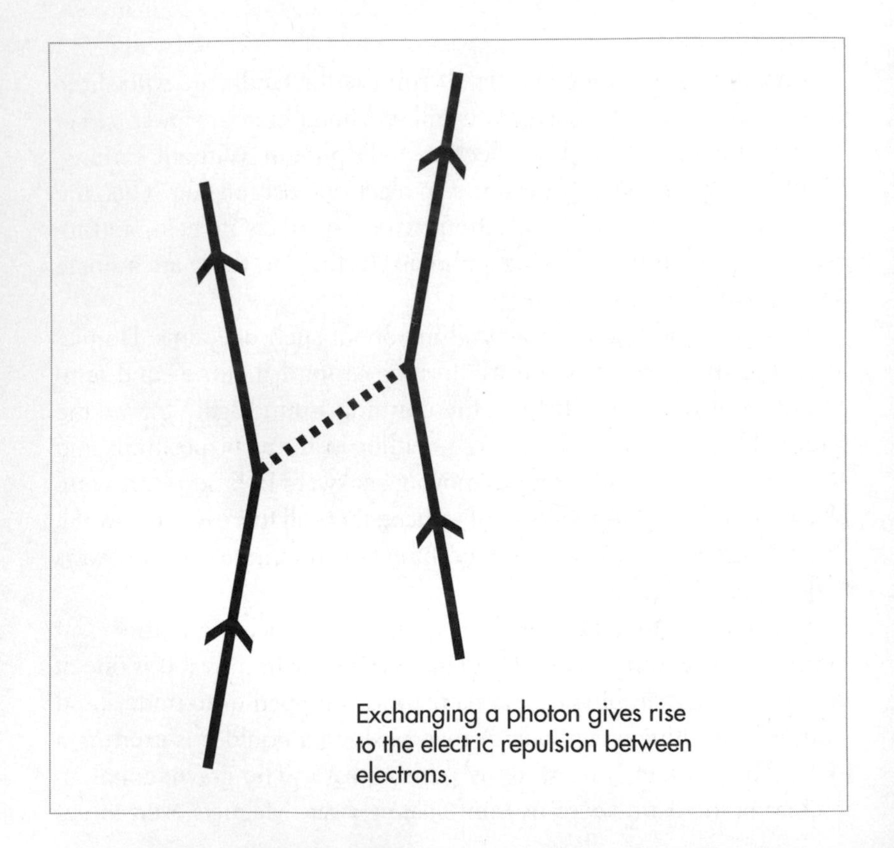

Exchanging a photon gives rise to the electric repulsion between electrons.

are at rest, the force is the usual electrostatic force that famously diminishes according to the square of the distance between the charges.[12] If the electrons happen to be moving, there is an additional magnetic force. The origin of both the electric and magnetic force is the same basic Feynman diagram.

Electrons are not the only particles that can emit photons. Any electrically charged particle can, including a proton. This means that photons can hop between two protons or even between a proton and an electron. This fact is of enormous importance to all of science and life in general. The continual exchange of photons between the nucleus and the atomic electrons provides the force that holds the atom together. Without those jumping photons, the atom would fly apart, and all matter would cease to exist.

Tremendously complicated Feynman diagrams — networks of vertices and propagators — represent complex processes involving any number of particles. In this way Feynman's theory describes all matter from the simplest to the most complicated objects.

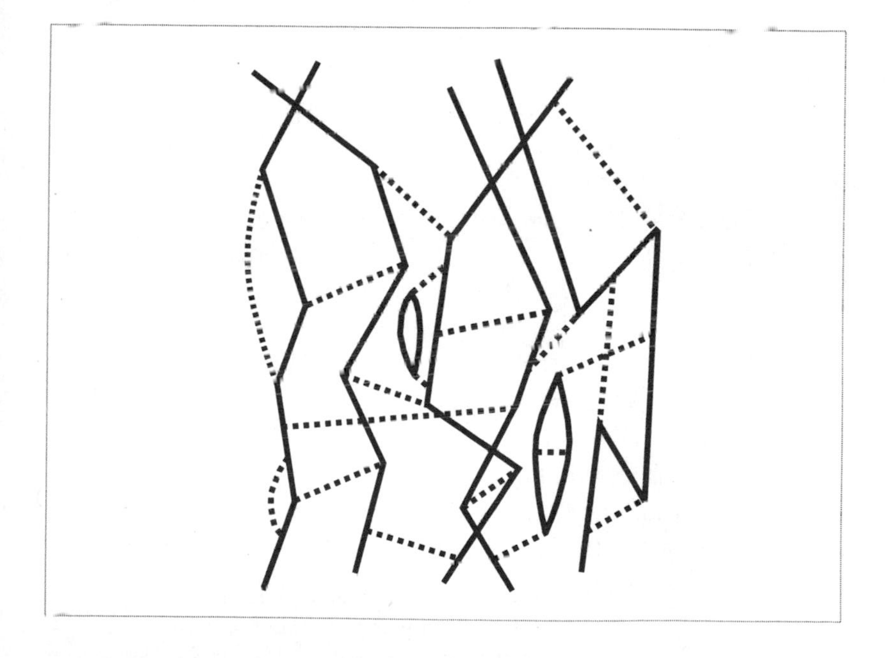

12. The electrostatic force is also known as the Coulomb force.

Feel free to add arrows to this picture in various ways to make the solid lines into electrons or positrons.

The Fine Structure Constant

The various equations and formulas of physics contain a variety of different numerical constants. Some of these constants are numbers derived from pure mathematics. An example is the number 3.14159 . . . , better known by its Greek name, π. We know the value of π to billions of decimal places, not by measuring it, but from its purely mathematical definition: π is defined to be the ratio of the circumference of a circle to its diameter. Other purely mathematical numbers, such as the square root of 2 and the number called e, may also be computed to endless precision if anyone were motivated to do it.

But other numbers that appear in physics equations have no special mathematical significance. We might call them empirical numbers. An example that is important in nuclear physics is the ratio of the proton mass to the neutron mass. Its numerical value is known to seven figures — 1.001378. The next digit cannot be obtained by mathematics alone. One must go into the lab and measure it. The most fundamental of these empirical numbers are crowned with the title "constants of nature." The fine structure constant is one of the most important constants of nature.[13] Like π, the fine structure is named for a Greek letter, in this case α (alpha). It is often approximated by the fraction $\frac{1}{137}$. Its precise value is known to a limited number of decimal places — 0.007297351 — but it is, nonetheless, one of the most accurately known physical constants.

The fine structure constant is an example of a quantity that physicists call *coupling constants*. Each coupling constant is associated with one of the basic events — the vertex diagrams — of quantum field theory. The coupling constant is a measure of the *strength*, or *potency*, of the event depicted by the vertex. In QED the sole vertex diagram is the emission of a photon by an electron. Let's consider more closely what happens when a photon is emitted.

13. The term *fine structure* has to do with the details of the atomic spectrum of hydrogen. The fine structure constant first made its appearance in the theory of the hydrogen spectrum.

One might begin by asking what determines the precise point at which an electron, as it moves through space-time, emits a photon. The answer is that nothing does — physics at the microscopic level is capricious. Nature has a random element that drove Einstein in his later years to distraction. He protested, "God does not play dice."[14] But whether Einstein liked it or not, nature is not deterministic. Nature has an element of randomness that is built into the Laws of Physics at the deepest level. Even Einstein couldn't change that. But if nature is not deterministic, neither is it completely chaotic. Here is where the principles of quantum mechanics enter. Unlike Newtonian physics, quantum mechanics never predicts the future in terms of the past. Instead, it provides very precise rules for computing the probability for various alternative outcomes of an experiment. Just as there is no way to predict the final location of a photon that has gone through a slit, there is also no way to predict exactly where along its path an electron will launch a photon or where another electron may absorb it. But there is a definite probability for these events.

The operation of a television screen provides a good illustration of such probabilities. The light coming from a TV screen is composed of photons that are created when electrons strike the screen. The electrons are ejected from an electrode at the rear of the set and are guided to the screen by electric and magnetic fields. But not every electron that hits the screen emits a photon. Some do. Most don't. Roughly speaking the probability that any particular electron will radiate a quantum of light is given by the fine structure constant α. In other words, only one lucky electron out of 137 emits a photon. That is the meaning of α: it is the probability that an electron, as it moves along its trajectory, will capriciously emit a photon.

Feynman didn't just draw pictures. He invented a set of rules for working out the probabilities of the complex processes depicted in the pictures. In other words, he discovered a precise mathematical calculus that predicts the probabilities for any process in terms of the simplest events — propagators and vertices. The probabilities for all processes in nature ultimately trace back to coupling constants like α.

14. Bohr's response was, "Einstein, don't tell God what to do."

The fine structure constant also controls the strength of the exchange diagrams, which in turn, determine the strength of electric forces among charged particles. It controls how tightly the atomic nucleus pulls the electrons toward it. As a consequence it determines how big the atom is, how fast the electrons move in their orbits, and ultimately controls the forces between different atoms that allow them to form molecules. But as important as it is, we don't know why its value is 0.007297351 and not something else. The Laws of Physics that were discovered during the twentieth century are very accurate and useful, but the underlying origin of those laws remains a mystery.

The theory of this simplified world of electrons, photons, and point nuclei is Quantum Electrodynamics, and Feynman's version of it was incredibly successful. Using his methods the properties of electrons, positrons, and photons were understood with astonishing accuracy. Moreover, if the simplified version of the nucleus was added, the properties of the simplest atom — hydrogen — could also be calculated to incredible precision. In 1965 Feynman, Julian Schwinger, and the Japanese physicist Sin-Itero Tomanaga won the Nobel Prize for their work on Quantum Electrodynamics. That was Act I.

If Act I was theater in the small, with only two characters, the play became an epic with hundreds of players in Act II. New particles were discovered throughout the 1950s and 1960s that eventually required an unruly cast including electrons, neutrinos, muons, tau particles, up-quarks, down-quarks, strange-quarks, charmed-quarks, bottom-quarks, top-quarks, photons, gluons, W- and Z-bosons, Higgs bosons, and a good deal more. Don't ever believe anyone who tells you that elementary-particle physics is elegant. This hodgepodge of particle names reflects an equally unwieldy jumble of masses, electric charges, spins, and other properties. But as messy as it is, we know how to describe it with enormous precision. The Standard Model is the name of the mathematical structure — a particular quantum field theory — that describes the modern theory of elementary particles. Although it is far more complicated than Quantum Electrodynamics, Feynman's methods are so powerful that, once again, they can be used to express everything in terms of simple pictures. The principles are exactly the same as those of QED: everything is built out of propagators, vertex diagrams, and coupling constants. But there are new actors and whole new plot lines, including one called QCD.

Quantum Chromodynamics

Many years ago I was invited to a famous university to give a series of lectures on a brand-new subject called Quantum Chromodynamics (QCD). As I was walking through the halls of the Physics Department on my way to the first lecture, I overheard a pair of graduate students discussing the title. One, who was looking at the lecture announcement on the bulletin board, said, "What's this all about? What is Quantum Chromodynamics?" The other fellow thought for a moment, then looked up and said, "Hmmm, it must be a new way to use quantum mechanics to develop your photographs."

Quantum Chromodynamics has nothing to do with photography or even light. QCD is the modern version of nuclear physics. Conventional nuclear physics begins with protons and neutrons (nucleons), but QCD goes a step deeper. It's been known for forty years that nucleons are not elementary particles. They are more like atoms or molecules except on a much smaller scale. If one could look into a proton with a sufficiently powerful microscope, one would see three quarks tied together by a string of particles called *gluons*. The theory of quarks and gluons — QCD — is a more complicated theory than QED, and I won't really be able to do it justice in a few pages. But the basic facts are not too difficult. Here is the cast of characters.

The Six Quarks

First there are the quarks: there are six distinct types of them. In order to distinguish one from another, physicists gave them meaningless, whimsical names: up-quark, down-quark, strange-quark, charmed-quark, bottom-quark, and top-quark, or more concisely, u-, d-, s-, c-, b-, and t-quarks. There is, of course, nothing stranger about the strange-quark or more charming about the charmed-quark, but the silly names give them a bit of personality.

Why are there six quark types and not four or two? Who knows? A theory with four or two types of quarks is every bit as consistent as one with six types. What we do know is that the mathematics of the Standard Model requires the quarks to come in pairs — up with down, charmed with strange, and top with bottom. But the reason for the threefold replication of the simplest theory — one with only up- and down-quarks — is a complete mystery. To make matters worse, only the up- and down-quarks play any essential role in ordinary nuclei.[15] If QCD were an engineering project, the rest of the quarks would be considered an extravagant squandering of resources.

Quarks are in some ways similar to electrons, although somewhat heavier, and they have peculiar electric charges. In order to have a basis for comparison, the charge of the proton is traditionally taken to be one (+1). The electron's charge is equally big but opposite in sign (−1). Quarks, on the other hand, have charges that are fractions of the proton charge. In particular, the charges of the u-, c-, and t-quarks are positive like the proton's but only two-thirds as big (⅔). The d-, s-, and b-quarks have negative charges equal to one-third the electron charge (−⅓).

Both protons and neutrons contain three quarks. In the case of the proton, there are two u-quarks and a single d-quark. Adding up the electric charges of these three quarks, the result is the charge of the proton:

$$\frac{2}{3}+\frac{2}{3}-\frac{1}{3}=1.$$

The neutron is very similar to the proton, the difference being that the up- and down-quarks are interchanged. Thus, the neutron contains two d-quarks and one u-quark. Again adding the three charges, we find the neutron has (as expected) no electric charge:

$$\frac{2}{3}-\frac{1}{3}-\frac{1}{3}=0.$$

What would happen if we tried to build a proton, or something like a proton, by substituting a strange-quark for the down-quark? Such objects do exist — they are called *strange particles* — but they don't exist anywhere except in physics laboratories. Even in such laboratories strange

15. The existence of the strange-quark has a minor effect on the properties of nucleons, but the others are of no importance.

particles are fleeting occurrences that last for no more than a tiny fraction of a second before disintegrating by a kind of radioactivity. The same is true of particles containing charmed-, bottom-, or top-quarks. Only up- and down-quarks can be assembled into stable, long-lasting objects. As I said, if the strange-, charmed-, bottom-, and top-quarks were suddenly removed from the list of elementary particles, hardly anyone would notice.

What about quarks that go backward in time? Like electrons, each kind of quark has its antiparticle. They can be assembled into antiprotons and antineutrons. At one time very early in the history of the universe, when the temperature was billions of degrees, antinucleons were about as abundant as ordinary nucleons. But as things cooled, the antiparticles almost entirely disappeared, leaving only the ordinary protons and neutrons to form the nuclei of atoms.

The Gluon

Nucleons are like tiny atoms made of quarks. But quarks alone would be powerless to bind themselves into nucleons. Like the atom, they require another ingredient to create the forces of attraction that "glue" them together. In the case of the atom, we know exactly what the glue is. The atom is held from flying apart by photons continuously being juggled back and forth between electrons and nuclei. But the force generated by photon exchange is far too weak to bind quarks into the tightly knit structure of a nucleon (remember that nucleons are 100,000 times smaller than atoms). Another particle with more potent properties is needed to pull quarks together that tightly. That particle is the aptly named gluon.

The basic events in any quantum field theory are always the same: the emission of particles by other particles. The Feynman diagrams describing these events always have the same form: vertex diagrams shaped like the letter Y. The basic vertex diagram for QCD looks exactly like the photon-emission vertex with a quark replacing the electron and a gluon taking the place of the photon.

Not surprisingly, the origin of the forces that bind quarks into protons and neutrons is the exchange of gluons. But there are two big differences between QED and QCD. The first is a quantitative difference:

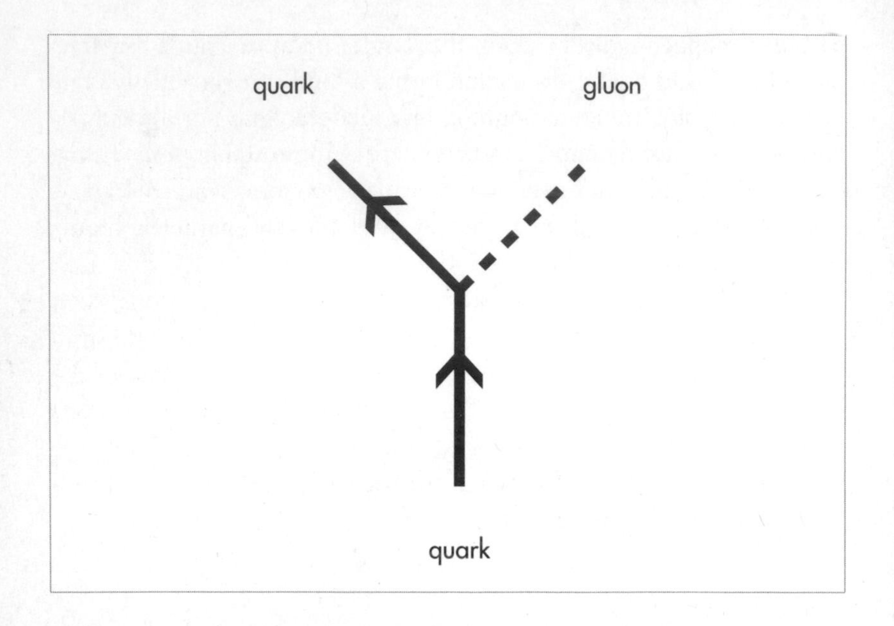

the numerical constant governing the emission of gluons is not as small as the fine structure constant. It is called α_{QCD} (alpha-QCD) and is about one hundred times larger than the fine structure constant. This is the reason that the force between quarks is much stronger than the electric force that acts in the atom. QCD is sometimes called the theory of the *strong interactions*.

The second difference is qualitative. It causes gluons to become a sticky substance that always reminds me of the tale of the tar baby. Once upon a time, according to southern folklore, Brer Rabbit[16] came upon a tar baby sitting in the road, minding his own business. Brer Rabbit said, "Mornin'." The tar baby said nothing. Brer Rabbit got offended. One thing led to another, and soon an altercation ensued — Brer Rabbit got mad and punched tar baby, but that was a huge mistake. Stuck with his fist in the tar, Brer Rabbit pulled and pulled, but the tar just stretched and pulled him back. As hard as he struggled to free him himself, tar baby just wouldn't let go.

Why the tar baby story? Because quarks are miniature tar babies, but only to other quarks. They are permanently glued together by a tarlike,

16. Brother Rabbit to Yankees.

stringy substance made of gluons. The origin of this strange behavior is one extra vertex that has no analog in QED. Any electrically charged particle can emit a photon. But photons themselves are not charged. They are electrically neutral and, therefore, will not emit another photon. In this respect gluons are very different from photons. The laws of QCD require a vertex in which a gluon splits into two gluons, each proceeding on one of the paths of the fork.

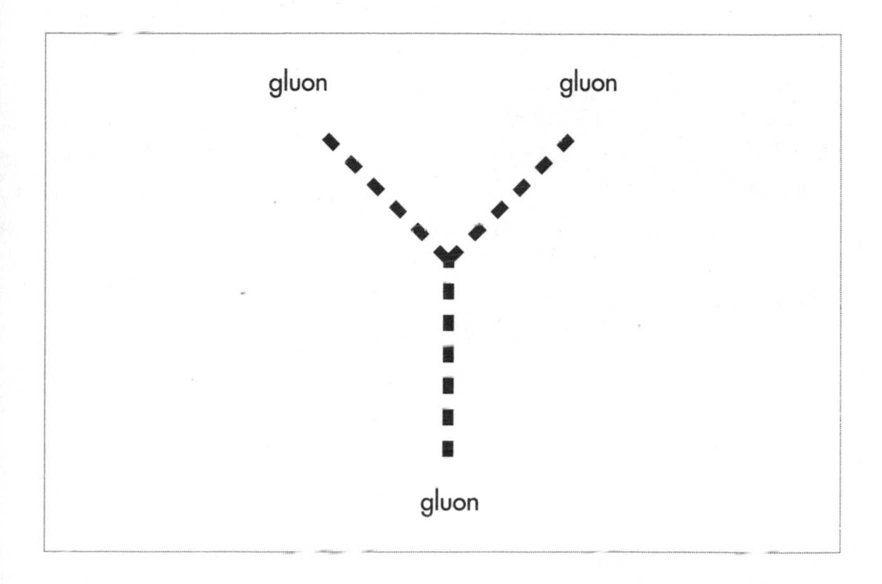

This is the big difference between QCD and QED that makes QCD a far more complicated theory than its electric counterpart. Among other things, it implies that gluons can exchange gluons and bind into objects called glueballs — particles with no quarks or electrons at all. In fact gluons don't stick together only in pairs. They can form long chains of stringy glue. Earlier I likened the electrons in an atom to balls being swung on ropes. In that case the rope was entirely metaphorical. But in the case of quarks, the strings holding them together are quite real. They are strings of gluons strung out between the quarks. In fact when a quark is forcibly ejected from a nucleon, a long string of gluons forms and eventually prevents the quark's escape.

The Weak Interactions

If you are getting weary of learning particle physics, I don't blame you. It's too complicated with too many things to remember. There are too many particles to keep track of and no good reason that we know of for their existence. QCD and QED hardly exhaust the pieces that make up the Standard Model. The whole thing is a very far cry from the elegant simple theory that physicists expected to find "at the bottom of it all." It's a good deal more like zoology or botany. But that's just the way it is. We can't change the facts.

I am going to guide you through one more piece of the Standard Model: that part known as the *weak interactions*. Like QED and QCD the weak interactions play an important role in explaining our own existence, although the reasons are subtler and won't become clear until later chapters.

The history of the weak interactions dates back to the very end of the nineteenth century, when the French physicist Antoine-Henri Becquerel discovered radioactivity. Becquerel's discovery preceded J. J. Thomson's discovery of the electron by one year.

Radioactivity comes in three different kinds, called alpha, beta, and gamma. They correspond to three very different phenomena, only one of which (beta) has to do with the weak interactions. Today we know that the *beta rays* coming from Becquerel's uranium sample were actually electrons emitted by neutrons in the uranium nucleus. Upon emitting the electron, the neutron immediately turns into a proton.

Nothing in either QED or QCD explains how a neutron can emit an electron and become a proton. The simplest explanation, which may already have occurred to you, is that there is an additional vertex diagram to add to our list of fundamental events. The vertex would involve an initial neutron coming to a fork in the road, whereupon a proton would continue on one path and an electron on the other. But this is not the correct explanation. The fact is that a new character is about to make its entrance — the neutrino. What Becquerel didn't know was that another particle flew off when the neutron decayed, namely, the antiparticle of the ghostly neutrino.

The Neutrino

The neutrino is similar to the electron but without electric charge. Think of it as an electron that has lost its electrical properties. In a way the relationship between the electron and the neutrino is similar to that between the proton and neutron.

What's left of the neutrino then? It has a tiny bit of mass but not much else. It doesn't emit photons. It doesn't emit gluons. This means it has none of the forces acting on it that electrically charged particles or quarks experience. It doesn't bind itself to other particles to form more complex objects. It hardly does anything. In fact the neutrino is such a loner that it will pass through light-years of lead without even getting deflected. But it's not a complete zero. To understand how the neutrino gets into the act, yet another actor has to be introduced — the W-boson.

The W-Boson

For the moment don't worry about the word *boson*. For now, it simply denotes another particle, similar in its properties to the photon or gluon but electrically charged. It comes in two versions, the positively charged W and the negatively charged W. They are, of course, each other's antiparticles.

The W-boson is the key to the neutrino's activities. Not only can electrons and quarks emit W-bosons — so, too, can the neutrino. Here is a (partial) list of the W-boson's activities:

- Electrons emit W-bosons and turn into neutrinos
- Up-quarks emit W-bosons and become down-quarks
- Up-quarks emit W-bosons and become strange-quarks
- Charmed-quarks emit W-bosons and become strange-quarks
- Top-quarks emit W-bosons and become bottom-quarks
- Higgs bosons emit Z-bosons

There is more, but it involves particles that we will meet only in later chapters.

As I have explained, protons and neutrons are not on the list of elementary particles because they are composed of the simpler quarks, but for some purposes it is useful to forget the quarks and think of nucleons as elementary particles. This will require us to add some additional vertices. For example, a proton can emit a photon. (In reality one of the hidden quarks produced the photon, but the net effect is as if the proton did it.) In a similar way, one of the two d-quarks in a neutron can emit a W-boson and become a u-quark, thus turning the neutron into a proton. In effect there is a vertex in which a neutron becomes a proton while emitting a W-boson.

Now we are ready to draw the Feynman diagram that explains the beta rays that Becquerel discovered emanating from his uranium. The diagram looks a good deal like a QED diagram except that the W-boson is exchanged, where in a QED diagram, the photon would be exchanged. Indeed, the weak interactions are very closely related to the electric forces due to photons.

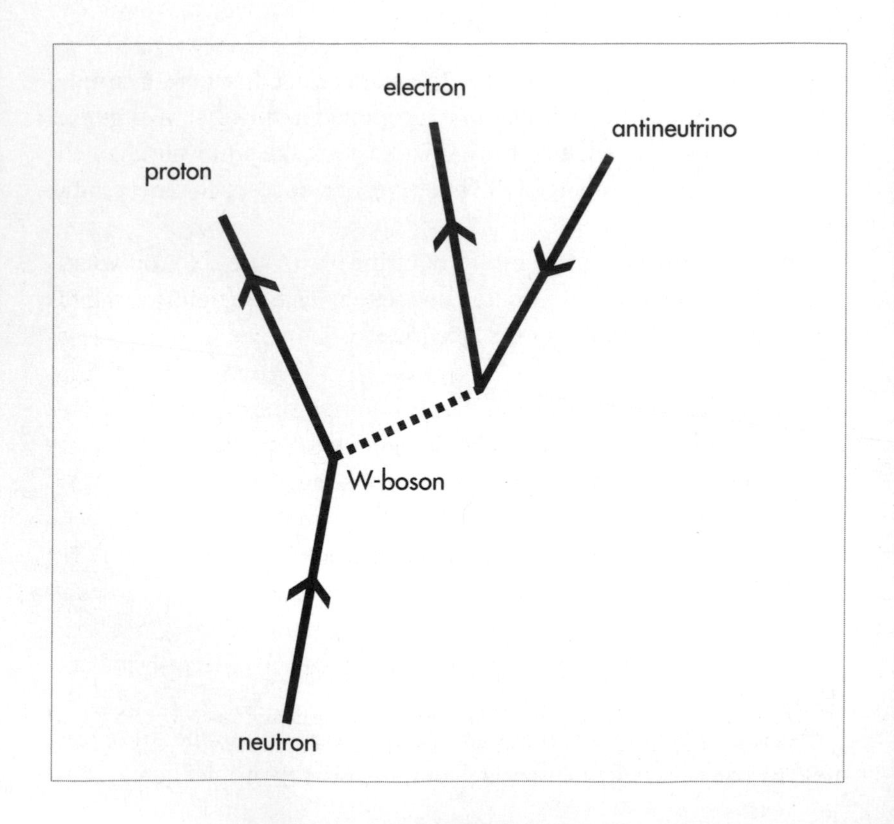

Take your slit cardboard square and start at the bottom. The neutron (which might be inside a nucleus) emits a negatively charged W-boson and becomes a proton. The W-boson goes a short way (about 10^{-16} centimeters) and splits into two particles: an electron and a neutrino "moving backward in time," or more mundanely, an antineutrino. That's what Becquerel would have seen in 1896 had he a powerful enough microscope. Later we will see the importance of this kind of process in creating the chemical elements of which we are made.

The Laws of Physics

You should now have a clear idea of what I mean by the Laws of Physics. I wish I could tell you that they are elegant, as some physicists would claim. But the unvarnished truth is that they are not. There are too many particles, too many vertex diagrams, and too many coupling constants. I haven't even told you yet about the random collection of masses that characterize the particles. The whole thing would be a very unappealing concoction if it were not for one thing: it describes the properties of elementary particles, nuclei, atoms, and molecules with incredible precision.

But there is a cost. It can be accomplished only by introducing about thirty "constants of nature" — masses and coupling constants — whose values have no other justification than that they "work."[17] Where do these numbers come from? Physicists don't pull the various numbers from thin air or even from mathematical calculation in some master theory. They are the results of many years of experimental particle physics done at accelerator laboratories in many countries. Many of them, like the fine structure constant, have been measured with great precision, but the bottom line is, as I said, that we don't understand why they are what they are.

The Standard Model is the culmination and distillation of more than half a century of particle physics. When combined with Feynman's graphical rules, it provides accurate descriptions of all elementary-

17. Thirty is an absolute minimum that does not include the numerical parameters that are required for cosmology or for various extensions of the Standard Model. Including these additional constants, their number easily grows to more than one hundred.

particle phenomena, including the ways particles combine to form nuclei, atoms, molecules, gases, liquids, and solids. But it is far too complicated to be the paragon of simplicity that we hope would be the hallmark of a truly fundamental theory — a final theory — of nature.

Unlike the laws of men, the Laws of Physics really are laws. We can choose to obey the law or disregard it, but an electron has no choice. These laws are not like the traffic laws or the tax laws that change from state to state and year to year. Perhaps the most important experimental fact, a fact that makes physics possible altogether, is that the constants of nature really are constant. Experiments at different times and places require exactly the same Feynman diagrams and give exactly the same values for each coupling constant and mass. When the fine structure constant was measured in Japan in the 1990s, it had exactly the same value as it had in Brookhaven, Long Island, in 1960 or at Stanford in the 1970s.

Indeed, when physicists are pursuing the study of cosmology, they tend to take completely for granted that the laws of nature are the same everywhere in the universe. But it needn't be so. One can certainly conceive of a world in which the fine structure constant changes as time goes on or in which some other constant varies from one location to another. From time to time physicists have questioned the assumption that the constants are absolutely constant, but thus far strong evidence suggests that they really are the same in every part of the *observed universe:* not the gigantic megaverse, but that part of the universe that we can see with the various kinds of telescopes at our disposal.

Someday we may be able to go to distant galaxies and measure the constants directly in those places. But even now we continually receive messages from remote regions of the universe. Astronomers routinely study the light from far-off sources and disentangle the spectral lines that were emitted or absorbed by distant atoms.[18] The relations between individual spectral lines are intricate, but they are always the same, no matter where and when the light originated. Just about any change in the local Laws of Physics would change the details, so we have excellent evidence that the laws are the same in all parts of the observed universe.

18. Spectral lines are discussed in chapter 4.

These rules — a list of particles, a list of masses and coupling constants, and Feynman's methods — that I call the Laws of Physics are extremely powerful. They govern almost every aspect of physics, chemistry, and ultimately, biology, but the rules do not explain themselves. We have no theory that tells us why the Standard Model is the right thing and not something else. Could other things have been the Laws of Physics? Might the list of elementary particles, the masses, and the coupling constants be different in other parts of the universe that we cannot observe? Might the Laws of Physics be different in very distant times and places? If so, what governs the way they change? Are there deeper laws that tell us what laws are possible and what are not? These are the questions that physicists are starting to grapple with at the beginning of the twenty-first century. They are the questions about which *The Cosmic Landscape* is concerned.

One thing may be puzzling you about this chapter. I haven't once mentioned the most important force in the universe — the force of gravity. Newton discovered the elementary theory of gravity that bears his name. Einstein also delved deeply into the meaning of gravity in the General Theory of Relativity. Even though the laws of gravity are far more important in determining the fate of the universe than all the others, gravity is not considered to be part of the Standard Model. The reason is not that gravity is unimportant. Of all the forces of nature, it will play the largest role in this book. My reason for separating it from the other laws is that the relationship between gravity and the microscopic world of quantum mechanical elementary particles is not yet understood. Feynman himself tried to apply his methods to gravity and gave up in disgust. In fact he once advised me to never get involved in the subject. That was like telling a small boy to stay out of the cookie jar.

In the next chapter I will tell you about the "mother of all physics problems." It is a grim tale of what goes wrong when gravity is combined with these Laws of Physics. It is also a tale of extreme violence. The Laws of Physics as we have understood them predict an extraordinarily lethal universe. Evidently we are missing something.

CHAPTER TWO

The Mother of All Physics Problems

New York City, 1967

I first learned about the "mother of all physics problems" one crisp fall day in New York City in an unlikely place: Washington Heights. Located three miles north of Columbia University, the Heights is part of Manhattan but in many ways resembles the South Bronx, where I grew up. Once, it had been a middle-class Jewish neighbourhood, but most of the Jews had left and were replaced by Latin Americans, especially working-class Cubans. It was a great neighborhood for inexpensive Cuban restaurants. My favorite was a Cuban-Chinese place.

People familiar with the area know that there is a group of unusual Byzantine-looking buildings on Amsterdam Avenue at about 187th Street. The streets nearby are filled with young Orthodox Jewish students and rabbis; the local student hangout at that time was a falafel joint called MacDovid's. The odd buildings are the campus of Yeshiva University, the oldest Jewish institution of higher learning in the United States. It specializes in the education of rabbis and Talmudic scholars, but in 1967, it also had a graduate school of physics and mathematics called the Belfer Graduate School of Science.

I had just come from a year of postdoctoral work in Berkeley to be an assistant professor in the Belfer School. The exotic Yeshiva buildings didn't look anything like the Berkeley or Harvard campus, or any other campus for that matter. Finding the Physics Department would be a challenge. A bearded fellow on the street directed me toward the top of one building, where there was a turret or onion dome of some kind.

This didn't look promising, but it was the only job I had, so I entered and climbed the spiral staircase. At the top was an open door to a very small, dark office containing a massive bookcase filled with large, leather-bound volumes, all of whose titles were written in Hebrew. In the office sat a rabbinical-looking, gray-bearded gentleman reading some ancient tome. The sign said:

Physics Department
Professor Posner

"Is this the Physics Department?" I asked, uncomprehendingly.

"Yes it is," he said, "and I am the physics professor. Who are you?"

"I'm the new guy, the new assistant professor, Susskind." A kindly but very puzzled look came over his face.

"Oy vey, they never tell me anything. What new guy?"

"Is the chairman here?" I sputtered.

"I am the chairman. In fact I am the only physics professor, and I don't know anything about a new guy." At that time I was twenty-six years old with a wife and two small children, and it was beginning to look like I might be unemployed.

Confused and embarrassed, I slunk out of the building and started to cross the street, when I saw a guy I knew from college named Gary Gruber. "Hey Gruber, what's going on here? I just came from the Physics Department. I thought it was full of physicists, but there seems to be only an old rabbi named Posner."

Gruber found this much more amusing than I did. He laughed and said, " I think you probably want the graduate department, not the undergraduate. It's around the corner on One hundred eighty-fourth Street. I'm a graduate student there." Sweet relief! I walked over to 184th and looked on the side of the street that Gruber had indicated, but I could see nothing that looked like a graduate school of science. The street was just a row of seedy-looking storefronts. One of them advertised: "Abogado — Bail Bonds." Another was empty and boarded up. The biggest storefront was an establishment of the kind that caters bar mitzvahs and Jewish weddings. It looked like it was no longer in business, but a small establishment that prepared kosher food was still in the basement. I passed it once, but on the second pass, I looked a little closer. Sure enough, a small sign next to the caterer's said:

Belfer Graduate School

and pointed up a broad flight of stairs. The stairs had an old stained and worn carpet, and from the lower floor the smell of food floated up. I wasn't sure if I liked the look of this place any better than the last. I climbed up to a big room that I could see was once a ballroom for weddings and bar mitzvahs. Now it was a large space with sofas, comfortable chairs, and much to my relief, blackboards. Blackboards meant physicists.

Surrounding the space were about twenty offices. The entire school was housed in this one-time ballroom. It would have been very depressing except that several people were having a lively physics conversation at one of the boards. What's more, I recognized some of them. I saw Dave Finkelstein, who had arranged my new job. Dave was a charismatic and brilliant theoretical physicist who had just written a paper on the use of topology in quantum field theory that was to become a classic of theoretical physics. I also saw P. A. M. Dirac, arguably the greatest theoretical physicist of the twentieth century after Einstein. Dave introduced me to Yakir Aharonov, whose discovery of the Aharonov-Bohm effect had made him famous. He was talking to Roger Penrose, who is now Sir Roger. Roger and Dave were two of the most important pioneers in the theory of black holes. I saw an open door with a sign that said Joel Lebowitz. Joel, a very well-known mathematical physicist, was arguing with Elliot Lieb, whose name was also familiar. It was the most brilliant collection of physicists that I had ever seen assembled in one place.

They were talking about vacuum energy. Dave was arguing that the vacuum was full of zero-point energy and that this energy ought to affect the gravitational field. Dirac didn't like vacuum energy because whenever physicists tried to calculate its magnitude, the answer would come out infinite. He thought that if it came out infinite, the mathematics must be wrong and that the right answer is that there is no vacuum energy. Dave pulled me into the conversation, explaining as he went. For me this conversation was a fateful turning point — my introduction to a problem that would obsess me for almost forty years and that eventually led to *The Cosmic Landscape.*

The Worst Prediction Ever

The part of the mind — I guess we call it the ego — that gets pleasure from being proved right is especially well developed in theoretical physicists. To make a theory of some phenomenon followed by a clever calculation and then finally to have the result confirmed by an experiment provides a tremendous source of satisfaction. In some instances the experiment takes place before the calculation, in which case it's not predicting but, rather, explaining a result, and it's almost as good. Even very good physicists now and then make wrong predictions. We tend to forget about them, but one wrong prediction just will not go away. It is by far the all-time worst calculation of a numerical result that any physicist ever made. It was not the work of any one person, and it was so wrong that no experiment was ever needed to prove it so. The problem is that the wrong result seems to be an inevitable consequence of our best theory of nature, quantum field theory.

Before I tell you what the quantity is, let me tell you how wrong the prediction is. If the result of a calculation disagrees with an experiment by being 10 times too large or too small, we say that it was off by one order of magnitude; if wrong by a factor of 100, then it's two orders of magnitude off; a factor of 1,000, three orders; and so on. Being wrong by one order of magnitude is bad; two orders, a disaster; three, a disgrace. Well, the best efforts of the best physicists, using our best theories, predict Einstein's cosmological constant incorrectly by 120 orders of magnitude! That's so bad that it's funny.

Einstein was the first to get burned by the cosmological constant. In 1917, one year after the completion of the General Theory of Relativity, Einstein wrote a paper that he later regretted as his worst mistake. The paper, titled "Cosmological Considerations on the General Theory of Relativity," was written a few years before astronomers understood that the faint smudges of light called nebulae were actually distant galaxies. It was still twelve years till the American astronomer Edwin Hubble would revolutionize astronomy and cosmology, demonstrating that the galaxies are all receding away from us with a velocity that grows with distance. Einstein didn't know in 1917 that the universe was expanding. As far as he or anyone else knew, the galaxies were stationary, occupying the same location for all eternity.

According to Einstein's theory the universe is *closed and bounded,* which first of all means that space is finite in extent. But it doesn't mean that it has an edge. The surface of the earth is an example of a closed-and-bounded space. No point on earth is more than twelve thousand miles from any other point. Moreover, there is no edge to the earth: no place that represents the boundary of the world. A sheet of paper is finite, but it has an edge: some people would say four edges. But on the earth's surface, if you keep going in any direction, you never come to the end of space. Like Magellan you will eventually come back to the same place.[1]

We often say that the earth is a sphere, but to be precise, the term *sphere* refers only to the surface. The correct mathematical term for the solid earth is a ball. To understand the analogy between the surface of the earth and the universe of Einstein, you must learn to think *only* of the surface and not the solid ball. Let's imagine creatures — call them flatbugs — that inhabit the surface of a sphere. Assume that they can never leave that surface: they can't fly, and they can't dig. Let's also assume that the only signals they can receive or emit travel along the surface. For example, they might communicate with their environment by emitting and detecting surface waves of some kind. These creatures would have no concept of the third dimension and no use for it. They would truly inhabit a two-dimensional closed-and-bounded world. A mathematician would call it a *2-sphere,* because it is two-dimensional.

We are not flatbugs living in a two-dimensional world. But according to Einstein's theory, we live in a three-dimensional analog of a sphere. A three-dimensional closed-and-bounded space is more difficult to picture, but it makes perfect sense. The mathematical term for such a space is a *3-sphere.* Just like the flatbugs, we would discover that we live in a 3-sphere by traveling out along any direction and eventually finding that we always return to the starting point. According to Einstein's theory, space is a 3-sphere.

In fact spheres come in every dimension. An ordinary circle is the simplest example. A circle is one-dimensional like a line: if you lived on

1. Actually, Magellan never did arrive back in Europe. He was killed in the Philippines. But some of his crew did manage to circumnavigate the globe, thus proving that it was a sphere.

it, you could move only along a single direction. Another name for a circle is a *1-sphere.* Moving along the circle is much like moving along a line except that you come back to the same place after a while. To define a circle, start with a two-dimensional plane and draw a closed curve. If every point on the curve is the same distance from a central point (the center), the curve is a circle. Notice that we began with a two-dimensional plane in order to define the 1-sphere.

The 2-sphere is similar except that you begin with three-dimensional space. A surface is a 2-sphere if every point is the same distance from the center. Perhaps you can see how to generalize this to a 3-sphere or, for that matter, a sphere of any dimension. For the 3-sphere we begin with a four-dimensional space. You can think of it as a space described by four coordinates instead of the usual three. Now just pick out all the points that are at a common distance from the origin. Those points all lie on a 3-sphere.

Just as the flatbugs living on the two-sphere had no interest in anything but the surface of the sphere, the geometer studying a 3-sphere has no interest in the 4-dimensional space in which it is embedded. We can throw it away and concentrate only on the 3-sphere.

Einstein's cosmology involved a space that has the overall shape of a 3-sphere, but like the earth's surface, the spherical shape is not perfect. In the General Theory of Relativity, the properties of space are not rigidly fixed. Space is more like the deformable surface of a rubber balloon than the surface of a rigid steel ball. Picture the universe as the surface of such a giant, deformable balloon. Flatbugs live on the rubber surface, and the only signals they receive propagate along that surface. They know nothing of the other dimension of space. They have no concept of the interior or exterior of the balloon. But now their space is flexible, and the distance between points can change with time as the rubber stretches.

On the balloon are markings indicating the galaxies, which more or less cover the balloon uniformly. If the balloon expands, the galaxies move apart. If it shrinks, the galaxies move closer. All of this is fairly easy to understand. The hard part is the jump from two to three dimensions. Einstein's theory describes a world in which space is flexible and stretchable but has the overall shape of a 3-sphere.

Now let's add the element of gravitational attraction. According to both Newton's and Einstein's theories of gravity, every object in the universe attracts every other object with a force proportional to the product of their masses and inversely proportional to the square of the distance between them. Unlike electric forces, which are sometimes attractive and sometimes repulsive, gravity is *always* attractive. The effect of gravitational attraction is to pull the galaxies together and shrink the universe. A similar effect exists on the surface of a real balloon, namely, the tension in the rubber that tries to shrink the balloon. If you want to see the effect of tension, just stick a pin in the rubber.

Unless some other force counteracts gravitational attraction, the galaxies should start to accelerate toward one another, collapsing the universe like a punctured balloon. But in 1917 the universe was thought to be static — unchanging. Astronomers, like ordinary people, looked at the sky and saw no motion of the distant stars (apart from that due to the earth's motion). Einstein knew that a static universe was impossible if gravity was universally attractive. A static universe is like a stone, hovering above the surface of the earth, completely motionless. If the stone were thrown vertically upward, then a momentary glance might see it ascending or descending. You might even catch it at the precise instant when it was turning around. What the stone cannot do is to just eternally hover at a fixed height. That is, not unless some other force is acting on the stone opposing the gravitational attraction to the earth. In exactly the same way, a static universe defies the universal law of gravitational attraction.

What Einstein needed was a modification of his theory that would provide a compensating force. In the case of the balloon, the air pressure from inside is the force that counteracts the tension in the rubber. But the real universe doesn't have an inside with air in it. There is only the surface. So Einstein reasoned that there must be some kind of repulsive force to counteract the gravitational pull. Could there be a hidden possibility of a repulsive force in the General Theory of Relativity?

Examining his equations, Einstein discovered an ambiguity. The equations could be modified without destroying their mathematical consistency by adding one more term. The meaning of the additional term was surprising: it represented an addition to the usual laws of gravity —

a gravitational force that became increasingly strong with distance. The strength of this new force was proportional to a new constant of nature that Einstein denoted by the Greek letter λ (lambda). Ever since, the new constant has been called the cosmological constant, and it continues to be denoted by λ.

What had especially caught Einstein's attention was that if λ were chosen to be a positive number, then the new term corresponded to a universal repulsion that increased in proportion to the distance. Einstein realized he could play off the new repulsive force against the usual gravitational attraction. The galaxies could be kept in equilibrium at a separation that could be controlled by choosing the magnitude of the new constant, λ. The way that it worked was simple. If the galaxies were closely spaced, their attraction would be strong and an equally strong repulsion would be needed to keep them in equilibrium. On the other hand, if the distance between galaxies were so large that they barely felt each other's gravitational fields, then only a weak repulsion would be needed. Thus Einstein argued that the size of the cosmological constant should be closely connected to the average distance between the galaxies. Although from a mathematical perspective the cosmological constant could be anything, it could be easily determined if one knew the average distance separating galaxies. In fact at that time, Hubble was busy measuring the distance between galaxies. Einstein believed that he had the secret of the universe. It was a world kept in balance by competing attractive and repulsive forces.

Many things are wrong with this theory. From the theoretical point of view, the universe that Einstein had built was unstable. It was in equilibrium but *unstable equilibrium*. The difference between stable and unstable equilibrium is not hard to understand. Think of a pendulum. When the pendulum is vertical and the bob is at its low point, the pendulum is in stable equilibrium. This means that if you disturb it a little, for example, by giving it a slight push, it will return to its original position.

Now imagine turning the pendulum upside down so that the bob is delicately balanced in the straight-up position. If it is disturbed ever so slightly, perhaps by nothing more than the breeze from a butterfly's wing, the disturbance will build up, and the pendulum will fall over. Moreover, the direction in which it falls is very unpredictable. Einstein's static universe was like the unstable upside-down pendulum. The slightest perturbation would either cause it to explosively grow or implode it like a popped balloon. I don't know whether Einstein missed this elementary point or if he just decided to ignore it.

But the worst thing about the theory was that it was trying to explain something that was just not true. Ironically, there was no need for the new term. Hubble, working with the hundred-inch telescope on Mount Wilson in Southern California, discovered that the universe was not standing still.[2] The galaxies were flying apart from one another, and the universe was expanding like an inflating balloon. The forces did not need to cancel, and the cosmological term, which added nothing to the beauty of the equations, could be discarded by setting it to zero.

2. One hundred inches sounds like a modest size for a telescope, but it refers only to the diameter of the light-gathering mirror not the overall size of the instrument. In fact the Mount Wilson telescope was the largest in the world until the 200-inch Mount Palomar telescope was completed, in 1949.

But Pandora's box, once opened, could not be closed so easily.

The cosmological constant is equivalent to another term that may be easier to picture: the *vacuum energy.*[3] You'll recall this term from the argument I first encountered at the Belfer School. Vacuum energy sounds like an oxymoron. The vacuum is empty space. By definition it is empty, so how can it have any energy? The answer lies in the weirdness brought to the world by quantum mechanics, the weird uncertainty, the weird granularity, and the weird incessant jitteriness. Even empty space has the "quantum jitters." Theoretical physicists are used to thinking of the vacuum as being full of particles flickering into and out of existence so quickly that we cannot detect them under normal circumstances. These vacuum fluctuations are like very high-frequency noise that is way beyond the ability of the human ear to detect. But vacuum fluctuations do have an effect on atoms, which like dogs, are much better tuned to the high frequencies. The precise energy levels of the hydrogen atom can be measured to exquisite accuracy, and the results are sensitive to the presence of the fluctuating sea of electrons and positrons in the vacuum.

These strange violent fluctuations of the vacuum are consequences of quantum field theory and can be visualized using Feynman's intuitive diagrams. Imagine completely empty space-time with not a single particle initially. Quantum fluctuations can create particles for a short period of time, as in the following figures.

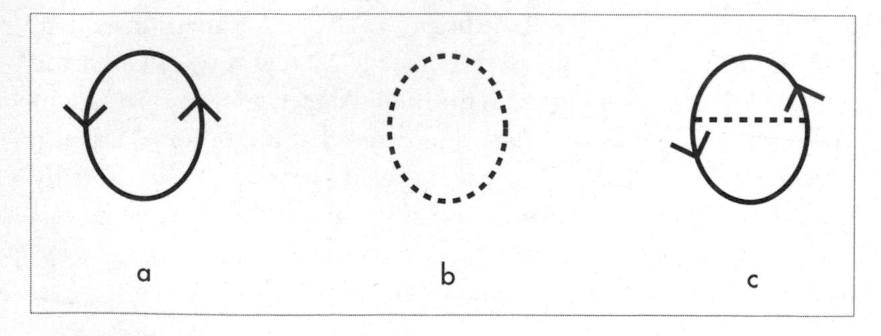

The first diagram shows an electron and a positron spontaneously created from nothing and then annihilating each other when they come together. You can also think of it as an electron going around a closed

3. The third term preferred by the press is *dark energy.*

loop in space-time, the positron being the electron moving backward in time. The second diagram shows two photons spontaneously created and then annihilated. The last diagram is like the first except that a photon hops between the electron and positron before they disappear. An infinite number of more and more complex "vacuum diagrams" are possible, but these three are more or less representative.

How long do the electrons and positrons last? About a billion-trillionth of a second. Next imagine these diagrams occurring all over space-time, filling it with a rapidly fluctuating population of elementary particles. These short-lived quantum particles that fill the vacuum are called *virtual particles*, but their effects can be quite real. In particular, they cause the vacuum to have energy. The vacuum is not the state of zero energy. It is merely a state of *minimum energy*.

Back to the Cosmological Constant

Now a clever reader might ask: "Who cares if the vacuum has energy? If that energy is always present, why don't we just readjust our definition of energy by subtracting it away?" The reason is that *energy gravitates*. To understand the meaning of this phrase, you need to remember two easy pieces of physics. The first is (I promised no equations, but I think I'll be excused for this one) $E = mc^2$. Even schoolchildren know this famous formula that expresses the equivalence between mass and energy. Mass and energy are really the same thing. They are just expressed in different units; to convert from mass to energy, you multiply by the square of the speed of light.

The second easy piece is Newton's law of gravity, slightly rephrased in this form: "Mass is the source of the gravitational field." This is a way of saying that the presence of a mass, like the sun, affects the motion of nearby objects. We can say that either the sun affects the motion of the earth or, in fancier terms, the sun creates a gravitational field, which in turn influences the motions of other objects like the planets.

Quantitatively, Newton's law tells us that the magnitude of the sun's field is proportional to the sun's mass. If the sun were one hundred times heavier, its field would be one hundred times stronger, and the force on the earth would be one hundred times greater. That's what it means to say, "Mass is the source of the gravitational field."

But if energy and mass are the same thing, then this sentence could also be read: "Energy is the source of the gravitational field." In other words, all forms of energy affect the gravitational field and, therefore, also influence the motion of nearby masses. The vacuum energy of quantum field theory is no exception. Even empty space will have a gravitational field if the energy density of the vacuum is not zero. Objects will move through empty space as if there were a force on them. The interesting thing is that if the vacuum energy is a positive number, then its effect is a universal repulsion, a kind of antigravity that would tend to drive galaxies apart. This is exactly what we said of the cosmological constant earlier, you will recall.

This point is so important that I want to stop and explain it again. If, in fact, empty space is filled with vacuum energy (or vacuum mass) it will exert forces on objects that are indistinguishable from the effects of Einstein's cosmological constant. Einstein's misbegotten child is nothing but the energy content of the fluctuating quantum vacuum. In deciding to eliminate the cosmological constant from his equations, Einstein was, in effect, claiming that there really is no vacuum energy. But from a modern perspective, we have every reason to believe that the quantum jitters inevitably give rise to energy in empty space.

If there really is a cosmological constant, or vacuum energy, then there are severe limits on its magnitude. If it were too big, it would lead to detectable distortions of the trajectories of astronomical bodies. The cosmological constant, if not zero, must be very small indeed. The problem is that once we identify the cosmological constant with vacuum energy, nobody has any idea why it should be zero or even small. Evidently, combining the theory of elementary particles with Einstein's theory of gravity is a very risky thing to do. It seems to lead to an unpromising universe with a cosmological constant many orders of magnitude too big.

Every kind of elementary particle is present in the violently fluctuating sea of virtual particles called the vacuum. In this sea are electrons, positrons, photons, quarks, neutrinos, gravitons, and many more. The energy of the vacuum is the sum total of the energies of all these virtual particles; each type of particle makes its contribution. Some of the virtual particles are moving slowly and have small energy, while others are moving faster and have higher energy. If we add up all the energy in this sea of particles using the technical mathematics of quantum field theory, we find a disaster. There are so many high-energy virtual particles that the total energy comes out infinite. Infinity is a senseless answer. It's what made Dirac skeptical of vacuum energy. But as Dirac's contemporary Wolfgang Pauli quipped, "Just because something is infinite doesn't mean it's zero."

The problem is that we have overestimated the effects of very energetic virtual particles. In order to make sense of the mathematical expressions, we somehow have to do a better job of accounting for their effects. But we don't understand much about the behavior of particles when their energy gets above a certain point. Physicists have used giant accelerators to study the properties of very high-energy particles, but every accelerator has a limit. Even theoretical ideas run out of steam at some point. Ultimately we reach a value of the energy so large that if two particles with that much energy collide, they create a black hole! By this point we are far beyond what we can understand with present tools. Even String Theory is not up to the task. So what we do is to make an agreement with one another. We will just ignore the contributions (to the vacuum energy) from all virtual particles whose energy is so large that they would make a black hole if they were to collide. We call it cutting off the divergences or regulating the theory. But whatever words we use, the meaning is the same: let's just agree to ignore the effects of very high-energy virtual particles that we don't yet understand.

It's a very unsatisfactory situation, but once we do this we can estimate the vacuum energy stored in electrons, photons, gravitons, and all the other known particles. The result is no longer infinite, but it is also not small. The joule is an ordinary unit of energy. It takes about four thousand joules to heat a liter of water 1 degree centigrade. A cubic centimeter is a common unit of volume. It is about as big as the tip of your pinky finger. In the ordinary world the joule per cubic centimeter is a useful unit of energy density. Well then, how many joules of vacuum energy are there in the form of virtual photons in a volume of space as big as the tip of your little finger? The estimate that quantum field theory gives is so big that it requires a 1 with 116 zeros after it: 10 to the 116th power! That many joules of vacuum energy are in your little finger in the form of virtual photons. That is far more energy than it would take to boil all the water in the universe. It's far more energy than the sun will radiate in a million or a billion years. It is far more energy than all the stars in the observable universe will ever radiate in their entire lives.

The gravitational repulsion due to that much vacuum energy would be disastrous. It would tear apart not only galaxies but also atoms, nuclei, and even the protons and neutrons that make up the galactic material. The cosmological constant, if it exists at all, must be very much smaller to avoid conflicting with all the things we know about physics and astronomy.

Now this was just the vacuum energy due to one type of particle, photons. What about virtual electrons, quarks, and all the others? They also fluctuate and create vacuum energy. The precise amount of energy from each type of particle is sensitive to the mass of that particle as well as the various coupling constants. We might expect that if we add the contribution from electrons, it would make the energy even bigger. But that's not necessarily correct. Photons and other similar particles contribute positive energy to the vacuum. It is one of those paradoxical quantum facts that virtual electrons in the vacuum have negative energy. The photon and electron belong to a class of particles that creates opposite energy in the vacuum.

These two kinds of particles are *bosons* and *fermions*. For our purposes it is not so important to know the detailed difference between these two, but I will take a paragraph or two to explain. Fermions are particles like the electron. If you know any chemistry, you will remember Pauli's exclusion principle. It says that no two electrons in the atom can occupy exactly the same quantum state. That's why the periodic table has the structure that it has. As electrons are added to an atom, they fill up ever-higher atomic shells. This is characteristic of all fermion particles. No two fermions of the same type can occupy the same quantum state. They are isolationist hermits.

Bosons are the opposite, the sociable particles. Photons are bosons. It is especially easy to have many bosons in the same state. In fact a laser beam is an intense collection of photons, all in the same quantum state. You can't build a laser to make a beam of fermions. On the other hand, you can't make atoms out of bosons, at least not atoms that have a periodic table.

What does all this have to do with vacuum energy? The answer is that virtual bosons in the vacuum have positive energy, but virtual fermions like the electron have negative energy. The reasons why are technical, but let's just accept it as a boon: fermion vacuum energy and boson vacuum energy can cancel because they have opposite signs.

So if we account for all the types of fermions and bosons in nature — photons, gravitons, gluons, W-bosons, Z-bosons, and Higgs particles on the boson side; neutrinos, electrons, muons, quarks on the fermion side — do they cancel? Not even approximately! The truth is that we have no idea why the vacuum energy is not enormous, why it is not big enough to tear apart the atoms, protons, and neutrons and all other known objects.

Nevertheless, physicists have been able to construct mathematical theories of imaginary worlds in which the positive contributions of bosons exactly cancel the negative vacuum energy of fermions. It's simple. All you have to do is make sure that the particles come in matched pairs: a fermion for each boson, a boson for each fermion, with each having exactly the same mass. In other words, the electron would have a twin, a boson, with precisely the same mass and charge as the electron. The photon would also have a twin, a massless fermion. In the arcane language of theoretical physics, a matching of that kind, between one thing and another, is called a *symmetry*. The matching between things and their mirror images is called reflection symmetry. The matching between particles and their antiparticles is called charge conjugation symmetry. It is in keeping with tradition to refer to the fermion-boson matching (in this fictitious world) of elementary particles as a symmetry. The most overworked word in the physicist's vocabulary is super: superconductors, superfluids, supercollider, supersaturated, superString Theory. Physicists are not often verbally challenged, but the only term they could think of for fermion-boson twinning was *supersymmetry*. Supersymmetric theories have no vacuum energy because the fermions and bosons exactly cancel.

But super or not, fermi-bose symmetry is not a feature of the real world. There is no superpartner of the electron or for any other elementary particle. The vacuum energies of fermions and bosons do not cancel, and the bottom line is that our best theory of elementary particles predicts vacuum energy whose gravitational effects would be vastly too large. We don't know what to make of it. Let me put the magnitude of the problem in perspective. Let's invent units in which 10^{116} joules per cubic centimeter is called one Unit. Then each kind of particle gives a vacuum energy of roughly a Unit. The exact value depends on the mass and other properties of the particle. Some of the particles give a positive number of Units, and some negative. They must all add up to some incredibly small energy density in Units. In fact a vacuum energy density bigger than .0000000000000000000000000000000000000000000000 000000000000000000000000000000000000000000000000000000000 0000000000000001 Units would conflict with astronomical data. For a bunch of numbers, none of them particularly small, to cancel one another to such precision would be a numerical coincidence so incredibly absurd that there must be some other answer.

Theoretical physicists and observational cosmologists have regarded this problem differently. The traditional cosmologists have generally kept an open mind about the possibility that there may be a tiny cosmological constant. In the spirit of experimental scientists, they have regarded it as a parameter to be measured. The physicists, myself included, looked at the absurdity of the required coincidence and said to themselves (and each other) that there must be some deep hidden mathematical reason why the cosmological constant must be *exactly* zero. This seemed more likely than a numerical cancellation of 119 decimal places for no good reason. We have sought after such an explanation for almost half a century with no luck. String theorists are a special breed of theoretical physicist with very strong opinions about this problem. The theory that they work on has often produced unexpected mathematical miracles, perfect cancellations for deep and mysterious reasons. Their view (and it was, until not too long ago, also my view) has been that String Theory is such a special theory that it must be the one true theory of nature. And being true, it must have some profound mathematical reason for the supposed fact that the vacuum energy is exactly zero. Finding the reason has been regarded as the biggest, most important, and most difficult problem of modern physics. No other phenomenon has puzzled physicists for as long as this one. Every attempt, be it in quantum field theory or in String Theory, has failed. It truly is the mother of all physics problems.

Weinberg Says the A Word

By the mid-1980s physicists had wracked their collective brains for decades about the cosmological constant and had come up completely empty-handed. Desperate situations require desperate measures, and in 1987 Steven Weinberg, one of the most eminent scientists in the world, acted in desperation. Throwing all caution to the wind, he suggested the unthinkable: perhaps the cosmological constant is so small for reasons having nothing to do with special properties of String Theory or any other mathematical theory. Maybe the reason is that if λ were any larger, our own existence would be in jeopardy. This kind of logic went by the name of the Anthropic Principle: some property of the universe or Laws of Physics must be true because if it wasn't, we couldn't exist. There are plenty of candidates for anthropic explanations:

Q: Why is the universe big?

Q: Why does the electron exist?

Q: Why is space three-dimensional?

A: At the very least the universe must be as big as the solar system in order for an earthlike planet warmed by a sunlike star to exist.

A: Without electrons there would be no atoms and no organic chemistry.

A: There are many special things that happen in three dimensions that don't happen in other dimensions. One example is that the stability of the solar system would be compromised in other dimensions. Solar systems in a world of four or more directions would be very chaotic and would not provide stable environments for the billions of years needed for biological evolution to do its work. Even worse is that the forces between electrons and nuclei would suck the electrons into the nucleus, ruining chemistry.

A small universe, a universe with no electrons, or a universe with some other number of dimensions would be a barren universe that could not sustain intelligent creatures to even ask these questions.

No doubt some legitimate applications of anthropic reasoning are justified. We live on the surface of a planet and not the surface of a star because life couldn't exist at 10,000-degree temperatures. But to use it to explain a fundamental constant of physics? The idea that a fundamental constant was determined by appeal to our own existence was anathema to most physicists. What possible mechanism could adjust a law of nature just so that the human race could exist? What mechanism, other than an appeal to supernatural forces? Physicists often refer to the Anthropic Principle as religion or superstition or "the A word" and claim that it is "giving up."

Steve Weinberg has been a friend of mine for longer than I care to remember. I first heard his booming baritone voice in a Mexican café in Berkeley. It was 1965: the heyday of Mario Savio's Free Speech Movement, Jefferson Poland and the Sexual Freedom Movement, LSD and the Vietnam peace protests. I tried all four, plus a few other things. My hair was long, and I was usually dressed in jeans and a tight black T-shirt. Twenty-five years old, I had just arrived there with a freshly minted PhD from Cornell University in upstate New York. Steve was in his early thirties. We both had grown up in the Bronx and gone to the same high school, but there the resemblance ended. When I met Steve, he was already a distinguished academic, the very model of a Berkeley professor. He even dressed like a Cambridge don.

That day in the café he held center stage, pontificating about something or other French and historical. Needless to say I was prepared to dislike him. But once I got to know him, I realized that Steve had that most winning of attributes, the capacity to laugh at himself. He loved being an important man but knew that his own self-importance had its ridiculous side. As you can probably tell, despite our difference in style, I like Weinberg very much.

I have always admired the clarity and depth of Steven Weinberg's physics. In my opinion he, more than anyone, can lay claim to being the father of the Standard Model. But in recent years I have come to admire him even more for his courage and intellectual integrity. He is one of the leading voices against creationism and other forms of antiscientific thinking. But on one occasion he was brave enough to express an opinion that went against the scientific prejudices of his colleagues. In fact it was evident from his own writings that he himself strongly disliked the Anthropic Principle. I imagine that it sounded much too close to what some people now call intelligent design. Nevertheless, given the state of desperation concerning the cosmological constant, he felt he could not ignore the possibility of an anthropic explanation. Characteristically, he took a practical course, asking if a cosmological constant bigger than the observed limit of 10^{-120} Units might be catastrophic for the development of life. If there was no way that a bigger λ could inhibit life, then the existence of life would not be significant, and string theorists could go on trying to find an elegant mathematical solution to the problem. But if a reason could be found why a slightly larger cosmological constant would prevent life, then the Anthropic Principle would have to be taken seriously. I've always wondered which way Weinberg wanted it to turn out.

To be fair many cosmologists were not only open-minded about the Anthropic Principle but even advocated it. The conjecture that the smallness of the cosmological constant might be anthropic had already appeared in a pioneering book by two cosmologists, John Barrow and Frank Tipler.[4] Among others who advocated at minimum an open mind were Sir Martin Rees, the British "Astronomer Royal," and Andrei Linde and Alex Vilenkin, both famous Russian cosmologists living in the United States. Perhaps cosmologists were more receptive to the idea than physicists because a look at the real universe, instead of abstract equations, is less suggestive of simplicity and elegance than of random and arbitrary numerical coincidences.

In any case Weinberg set out to see if he could find a reason why a cosmological constant much bigger than 10^{-120} Units would prevent life. To give some idea of the challenge he faced, we can ask how big the effects of such a cosmological constant would be on ordinary terrestrial phenomena. Remember that the cosmological constant manifests itself as a universal repulsion. A repulsive force between the electrons and the nuclei of an atom would change the properties of atoms. But if you plug

4. John D. Barrow and Frank J. Tipler, *The Anthropic Cosmological Principle* (Oxford: Oxford University Press, 1986).

the numbers in, the repulsion due to such a small cosmological constant would be far smaller than anything that can ever be detected from the properties of atoms or molecules. A cosmological constant many orders of magnitude bigger than 10^{-120} Units would still be far too small to have any effect on molecular chemistry. Might a small cosmological constant affect the stability of the solar system? Again, the effects are too small for it to do so by many orders of magnitude. There does not seem to be any way that a small cosmological constant could affect life.

Nevertheless, Weinberg found his quarry. It didn't have to do with physics, chemistry, or astronomy today but rather at the time when galaxies were first beginning to form from the primordial stuff of the early universe. At that time the hydrogen and helium that made up the mass of the universe were spread out in an almost perfectly smooth or homogeneous distribution. The variations of density from one point to another were almost nonexistent.

Today, the universe is full of lumps of many different sizes: everything from small planets and asteroids to giant superclusters of galaxies. If conditions in the past were perfectly homogeneous, then no clumps could ever have formed. The perfect symmetry of an exactly spherical universe would be maintained for all time. But the universe was not exactly homogeneous. At the earliest times that we can see, slight variations in the density and pressure amounted to a few parts in 100,000. In other words, the variations in density were 100,000 times smaller than the density itself. The tendency for gravity to cause clumping is not measured by the overall density of matter but by these small variations.

Even those infinitesimal irregularities were enough to get the process of galaxy formation started. As time progressed, regions with a slight overdensity attracted the matter from the less dense regions. This had the effect of magnifying the slight density contrasts. Eventually the process speeded up, and the galaxies were formed.

But because these density contrasts were initially so small, even a very tiny amount of repulsion could reverse the tendency to cluster. Weinberg found that if the cosmological constant were just an order of magnitude or two bigger than the empirical bound, no galaxies, stars, or planets would ever have formed!

The Case of Negative λ

So far I have told you about the repulsive effects that accompany positive vacuum energy. But suppose that the contribution of fermions outweighed that of bosons: then the net vacuum energy would be a negative number. Is this possible? If so, how does it affect Weinberg's arguments?

The answer to the first question is yes, it can happen very easily. All you need is a few more fermion-type particles than bosons, and the cosmological constant can be made negative. The second question has an equally simple answer — changing the sign of λ switches the repulsive effects of a cosmological constant to a universal attraction: not the usual gravitational attractive force but a force that increases with distance. To argue convincingly that a large cosmological constant would automatically render the universe uninhabitable, we need to show that life could not form if the cosmological constant were large and negative.

What would the universe be like if the laws of nature were unaltered except for a negative cosmological constant? The answer is even easier than the case of positive λ. The additional attractive force would eventually overwhelm the outward motion of the Hubble expansion: the universe would reverse its motion and start to collapse like a punctured balloon. Galaxies, stars, planets, and all life would be crushed in an ultimate "big crunch." If the negative cosmological constant were too large, the crunch would not allow the billions of years necessary for life like ours to evolve. Thus, there is an anthropic bound on negative λ to match Weinberg's positive bound. In fact the numbers are fairly similar. If the cosmological constant is negative, it must also not be much bigger than 10^{-120} Units if life is to have any possibility of evolving.

Nothing we have said precludes there being pocket universes far from our own with either a large positive or large negative cosmological constant. But they are not places where life is possible. In the ones with large positive λ, everything flies apart so quickly that there is no chance for matter to assemble itself into structures like galaxies, stars, planets, atoms, or even nuclei. In the pockets with large negative λ, the expanding universe quickly turns around and crushes any hope of life.

The Anthropic Principle had passed the first test. Nevertheless, the general attitude of theoretical physicists to Weinberg's work was to ignore it. Traditional theoretical physicists wanted no part of the Anthropic

Principle. Part of this negative attitude stemmed from lack of any agreement about what the principle meant. To some it smacked of creationism and the need for a supernatural agent to fine-tune the laws of nature for man's benefit: a threatening, antiscientific idea. But even more, theorists' discomfort with the idea had to do with their hopes for a unique consistent system of physical laws in which every constant of nature, including the cosmological constant, was predictable from some elegant mathematical principle.

But Weinberg took the practical route a little further. He said that whatever the meaning of the Anthropic Principle and the mechanism that enforces it, one thing was clear. The principle may tell us that λ is small enough to not kill us, but there is no reason why it should be exactly zero. In fact there is no reason for it to be very much smaller than what is needed to ensure life. Without worrying about the deeper meaning of the principle, Weinberg was, in effect, making a prediction. If the Anthropic Principle is correct, then astronomers would discover that the vacuum energy was nonzero and probably not much smaller than 10^{-120} Units.

The Planck Length

The process of discovery has always fascinated me. I'm referring to the mental process; what was the line of reasoning — the insight — that led to the "eureka" moment? One of my favorite daydreams is to put myself in the mind of a great scientist and imagine how I might have made a crucial discovery.

Let me share with you how I would have made the first great contribution to the quantum theory of gravity. It was a full sixteen years before young Einstein would invent the modern theory of gravity and twenty-six years before those upstarts Werner Heisenberg and Schrödinger invented modern quantum mechanics. As a matter of fact I, Max Planck, did it without even realizing it.

Berlin 1900, The Kaiser Wilhelm Institute

Recently I made the most wonderful discovery of a completely new fundamental constant of nature. People are calling it my constant, *Planck's constant.* I was sitting in my office thinking to myself: why is it that the fundamental constants like the speed of light, Newton's gravitational constant, and my new constant have such awkward values? The speed of light is 2.99×10^8 meters per second. Newton's constant is 6.7×10^{-11} square meters per second-kilogram. And my constant is even worse, 6.626×10^{-34} kilogram-square meters per second. Why are they always so big or so small? Life for a physicist would be so much easier if they were ordinary-size numbers.

Then it hit me! There are three basic units describing length, mass, and time: the meter, kilogram, and second. There are also three fundamental constants. If I change the units, say, to centimeters, grams, and hours, the numerical value of all three constants will change. For example, the speed of light will get worse. It will become 1.08×10^{14} centimeters per hour. But if I use years for time and light-years for distance, then the speed of light will be exactly one since light travels one light-year per year. Doesn't that mean that I can invent some new units and make the three fundamental constants anything I want? I can even find units in which all three fundamental constants of physics are equal to one! That will simplify so many formulas. I'll call the new units natural units since they're based on the constants of nature. Maybe if I'm lucky, people will start calling them Planck units.

Calculate, Calculate, Calculate . . .

Ah, here's my result: the natural unit of length is about 10^{-33} centimeters. Holy Bernoulli! That's far smaller than anything I've ever thought about. Some of those people who think about atoms say that they may be about 10^{-8} centimeters in diameter. That means my new natural unit is as much smaller than an atom as an atom is smaller than the galaxy![5]

How about the natural unit of time? That comes out to be about 10^{-42} seconds! That's unimaginably small. Even the time of oscillation of a high-frequency light wave is vastly longer than a natural time unit.

And now for mass: ah, the unit of mass is not so strange. The natural unit of mass is small but not very. It's 10^{-5} grams: about the same as a

5. Sorry for the anachronism; the idea of a galaxy didn't exist yet in 1900.

dust mote. These units must have some special meaning. All the formulas of physics are so much simpler if I work in natural units. I wonder what it means?

That's how Planck made one of the great discoveries about quantum gravity without realizing it.

Planck lived forty-seven more years, to the age of eighty-nine. But I don't think he ever imagined the profound impact that his discovery of Planck units would have for later generations of physicists. By 1947 the General Theory of Relativity and quantum mechanics were part of the basic foundation of physics, but hardly anyone had started to think about the synthesis of the two: *quantum gravity*. The three Planck units of length, mass, and time were critical in the development of the discipline, but even now, we are only beginning to understand the depth of their significance. I'll give some examples of their importance.

Earlier we discussed the fact that in Einstein's theory, space is stretchable and deformable like the surface of a balloon. It can be stretched flat and smooth or it can be all wrinkled and bumpy. Combine this idea with quantum mechanics, and space becomes very unfamiliar. According to the principles of quantum mechanics, everything that can fluctuate does fluctuate. If space is deformable, then even it has the "quantum jitters." If we could look through a very high-powered microscope, we would see space fluctuating, shaking and shimmering, bulging out in knots, and forming donut holes. It would be like a piece of cloth or paper. On the whole it looks flat and smooth, but if you look at it microscopically, the surface is full of pits, bumps, fibers, and holes. Space is like that but worse. It would appear not only full of texture but of texture that fluctuates incredibly rapidly.

How powerful does the microscope have to be in order to see the fluctuating texture of space? You guessed it. The microscope would have to discern features whose size is the Planck length, i.e., 10^{-33} centimeters. That's the scale of the quantum-texture of space.

And how long do the features last before fluctuating to something new? Again you can guess the answer; the time scale of these fluctuations is the Planck time, 10^{-42} seconds! Many physicists think there is a

sense in which the Planck length is the smallest distance that can ever be resolved. Likewise, the Planck time may be the shortest interval of time.

Let's not leave out the Planck mass. To understand its importance, imagine two particles colliding so hard that they create a black hole at the collision point. Yes, it can happen; two colliding particles, if they have enough energy, will disappear and leave behind a black hole, one of those mysterious objects that will occupy chapter 11 of this book. The energy needed to form such a black hole played a role in our earlier discussion about vacuum energy. Just how large must that energy be (remembering that energy and mass are the same thing)? The answer, of course, is the Planck mass. The Planck mass is neither the smallest nor the largest possible mass, but it is the smallest possible mass of a black hole. By the way, a Planck mass black hole would be about one Planck length in size and it would last for about one Planck unit of time before exploding into photons and other debris.

As Planck discovered, his mass is about a hundred-thousandth of a gram. By ordinary standards that's not much mass, and even if we multiply it by the speed of light squared, it's not a huge amount of energy. It more or less corresponds to a tank full of gasoline. But to concentrate that much energy in two colliding elementary particles — that would be a feat. It would take an accelerator many light-years in size to do the job.

Recall that we estimated the vacuum energy density due to virtual particles. Not surprisingly, the answer translates to about one Planck mass per cubic Planck length. In other words, the unit of energy density that I defined as one Unit was nothing but the natural Planck unit of energy density.

The world at the Planck scale is a very unfamiliar place, where geometry is constantly changing, space and time are barely recognizable, and high-energy virtual particles are perpetually colliding and forming tiny black holes that last no longer than a single Planck time. But it's the world in which string theorists spend their working days.

Let me take a bit of space and time to summarize the two difficult chapters that you've worked your way through and the dilemma they lead to. The microscopic laws of elementary particles in the form of the Standard Model are a spectacularly successful basis for calculating the properties not only of the particles themselves, but of nuclei, atoms, and simple molecules. Presumably, with a big enough computer and enough time, we could calculate all molecules and move on to even more complex objects. But the Standard Model is enormously complicated and arbitrary. In no way does it explain itself. There are many other imaginable lists of particles and lists of coupling constants that are every bit as mathematically consistent as those found in nature.

But things get worse. When we combine the theory of elementary particles with the theory of gravity, we discover the horror of a cosmological constant big enough to not only destroy galaxies, stars, and planets but also atoms, and even protons and neutrons — *unless*. Unless what? Unless the various bosons, fermions, masses, and coupling constants that go into calculating the vacuum energy conspire to cancel the first 119 decimal places. But what natural mechanism could ever account for such an unlikely state of affairs? Are the Laws of Physics balanced on an incredibly sharp knife-edge, and if so, why? Those are the big questions.

In the next chapter we will discuss what determines the Laws of Physics and just how unique they are. What we will find is that these laws are not at all unique! They can even vary from place to place in the megaverse. Could it be that there are special rare places in the megaverse where the constants conspire in just the right way to cancel the vacuum energy with sufficient precision for life to exist? The basic idea of a Landscape of possibilities that allows such variation is the subject of chapter 3.

CHAPTER THREE

The Lay of the Land

"Navigator, is it gaining on us?" The captain's face was grim as sweat beads rolled down his bald dome and dropped from his chin. The veins in his forearm bulged as his hand clenched the control stick.

"Yes Captain, I'm afraid there is no way to outrun it. The bubble is growing and unless my calculation is way off, it's going to engulf us."

The captain winced and punched the desktop in front of him. "So this is how it ends. Swallowed by a bubble of alternate vacuum. Can you tell what the laws of physics are like inside it? Any chance we can survive?"

"Not likely. I compute that our chances are about one in ten to the one-hundredth power — one in a googol. My guess is the vacuum inside the bubble can support electrons and quarks, but the fine structure constant is probably way too large. That'll blow the hell out of our nuclei." The navigator looked up from his equations and smiled ruefully. "Even if the fine structure constant is okay, the chances are overwhelming that there is a big CC."

"CC?"

"Yeah, you know — cosmological constant. It's probably negative and big enough to squash our molecules like that." The navigator snapped his fingers. "Here it comes now! Oh god, no, it's supersymmetric.[1] No chance. . . ." Silence.

1. See chapter 8 for an explanation of the navigator's fear of supersymmetry.

. . .

That was the beginning of a very bad science-fiction story that I started to write. After a few more paragraphs, I concluded that I am a sadly untalented sci-fi author and abandoned the project. But the science may be a good deal better than the fiction.

It is gradually becoming accepted, by many theoretical physicists, that the Laws of Physics may not only be variable but are almost always deadly. In a sense the laws of nature are like East Coast weather: tremendously variable, almost always awful, but on rare occasions, perfectly lovely. Like deadly storms, bubbles of extremely hostile environments may propagate through the universe causing destruction in their wake. But in rare and special places, we find Laws of Physics perfectly suited to our existence. In order to understand how it came to pass that we find ourselves in such an exceptional place, we have to understand the reasons for the variability of the Laws of Physics, just how large the range of possibilities is, and how a region of space can suddenly change its character from lethal to benign. This brings us to the central concern of this book, the Landscape.

As I have said, the Landscape is a space of possibilities. It has geography and topography with hills, valleys, flat plains, deep trenches, mountains, and mountain passes. But unlike an ordinary landscape, it isn't three-dimensional. The Landscape has hundreds, maybe thousands, of dimensions. Almost all of the Landscape describes environments that are lethal to life, but a few of the low-lying valleys are habitable. The Landscape is *not* a real place. It doesn't exist as a real location on the earth or anywhere else. It doesn't exist in space and time at all. It's a mathematical construct, each of whose points represents a possible environment or, as a physicist would say, a possible *vacuum*.

In common usage the word *vacuum* means empty space, space from which all air, water vapor, and other material has been sucked out. That's also what it means to an experimental physicist who deals in vacuum tubes, vacuum chambers, and vacuum pumps. But to a theoretical physicist, the term *vacuum* connotes much more. It means a kind of background in which the rest of physics takes place. The vacuum represents potential for all the things that can happen in that background. It means a list of all the elementary particles as well as the constants of nature that would be revealed by experiments in that vacuum. In short, it means an environment in which the Laws of Physics take a particular form. We say of our vacuum that it can contain electrons, positrons, photons, and the rest of the usual elementary particles. In our vacuum the electron has a mass of .51 MeV,[2] the photon's mass is zero, and the fine structure constant is 0.007297351. Some other vacuum might have electrons with no mass, a photon with mass 10 MeV, and no quarks but forty different kinds of neutrinos and a fine structure constant equal to 15.003571. A different vacuum means different Laws of Physics; each point on the Landscape represents a set of laws that are, most likely, very different from our own but which are, nonetheless, entirely consistent possibilities. The Standard Model is merely one point in the Landscape of possibilities.

And if the Laws of Physics can be different in other vacuums, so can all of science. A world with much lighter electrons but heavier photons would have no atoms. No atoms means no chemistry, no periodic table, no molecules, no acids, no bases, no organic substances, and of course, no biology.

The idea of universes with alternative laws of nature seems like the stuff of science fiction. But the truth is more mundane than it sounds. Modern medical technology routinely produces alternative universes inside MRI machines. The abbreviation MRI was not the original name for this technology: it replaced NMR, which stands for Nuclear Magnetic Resonance. But patients got scared by the word *nuclear* and wouldn't go near the thing. So the name was changed to Magnetic Resonance Imaging to emphasize the magnetic aspects of the technology instead of the nuclear. In fact the nuclei that are involved in NMR are not uranium or plutonium nuclei as in nuclear warheads, they are the patient's own nuclei that are ever so gently tickled by the magnetic field of the machine.

An MRI machine is basically a cylinder of empty space with a coil of wire surrounding it. An electric current through the coil creates a powerful magnetic field in the cylinder. It's essentially a very strong electromagnet. The patient in the interior of the MRI machine is in a small

2. An MeV is a tiny unit of mass used by elementary-particle physicists. Approximately five times ten to the twenty-ninth (5×10^{29}) MeVs is equal to one kilogram.

private universe, where as we will see, the properties of the vacuum are slightly different from those on the outside. Imagine waking up one morning inside the machine, not knowing where you were. Something would seem amiss about the Laws of Physics. The most obvious thing you would notice is that iron objects would move in very odd ways, even presenting serious danger. If you happened to have a compass, it would rigidly lock into place along some particular direction.

It probably wouldn't be a good idea to have a TV in the MRI machine, but let's suppose you did. The picture would be distorted in bizarre ways. If you know how a television operates, you would trace the strange distortion to the motion of electrons. The strong magnetic field inside the cylinder exerts forces on the electrons that curve their trajectories from straight lines to corkscrew spirals. A theoretical physicist who knew about Feynman diagrams would say that something was different about the electron propagator. The propagator is not just a picture of an electron moving from one point to another, it's also a mathematical expression that describes the motion.

The constants of nature would also be slightly unusual. The strong magnetic field interacts with an electron's spin and even modifies the electron's mass. Funny things happen to atoms in strong magnetic fields. The magnetic forces on the atomic electrons cause the atom to be slightly squashed in directions perpendicular to the field. The effects in a real MRI machine would be tiny, but if the magnetic field could be made much stronger, atoms would get squeezed into strands resembling spaghetti along the magnetic field lines.

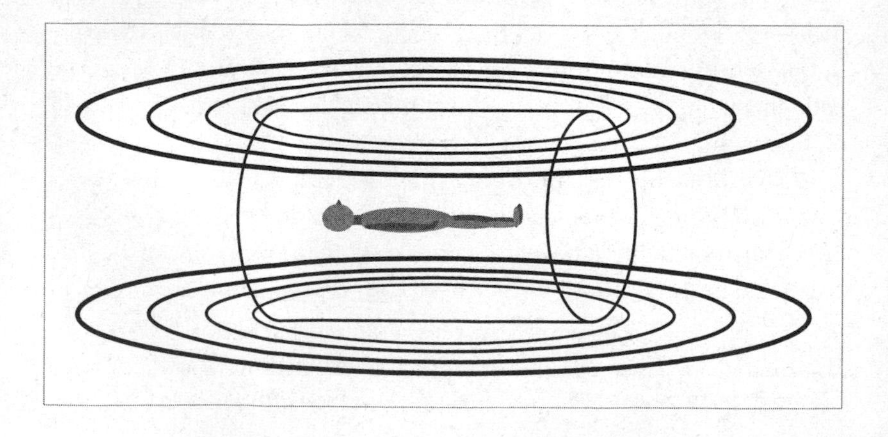

The effects of the magnetic field can also be detected from small changes in the energy levels of atoms and, consequently, the spectrum of light they emit. There are changes in the precise manner in which electrons, positrons, and photons interact with one another. If the field were made strong enough, even the vertex diagrams would be affected. The fine structure constant would be a little different and depend on which way the electron moves.

Of course the field in the MRI machine is very weak, and the effects on the laws regulating charged particles are minute. If the field were very much stronger, the patient would feel funny. A field strong enough to seriously affect those laws would be absolutely fatal. The effects on atoms would have terrible consequences for chemical and biological processes.

There are two ways to view this, both of which are right. One is conventional: the Laws of Physics are exactly what they always were, but the environment is modified by the presence of the magnetic field. The other way to think about it is that the rules for Feynman diagrams have been changed and that something has happened to the Laws of Physics. Perhaps the most precise thing to say is:

The Laws of Physics are determined by the environment.

Fields

Fields, as we've seen, are invisible properties of space that influence objects moving through them. The magnetic field is a familiar example. Everyone who has played with magnets has discovered the mysterious action at-a-distance forces they exert on paper clips, pins, and steel nails. Most people who have had a science course in school have seen the effect of a magnetic field on iron filings — tiny bits of iron — sprinkled on a surface in the vicinity of a magnet. The field assembles the filings into long filaments that look like hairy threads, lined up along the direction of the field. The filaments follow mathematical lines called *magnet lines of force* or *magnetic field lines*. The magnetic field has a direction at every point, but it also has a strength that determines how forceful the field is in pushing pieces of iron. In the MRI machine the field is more than ten thousand times stronger than the earth's magnetic field.

The electric field is a slightly less familiar close relative of the magnetic field. It has no observable effects on iron filings, but it causes small bits of paper to move when there is some static electricity on them. Electric fields aren't caused by electric current but by accumulations of static electric charge. For example, rubbing one material on another — your rubber shoe soles on the carpet, say — causes the transfer of electrons. One material becomes charged negatively, and the other positively. The charged objects create an electric field around them that, like magnetic fields, have both direction and strength.

Ultimately the Laws of Physics are variable because they are determined by fields, and fields can vary. Switching on magnetic and electric fields is one way to change the laws, but it is by no means the only way to modify the vacuum, or even the most interesting way. The second half of the twentieth century was a time of discovery of new elementary particles, new forces, and above all, new fields. Einstein's gravitational field was one, but there were many others. Space can be filled with a wide variety of invisible influences that have all sorts of effects on ordinary matter. Of all the new fields that were discovered, the one that has the most to teach us about the Landscape is the *Higgs field.*

The discovery of the Higgs field wasn't an experimental discovery in the usual sense.[3] Theoretical physicists discovered that the Standard Model, without the Higgs field, is mathematically inconsistent. Without it the Feynman rules would lead to nonsensical results like infinite and even negative probabilities. But theorists in the late 1960s and early 1970s figured out a way to fix all the problems by adding one additional elementary particle: the Higgs particle.

Higgs particle, Higgs field — what's the connection between particles and fields that leads us to call them by the same name? The field idea first appeared in the mid-nineteenth century in the form of the electromagnetic field. Michael Faraday imagined a field to be a smooth disturbance in space that affects the motions of electrically charged particles, but the field itself was *not* supposed to be made of particles. For Faraday, and Maxwell who followed him, the world was composed of

3. A number of people were involved in the theoretical discovery of the Higgs field. In addition to Peter Higgs of Great Britain, Robert Brout and François Englert of Belgium were among the earliest to realize its necessity.

particles and fields, and there was no doubt whatsoever about which was which. But in 1905 Albert Einstein, in order to explain Planck's formula for heat radiation, proposed an outlandish theory. Einstein claimed that the electromagnetic field was really composed of a very large number of indivisible particles that he called photons. In small numbers, photons, or what are the same thing, light quanta behave like particles, but when many of them move in a coordinated way, the whole collection behaves like a field — a *quantum field.* This relation between particles and fields is very general. For each type of particle in nature, there is a field, and for each type of field there is a particle. Thus, fields and particles often go by the same name. The electromagnetic field (the collective name for electric and magnetic fields) could be called the *photon field.* The electron has a field. So, too, does the quark, the gluon, and each member of the cast of characters of the Standard Model, including the Higgs particle.

When I say that the Standard Model is mathematical nonsense without the Higgs field, I should qualify the statement. The theory without the Higgs field is mathematically consistent, but *only* if all particles move with the speed of light, like the photon. Particles that move with the speed of light cannot have any mass, so physicists say that the Higgs field is necessary in order to "give the elementary particles their mass." In my opinion this is a poor choice of words, but I can't think of a better one. In any case this is an important example of how the value of a field can influence the constants of nature.

Nobody has ever seen a Higgs particle, even in the indirect way that experimental physicists "see" particles. The difficulty isn't in detecting them but is in producing them in the first place. The problem is not a fundamental one; to produce a particle as heavy as the Higgs, you simply need a bigger accelerator. But both the Higgs particle and Higgs field are so important to the success of the Standard Model that no one seriously questions their existence.[4] As I write this book, the construction of an accelerator is nearing completion in the European Organization for Nuclear Research (CERN) that should easily do the job of

4. So important that the physicist Leon Lederman wrote a book about it, which in a moment of overenthusiasm he titled: *The God Particle.* (Lederman might have called his book *The God Field.* I suppose *The God Particle* has a better ring to it.)

producing the Higgs particle.[5] Just about forty years will have passed from the time the Higgs particle was first discovered by theorists to the time of its detection.

If it were as easy to "switch on" the Higgs field as it is to switch on the magnetic field, we could change the mass of the electron at will. Increasing the mass would cause the atomic electrons to be pulled closer to the nucleus and would dramatically change chemistry. The masses of quarks that comprise the proton and neutron would increase and modify the properties of nuclei, at some point destroying them entirely. Even more disruptive, shifting the Higgs field in the other direction would eliminate the mass of the electron altogether. The electron would become so light that it couldn't be contained within the atom. Again, this is not something we would want to do where we live. The changes would have disastrous effects and render the world uninhabitable. Most significant changes in the Laws of Physics would be fatal and therein lies a tale that we will return to repeatedly.

By varying the Higgs field, we can add diversity to the world; the laws of nuclear and atomic physics will also vary. A physicist from one region would not entirely recognize the Laws of Physics in another. But the variety inherent in the variations of the Higgs field is very modest. What if the number of variable fields were many hundreds instead of just one? This would imply a multidimensional Landscape, so diverse that almost anything could be found. Then we might begin to wonder what is not possible instead of what is. As we will see this is not idle speculation.

Whenever mathematicians or physicists have a problem that involves multiple variables, they think of a space representing the possibilities. A simple example is the temperature of the air. Imagine a line with a mark representing 0 degrees Fahrenheit, next to it a point representing 1 degree, another point at 2 degrees, and so on. The line is a one-dimensional space representing the possible values of the temperature. A point at 70 degrees would represent a beautiful, mild day; the point at 32 degrees, a freezing winter day. The temperature indicator on an ordinary household thermometer is exactly this kind of abstract space made concrete.

5. The name CERN is an acronym for "Conseil Europeen pour la Recherche Nucléaire."

Suppose that in addition to a thermometer outside the kitchen window, we also have a barometer to measure the air pressure. Then we might draw two axes, one to represent temperature and one to represent atmospheric pressure. Again, each point, now in a two-dimensional space, represents a possible weather condition. If we wanted even more information — for example, how moist the air is — we might add yet a third dimension to the space of possibilities: humidity.

The temperature, pressure, and humidity combined tell us more than just the temperature, pressure, and humidity. They tell us something about the kinds of particles that can exist: in this case not elementary particles but droplets of water. Depending on the conditions either snowflakes, liquid drops, or sleet particles can move through the atmosphere.

The Laws of Physics are like the "weather of the vacuum," except instead of the temperature, pressure, and humidity, the weather is determined by the values of fields. And just as the weather determines the kinds of droplets that can exist, the vacuum environment determines the list of elementary particles and their properties. How many controlling fields are there, and how do they affect the list of elementary particles, their masses, and coupling constants? Some of the fields we already know — the electric field, the magnetic field, and the Higgs field. The rest will be known only when we discover more about the overarching laws of nature than just the Standard Model. At the present time our best bet for these higher-level laws — our only bet — is String Theory. In chapters 7 and 8, we will see that String Theory has an unexpected answer to the question of how many fields control the local vacuum weather. From the current state of knowledge, it seems that it is in the hundreds or even thousands.

Whatever the number of fields, the principle is the same. Imagine a mathematical space with a direction for each field. If there are ten fields, then the space must be ten-dimensional. If there are one thousand fields, then the number of dimensions should be one thousand. This space is the Landscape. A point in the Landscape specifies a value for each field: a vacuum-weather condition. It also defines a particular set of elementary particles, their masses, and laws of interaction. If one could gradually move the universe from one point on the Landscape to another, all these things would gradually change. Responding to these changes, the properties of atoms and molecules would also change.

Hills and Valleys

A map of real terrain would not be complete unless it indicated how the altitude varies from point to point. That's the purpose of a topographical map, with its curved contours representing elevation. Even better than a topographical map would be a three-dimensional plaster model showing the mountains, valleys, and plains in miniature. Imagine we had such a model in front of us and add a little smooth ball, maybe a BB, to roll around on the model landscape. Place the ball anywhere and give it a little push: it will start to roll downhill until it eventually rolls to a stop at the bottom of some valley. Why does it do so? Many answers have been given to this question. The ancient Greeks believed that everything — every material — had its own place in the world and would always seek its correct level. I'm not sure what your answer would be, but as a physics professor, I would explain that the BB has potential energy, which depends on its elevation: the higher the altitude, the higher the potential energy. The BB rolls to a place of lowest energy or at least the lowest energy it can locate without having to climb over a hill to find an even lower valley. For a physicist studying the rolling BB, the contour map and the model landscape depict the variations of potential energy as the BB rolls along the landscape.

The Landscape of this book also has its highlands, lowlands, mountains, and valleys. It's not little balls that roll around on it: whole pocket universes occupy locations on the Landscape! What do I mean when I say that a pocket universe occupies a place in the Landscape? It's more or less the same as reporting the winter weather in Denver by saying that "the city occupies the point twenty-five on the thermometer scale."

It sounds odd, but it makes perfect sense to say that the major cities of the world populate the thermometer and crawl about, on a daily basis or even from moment to moment.

But what is the meaning of the altitude of a point on the Landscape? Obviously it has nothing to do with elevation above sea level. But it does have to do with potential energy, not the energy of a BB, but the

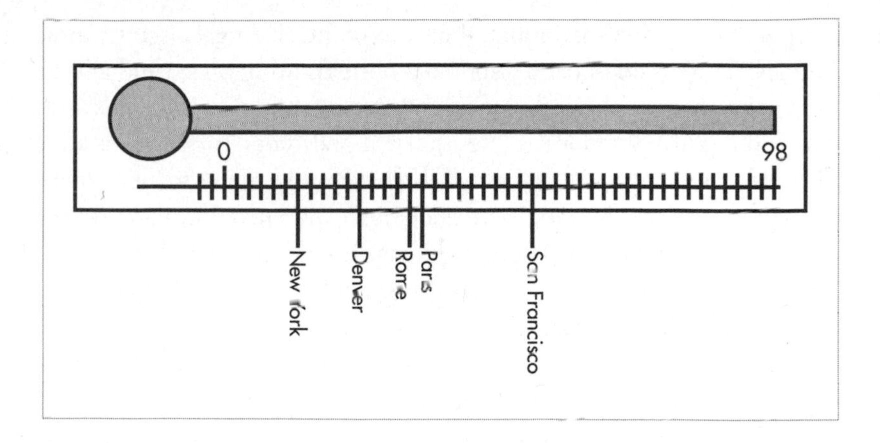

energy of a (pocket) universe.[6] And, as in the case of the BB, rolling toward the bottom of a valley represents the tendency of a universe to evolve toward lower potential energy. I will return to this point.

With this in mind let's return to the Laws of Physics in the MRI machine. If the magnetic field were the only field, the Landscape would be one-dimensional, like the temperature scale, with the single axis labeling the strength of the magnetic field.

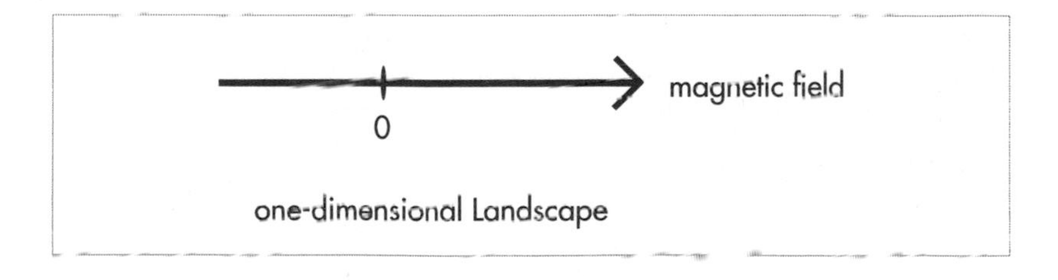

Magnetic fields do not come for free. It takes energy to create the field. In the early days of electromagnetic theory, before Michael Faraday introduced the field concept, it was thought that the energy was in the

6. Even in the case of the weather, there is a notion of energy. The energy of all the molecules in a cubic meter of atmosphere depends on the temperature and pressure. Thus, we could add a concept of altitude to the weather Landscape. It would of course have nothing to do with ordinary altitude.

electric currents flowing through the wires of the electrical circuitry. But Faraday's new vision of nature, in terms of fields filling space and affecting the behavior of charged objects, focused attention on the fields and away from wires, transformers, resistances, and other circuit elements. Physicists soon realized the great value of attributing energy to the field itself: wherever there is field, there is energy. The energy in the electromagnetic field of a beam of light will heat the cold object it illuminates.

There is also energy in the magnetic field of the MRI machine. Later we will encounter fields with vastly more energy content than the meager field in the MRI machine. The energy in the whole MRI field would be about enough to boil a few ounces of water.

Adding a vertical axis to the one-dimensional Landscape, we can plot the energy of every point. The energy contained in a magnetic field is proportional to the square of the field, which implies that the Landscape is a deep, parabolic valley with steep inclines on either side.

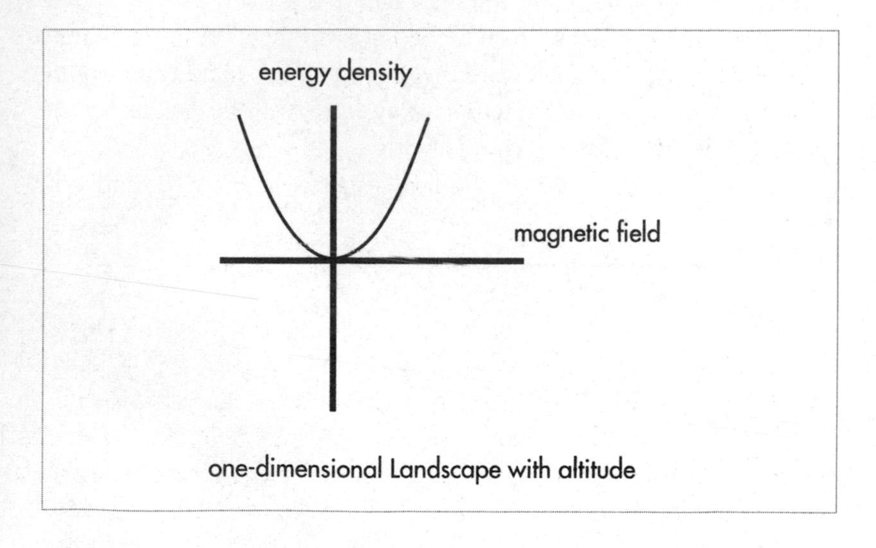

one-dimensional Landscape with altitude

The magnetic field is only one of two fields that Faraday introduced, the electric field being the other. The electric field doesn't affect a compass needle as a magnetic field would, but the electric field can make one's hair stand on end. Strong electric fields deform atoms, pushing the negatively charged electrons in one direction and the positively charged nucleus in the other. The lopsided atoms can form long chains, positive nuclei adjacent to negative electron clouds (the group of atomic electrons bound to the nuclei), right down the line. If the electric field is strong enough, atoms will be torn apart. In such regions of the Landscape, atoms couldn't exist at all. Neither, of course, could life.

Having both electric and magnetic fields adds diversity to the Landscape. It becomes two-dimensional. Since the electric field also has energy, the "altitude" varies in the two horizontal directions. This Landscape looks like a deep bowl surrounded by high, steep walls.

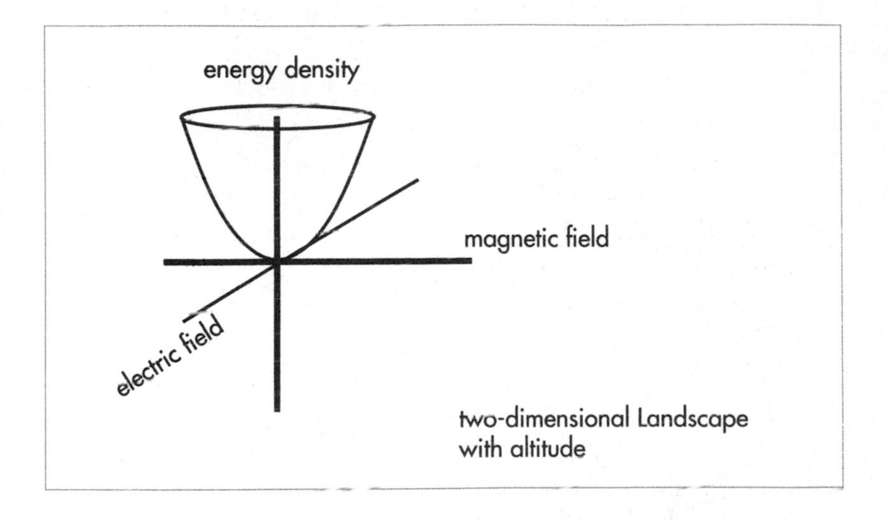

two-dimensional Landscape with altitude

Because electric and magnetic fields affect the properties of electrons in different ways, there is greater potential for variety in the Laws of Physics. Electrons in the combined field move along more complicated trajectories than in the simpler pure magnetic field. Atomic energy levels show a greater degree of complexity, and the Landscape is more varied. If all space were uniformly filled with electric and magnetic fields, we could say that the Laws of Physics depend on where, in the two-dimensional Landscape, the universe is "located." There are many more fields in nature than just the electric and magnetic fields, but the general principle is always the same. Every point on the Landscape — in other words, every value of the fields — has a corresponding value of energy density. If the fields are thought of as horizontal directions in the Landscape, one more vertical axis can be added to represent energy. If we call the vertical axis altitude, the Landscape can have plains, hills, mountain passes, and valleys.

Electric and magnetic fields are *vector fields*, which means that they not only have a size or magnitude at every point in space, but they also point in a direction. A compass located near a magnet points in the direction of the magnetic field. In a perfect world, the magnetic field would point precisely along the north-south axis everywhere on the surface of the earth. In the real world there are all sorts of variations from the ideal. A large iron deposit will influence the field in complicated ways. It's exactly the ability to vary from point to point that makes magnetism a field.

The fields making up the Landscape are, for the most part, simpler than electric and magnetic fields. Most are *scalar fields*. A scalar is a quantity that has a size or magnitude but no direction. For example, temperature is a scalar. You won't hear the weatherperson quote the temperature as "seventy-five degrees north by northwest." Temperature has a magnitude but no direction. Air pressure and humidity are also scalars. But the weatherperson also reports a vector field: the wind velocity. Wind velocity is a perfect example of a vector field: it has magnitude and direction. Temperature, pressure, humidity, and wind velocity all share the property that they can vary from place to place. That's what makes them fields. Of course they are just analogies. They have nothing to do with the fields that make up the Landscape.

The Higgs field is much like the magnetic field (except that it is scalar), but it is far more difficult to manipulate. It takes tremendous amounts of energy to change the Higgs field even a little bit. But if we could change it, the mass of each elementary particle (with the exception of the photon) would change.

Locomotives, cannon balls, and elementary particles all have mass. Mass is inertia; the more massive a body is, the harder it is to get it moving or to stop it. To determine the mass of an object, you subject it to a known force and measure its acceleration: the ratio of force to acceleration is its mass. If the object is at rest when the experiment begins, the measured mass is called the rest mass. In times past, a distinction was drawn between mass and rest mass, but today whenever the term *mass* is used it *always means rest mass*.

It's an empirical fact that all electrons have the same mass. So do all protons or any other type of particle. It's because all electrons do have the same mass that we can speak of *the mass of the electron.* It's the same for each kind of particle but, of course, the mass of one type of particle is not the same as another type. For example, the mass of the proton is about eighteen hundred times bigger than the electron mass.

Photons are eccentric oddballs when it comes to assigning them mass. The mass can't be defined by accelerating them from *rest* because photons always move with the same velocity. Photons are particles of light, and as Einstein explained, light always moves at the speed of light. A photon can never be brought to rest; instead of slowing down, it would just disappear. Thus, the mass of a photon is zero. Any particle that travels with the speed of light is said to be *massless.*

Of all the particles that are experimentally observed, only the photon is massless. But there is little doubt that at least one more massless particle exists. In the same way that an electron moving in an outer orbit of an atom radiates electromagnetic waves, a planet moving around the sun disturbs the gravitational field, thereby emitting gravitational waves. These gravitational waves are much too feeble to be detected on earth, but from time to time, tremendously violent events take place that produce very strong gravitational radiation. The collision of two black holes would release prodigious amounts of energy in the form of gravity waves that detectors on earth are being built to detect. Unless theoretical physicists are very badly mistaken, these waves will move with the speed of light. The reasonable presumption is that gravitational waves are made of massless quanta — gravitons.

Although I said that all electrons have the same mass, there is a qualification which you might have guessed. An electron's mass depends on the value of the Higgs field at the position of the electron. If we had the technology to vary the Higgs field, the electron's mass would depend on its location. This is true of the mass of every elementary particle, with the exception of the photon and graviton.

In our ordinary vacuum state, most of the known fields are zero. They may fluctuate due to quantum mechanics, but they fluctuate positively for a brief time and then negatively. If we ignore this rapid jittering, the fields average to zero. Changing the field away from zero costs energy.

The Higgs field, however, is somewhat different. Its average value in empty space is not zero. It is as if, in addition to the fluctuating sea of virtual particles, space is filled with an additional steady fluid made of Higgs particles. Why don't we notice the fluid? In a sense I suppose we could say we have gotten used to it. But if it were removed, we would certainly notice its absence! More precisely, we wouldn't exist to notice anything.

"The Higgs field gives particles their mass." What on earth does that mean? The answer is buried deep in the mathematics of the Standard Model, but I will try to give you an idea. As I mentioned earlier (on page 95), if the Higgs field (or particle) were left out of the cast of characters, the mathematical quantum field theory describing the Standard Model would be mathematically consistent only if all the other elementary particles were massless, like the photon. The actual masses of particles like electrons, quarks, W-bosons, and Z-bosons are due to their motion through the fluid of Higgs particles. I don't want to mislead you with false analogies, but there is a sense in which the Higgs fluid creates a resistance to the motion of particles. It's not a form of friction, which would slow moving particles and cause them to come to rest. Instead, it is a resistance to changes of velocity, in other words, inertia or mass. Once again a Feynman diagram is worth a thousand words.

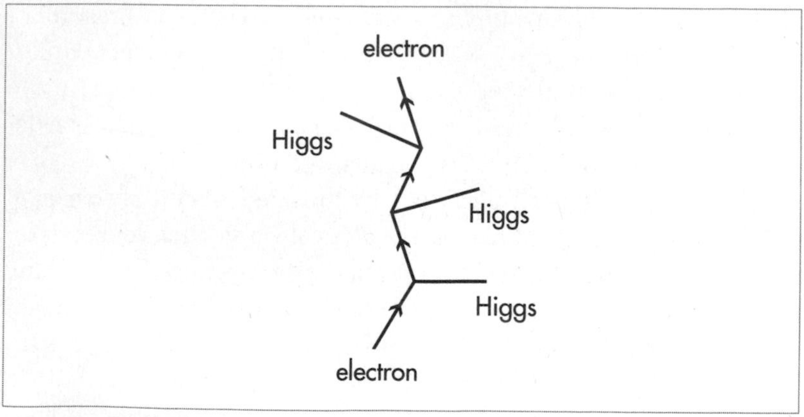

If we could create a region where the Higgs field was zero, the most singular thing we would notice (assuming our own survival) is that the electron mass would be zero. The effects on atoms would be devastating. The electron would be so light that it could not be contained within the atom. Neither atoms nor molecules would exist. Life of our kind would almost certainly not exist in such a region of space.

It would be very interesting to test these predictions the same way that we can test physics in a magnetic field. But manipulating the Higgs field is vastly more difficult than manipulating the magnetic field. Creating a region of space where the Higgs field is zero would cost an enormous quantity of energy. Just a single cubic centimeter of Higgs-free space would require energy of about 10^{40} joules. That's about the total amount of energy that the sun radiates in a million years. This experiment will have to wait a while.

Why is the Higgs field so different from the magnetic field? The answer lies in the Landscape. Let's simplify the Landscape to one dimension by ignoring the electric and magnetic fields and include only the Higgs field. The resulting "Higgs-scape" would be more interesting than the simple parabola that represents the magnetic field Landscape. It has two deep valleys separated by an extremely high mountain.

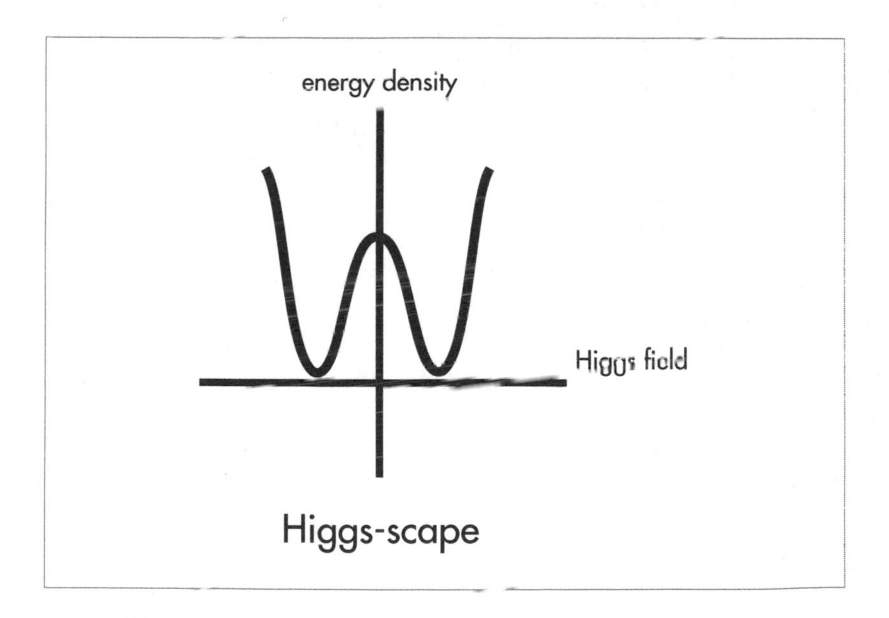

Higgs-scape

Don't worry if you don't understand why the Higgs-scape looks so different. No one completely understands it. It is another empirical fact that we have to accept for now. The top of the hill is the point on the

Landscape where the Higgs field is zero. Imagine that some superpowerful cosmic vacuum cleaner has sucked the vacuum clean of Higgs field. Here is the place in the Higgs-scape where all the particles of the Standard Model are massless and move with the speed of light. From the graph you can see that the top of the mountain represents an environment with a large amount of energy. It is also a deadly environment.

By contrast, our corner of the universe is safely nestled in one of the valleys where the energy is lowest. In these valleys the Higgs field is not zero, the vacuum is full of Higgs fluid, and the particles are massive. Atoms behave like atoms, and life is possible. The full Landscape of String Theory is much like these examples but infinitely richer in mostly unpleasant possibilities. Friendly, habitable valleys are very rare exceptions. But that's a later story.

Why, in each example, do we live at the bottom of a valley? Is it a general principle? Indeed, it is.

Rolling along the Landscape

Hermann Minkowski was a mathematician with a flair for words. Here is what he had to say about space and time: "Henceforth space by itself, and time by itself, are doomed to fade away into mere shadows, and only a kind of union of the two will preserve an independent identity." Minkowski was talking about Einstein's two-year-old child, the Special Theory of Relativity. It was Minkowski who announced to the world that space and time must be joined together into a single, four-dimensional space-time. It follows from the four-dimensional perspective that if the Laws of Physics can vary from one point of space to another, then it must also be possible for them to vary with time. There are things that can make all the normal rules — even the law of gravity — change, suddenly or gradually.

Imagine a very long-wavelength radio wave passing through a physics laboratory. A radio wave is an electromagnetic disturbance consisting of oscillating electric and magnetic fields. If the wavelength is long enough, a single oscillation will take a long time to pass through the lab. For argument's sake let's say the wavelength is two light-years. The fields in the lab will take one full year to go from zero to a maximum and back to zero.[7] If in our laboratory the field was zero in December, it will be maximum in June.

The slowly changing fields will mean that the behavior of electrons will slowly change with time. For the winter months, when the fields are smallest, the electrons, atoms, and molecules will behave normally. In the summer, when the fields are at their maximum, the electrons will move in strange orbits, and atoms will be squashed in directions perpendicular to the magnetic field. The electric field will also distort the shapes of atoms by pulling the electrons and nuclei in opposite directions. The Laws of Physics will appear to change with the seasons!

What about the Higgs field? Can it change with time? Remember that normal empty space is full of Higgs field. Imagine that an evil physicist invented a machine — a "vacuum cleaner" — that could sweep away the Higgs field. The machine would be so powerful that it could push the universe, or part of it, up the hill to the top of the mountain in the middle of the Higgs-scape. Bad things would happen; atoms would disintegrate, and all life would terminate. What happens next is surprisingly simple. Pretend the Higgs-scape really is a Landscape with a high hill separating two valleys. The universe would act like a small, round BB ball, balancing precariously on the knife-edge between falling to the left and falling to the right. Obviously the situation is unstable. Just a tiny tap one way or the other would send the ball plummeting toward a valley.

If the surface of the Landscape were perfectly smooth, without any friction, the ball would overshoot the valley, climb up the other side, and then roll back past the valley, up the hill, over and over. But if there is the smallest amount of friction, the ball will eventually come to rest at the lowest point of one of the valleys.[8]

That is how the Higgs field behaves. The universe "rolls" around on the Landscape and eventually comes to rest in a valley representing the usual vacuum.

7. Keep in mind that in one full cycle an oscillation starts at zero, increases, decreases through zero to negative values, and then increases back to zero. The electric and magnetic fields will pass through zero twice in a single wavelength.
8. While the universe is at the top of the mountain, the vacuum energy causes the universe to expand as if there were a cosmological constant. The expansion causes a kind of friction called cosmic friction.

The bottoms of valleys are the only places where an imaginary ball can stand still. Placed on a slope, it will roll down. Placed at the top of a hill, it will be unstable. In the same way, the only possible vacuum with stable, unchanging Laws of Physics is at the bottom of a valley in the Landscape.

A valley does not necessarily have to be the absolute lowest point on the Landscape. In a mountain range with many valleys, each surrounded by peaks, some of the valleys may be quite high, higher in fact than some of the summits. But as long as the rolling universe arrives at the bottom of a valley, it will remain there. The mathematical term for the lowest point of a valley is a *local minimum*. At a local minimum, any direction will be uphill. Thus, we arrive at a fundamental fact: the possible stable vacuums — or equivalently, the possible stable Laws of Physics — correspond to the local minima of the Landscape.

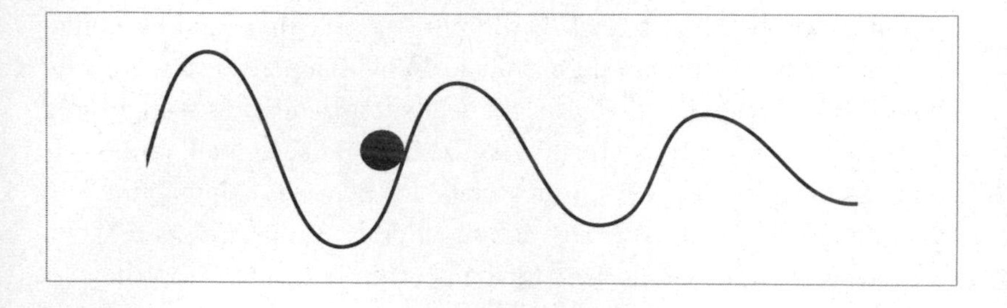

No mad scientist is ever going to sweep the Higgs field away. As I mentioned earlier, just to sweep out one cubic centimeter of space would require all of the energy radiated by the sun in a million years. But there was a time roughly fourteen billion years ago when the temperature of the world was so high that there was more than enough energy to sweep away the Higgs field from the entire known universe. I am referring to the very early universe, just after the Big Bang, when the temperature and pressure were tremendously large. Physicists believe that the universe began with the Higgs field equal to zero, i.e., up at the top of the hill. As the universe cooled, it rolled down the slope to the valley that we now "inhabit." Rolling on the Landscape plays a central role in all modern theories of cosmology.

The Higgs-scape has a small number of local minima. That one of the minima should have vacuum energy as small as 10^{-120} is incredibly improbable. But as we will see in chapter 10, the real Landscape of String Theory is far more complex, diverse, and interesting. Try to imagine a space of five hundred dimensions with a topography that includes 10^{500} local minima, each with its own Laws of Physics and constants of nature. Never mind. Unless your brain is very different from mine, 10^{500} is far beyond imagining. But one thing seems certain. With that many possibilities to choose from, it is overwhelmingly likely that the energy of many vacuums will cancel to the accuracy required by Weinberg's anthropic argument, namely 119 decimal places.

In the next chapter I want to take a break from technical aspects of physics and discuss an issue having to do with the hopes and aspirations of physicists. We will come back to "hard science" in chapter 5, but paradigm shifts involve more than facts and figures. They involve esthetic and emotional issues and fixations on paradigms that may have to be abandoned. That the Laws of Physics may be contingent on the local environment, somewhat like the weather, represents a devastating disappointment to many physicists, who have an almost spiritual feeling that nature must be "beautiful" in a certain special mathematical sense.

CHAPTER FOUR

The Myth of Uniqueness and Elegance

> "God used beautiful mathematics in creating the world."
>
> — PAUL DIRAC

> "If you are out to describe the truth, leave elegance to the tailor."
>
> — ALBERT EINSTEIN

> "Beauty is worse than wine, it intoxicates both the holder and beholder."
>
> — ALDOUS HUXLEY

What Physicists Mean by Beautiful

The anthropic controversy is about more than scientific facts and philosophical principles. It is about what constitutes good taste in science. And like all arguments about taste, it involves people's esthetic sensibilities. The resistance to anthropic explanations of natural facts derives, in part, from the special esthetic criteria that have influenced all great theoretical physicists — Newton, Einstein, Dirac, Feynman — right down to the current generation. To understand the strong feelings involved, we must first comprehend the esthetic paradigm that is being challenged and threatened by dangerous new ideas.

Having spent a good part of a lifetime doing theoretical physics, I am personally convinced that it is the most beautiful and elegant of all the sciences. I'm pretty sure that my physicist friends all think the same. But most of us have no clear idea of what we mean by beauty in physics. When I have raised the question in the past, the answers varied. The most common was that the equations are elegant. A few answered that the actual physical phenomena are beautiful.

Physicists no doubt have esthetic criteria by which they judge theories. Conversations are peppered with words like *elegant, beautiful, simple, powerful, unique,* and so on. Probably no two people mean exactly the same thing by these words, but I think we can give broad definitions on which physicists will more or less agree.

If there is a difference between elegance and simplicity, it is too subtle for me. Mathematicians and engineers also use these terms more or less interchangeably, and they mean roughly the same thing as they do to physicists. An elegant solution to an engineering problem means one that uses the minimal amount of technology to accomplish the task at hand. Making one component serve two purposes is elegant. The minimal solution is the most elegant.

In the 1940s the cartoonist Rube Goldberg specialized in designing "Rube Goldberg machines," which were fanciful, silly solutions to engineering problems. A Rube Goldberg alarm clock would have balls rolling down roller-coaster tracks, hammers tapping birds who pulled strings, and the whole thing ending with a bucket of water being poured on the sleeper. A Rube Goldberg machine was a decidedly inelegant solution to a problem.

Rube Goldberg's Picture-Taking Machine

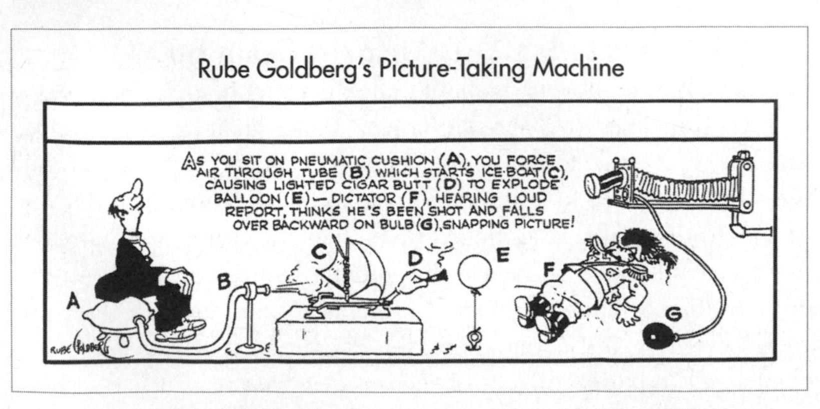

AS YOU SIT ON PNEUMATIC CUSHION (A), YOU FORCE AIR THROUGH TUBE (B) WHICH STARTS ICE-BOAT (C), CAUSING LIGHTED CIGAR BUTT (D) TO EXPLODE BALLOON (E)— DICTATOR (F), HEARING LOUD REPORT, THINKS HE'S BEEN SHOT AND FALLS OVER BACKWARD ON BULB (G), SNAPPING PICTURE!
A
B
C
D
E
F
G
RUBE GOLDBERG

Solutions to mathematical problems can similarly be evaluated in terms of elegance. A proof of a theorem should be as lean as possible, meaning that the number of assumptions, as well as the number of steps, should be kept to the minimum. A mathematical system such as Euclidean geometry should be based on the minimal number of axioms. Mathematicians love to streamline their arguments, sometimes to the point of incomprehensibility.

The theoretical physicist's idea of elegance is fundamentally the same as the engineer's or mathematician's. The General Theory of Relativity is elegant because so much flows out of so little. Physicists also like their axioms simple and few in number. Any more than is absolutely essential is inelegant. An elegant theory should be expressible in terms of a small number of equations, each of which is simple to write. Long equations with too many symbols crowded together are a sign of an inelegant theory or perhaps a theory that is expressed in a clumsy way.

Where did this esthetic taste for simplicity come from?[1] It's not just engineers, mathematicians, and physicists who derive a sense of satisfaction from a neat solution to a problem. My father was a plumber with a fifth-grade education. But he relished the symmetry and geometry of well-placed piping. He took deep professional pride in finding clever ways to minimize the pipe needed to run a water line from one point to another — without violating the esthetic rules of parallelism, rectangularity, and symmetry. It wasn't because of the money he could save by cutting down on materials. That was trivial. His pleasure at an ingenious simplification and an elegant geometry was not so different from my own when I find a neat way to write an equation.

Uniqueness is another property that is especially highly valued by theoretical physicists. The best theories are ones that are unique in two senses. First of all, there should be no uncertainty about their consequences. The theory should predict all that is possible to predict and no more. But there is also a second kind of uniqueness that would be especially treasured in what Steven Weinberg calls a final theory. It is a

1. These remarks about simplicity are intended to apply only to the engineer/mathematician/physicist's kind of simplicity. I hold no opinions about simplicity and complexity in music, poetry, or any other artform.

kind of inevitability — a sense that the theory could not be any other way. The best theory would be not only a theory of everything, but it would be the *only* possible theory of everything.

The combination of elegance, uniqueness, and the power to answer all answerable questions is what makes a theory beautiful. But I think physicists would generally agree that no theory yet devised has fully lived up to these criteria. Indeed, there is no reason why any but the final theory of nature should be perfect in its beauty.

If you asked theoretical physicists to rank all theories esthetically, the clear winner would be the General Theory of Relativity. Einstein's ideas were motivated by an elementary fact about gravity that every child can understand: the force of gravity feels the same as the force due to acceleration. Einstein performed a thought experiment in an imaginary elevator. His point of departure was the fact that in an elevator it is impossible to distinguish between the effects of a gravitational field and the effects of upward acceleration. Anyone who has been on a high-speed elevator knows that for the brief period of upward acceleration, you feel heavier: the pressure on the bottoms of your feet, the pull on your arms and shoulders feel exactly the same whether caused by gravity or the elevator's increasing velocity. And during the deceleration you feel lighter. Einstein turned this trivial observation into one of the most far-reaching principles of physics: the principle of equivalence between gravity and acceleration, or more simply, the *equivalence principle.* From it he derived the rules that govern all phenomena in a gravitational field as well as the equations for the non-Euclidean geometry of space-time. It is all summarized in a few equations, the Einstein equations, with universal validity. I find that beautiful.

This brings up another facet of what beauty means to some physicists. It's not just the final product of Einstein's work on gravity that I find pleasing. For me a great deal of the beauty lies in the way he made the discovery: how it evolved from a thought experiment that even a child can understand. And yet I have heard physicists claim that if Einstein had not discovered the General Theory of Relativity, they or someone else would have soon discovered it in a more modern, more technical, but in my opinion, much less beautiful way. It's interesting to compare the two routes to Einstein's equations. According to these alternate-world historians, they would have attempted to build a theory along the lines of Maxwell's electrodynamics. Maxwell's theory consists of a set of eight equations whose solutions describe wavelike motions of the electromagnetic field. Those same equations also imply the ordinary forces between magnets and between electric charges. It is not the phenomena but the form of the equations that would have been the inspiration for modern-day theorists. The starting point would have been an equation for gravitational waves, similar in form to the equations describing light or sound waves.[2]

Just as light is emitted from a vibrating charge, or sound from an oscillating tuning fork, waves of gravity are emitted by rapidly moving masses. While the equations describing the waves are mathematically consistent, trouble would arise when the waves were allowed to interact with massive objects. Inconsistencies would arise that do not occur in Maxwell's theory. Undaunted, the theorists would have searched for extra terms to add to the equations to make them consistent. By trial and error they would find a series of successive approximations, each being better than the last. But at any given stage, the equations would still be inconsistent.

Consistency would be achieved only when an infinite number of terms were summed together. Moreover, when all the terms were added, the result would be exactly equivalent to Einstein's equations! By a series of successive approximations, a route would have been found to a unique theory that would be equivalent to general relativity. There would be no need to ever think about accelerating elevators. The mathematical requirement of consistency, together with the method of successive approximations, would suffice. For some this is beautiful. It could hardly be called simple.

As for the elegance of the equations, I will display them in the wonderfully simple form that Einstein derived.

$$R_{mn} - \tfrac{1}{2} g_{mn} R = T_{mn}$$

2. Very similar equations describe the electromagnetic disturbances such as light, the pressure disturbances called sound waves, and the waves traveling up and down a long rope when you shake one end of it. Collectively, the equations for these types of phenomena are called wave equations.

This small box with a few simply placed symbols contains the entire theory of gravitational phenomena: the falling of stones, the motion of the moon and earth, the formation of galaxies, the expansion of the universe, and much more.

The approach advocated by the modernists, although yielding the same content, would lead to an open-ended infinity of successive approximations. In that form the equations are distinctly inelegant.

Nevertheless, I have to admit that while the "modern derivation" may have missed the elegance of the Einstein equations, it did one thing rather well. It demonstrated the uniqueness of the theory. At each level of approximation, the extra terms needed to restore consistency are uniquely determined; the theory is unambiguous. Not only does it describe how gravity works, but it also shows that it could not have been otherwise.

The theory is also powerful. It can describe with great accuracy a very wide variety of gravitational phenomena from the earth's ability to hold us to its surface to black holes at the center of quasars and gravitational waves from violent collisions of such black holes. With its elegant equations, an element of uniqueness, and the power to describe many phenomena, the General Theory of Relativity is the most beautiful physical theory yet devised. But as we have seen, it's not only the content of a theory — what it says about the world — that makes it beautiful but also the form in which the equations are written and even the reasoning that went into its discovery.

If the physics beauty contest would be won by general relativity, the ugly-man prize would have to go to nuclear physics. The problem with nuclear physics is not that it leads to the ugliness of nuclear reactors and mushroom clouds. That's technology, not physics. The problem is that the laws of nuclear physics are not clear or concise. As a result no elegant equations can capture its content, and no simple inevitable reasoning led to the discovery of its rules. If the rules simply declared that protons and neutrons attract each other according to some very simple force law, then the theory would be as elegant as atomic physics. But like the revisionist relativity, each approximation to the truth is defective. However, instead of appealing to mathematical consistency to improve it, various ad hoc rules of thumb must be introduced to make the theory agree with the properties of nuclei. Moreover, the rules of thumb that work for some nuclei don't work for others. There is a boatload of different approximation schemes and successive trial-and-error strategies, but unlike the case of general relativity, it doesn't add up to something simple, unique, and universally valid. Most theoretical physicists agree that the equations of nuclear physics are not in the least elegant, nor is their logic especially compelling.

Some physicists would claim that chemistry is ugly. Chemistry is also full of ad hoc recipes that have no universal validity. The first few lines of the periodic table are simple enough, but as you proceed down the table, more and more assumptions have to be added. The rules for molecular bonding are approximate and have plenty of exceptions. Sometimes they predict correctly and sometimes not. Whenever physicists want to disparage something as being unmotivated or overly complicated, they dismiss it as chemistry or even cookbook chemistry.

But a chemist might answer that physics is boring and impoverished. Chemistry is the subject that describes and explains the beauty and variety of the natural world. A flower, after all, is a collection of chemicals undergoing chemical reactions. To a scientific mind, understanding these processes adds to their esthetic value.[3] Many physicists and chemists look for beauty in the way in which very simple structures, like atoms, can combine themselves into macroscopic patterns. These phenomena, which are meaningful only for large numbers of atoms, are called *collective* or *emergent*. Phenomena that involve the collective behavior of many constituents *emerge* from the laws of simple elements such as

3. Apparently not everyone agrees. Walt Whitman wrote:
"When I heard the learn'd astronomer,
When the proofs, the figures, were ranged in columns before me,
When I was shown the charts and diagrams, to add, divide, and measure them,
When I sitting heard the astronomer where he lectured with much applause in the lecture room,
How soon unaccountable I became tired and sick,
Till rising and gliding out I wander'd off by myself,
In the mystical moist night-air, and from time to time,
Look'd up in perfect silence at the stars."

— *When I Heard the Learn'd Astronomer*

Personally, I prefer Alexander Pope's sentiment:
"Nature and nature's laws lay hid in night;
God said 'Let Newton be' and all was light."

— *Epitaph for Isaac Newton*

atoms. Life is a collective phenomenon. So is the formation of a snowflake or the way atoms line up, side by side, to form a lovely crystal like a diamond. Yet another example is the cooperative behavior that allows many atoms to move without friction in a superconducting material. This is emergent beauty.

Who's to say that beauty of this type has less merit than the reductionist-particle physicist's version? Not I. But the kind of beauty that I am speaking about is different. Elementary-particle physicists search for beauty in the underlying laws and equations. Most have had a kind of quasireligious belief in the gods of uniqueness and simplicity. As far as I can tell, they believe that at the "bottom of it all" lies a beautiful theory, a single unique powerful and compelling set of equations describing all phenomena, at least in principle, even if the equations are too hard to solve. These master equations should be simple and symmetric. Simplicity, roughly speaking, means it should be possible to write the equations in a box about this big:

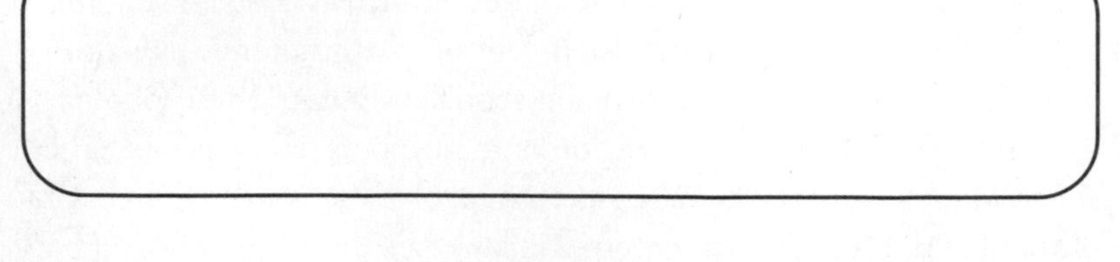

But above all, the equations should uniquely predict the Laws of Physics that have been discovered over the last few centuries, including the Standard Model of particle physics: the list of elementary particles, their masses, coupling constants, and the forces among them. No other alternative rules should be possible.

Origins of the Myth

The myth of uniqueness and elegance probably originated with our Greek intellectual forebears. Pythagoras and Euclid both believed in a mystical mathematical harmony of the universe. Pythagoras believed that the world operated according to mathematical principles that paralleled those governing music. While the connection between music and physics may seem to us naive and even silly, it's not hard to see in the Pythagorean creed the same love of symmetry and simplicity that so motivates modern physicists.

Euclidean geometry also has a strong esthetic flavor to it. Proofs should be as simple and elegant as possible, and the number of unproved axioms should be as small as possible: no more than five of them are needed. Euclidean geometry is usually considered a branch of mathematics. But the Greeks made no distinction between mathematics and physics. For them, Euclidean geometry was the theory of how real physical space behaves. You could not only prove theorems, but you could also go out and measure the properties of space, and they would necessarily (according to the Greeks) agree with the theorems. For example, you may draw a triangle using a ruler and pencil and then with a protractor measure the three interior angles. One of the theorems of Euclidean geometry states that the sum of these angles is exactly 180 degrees. The Greeks believed that any real triangle, drawn in real space, would agree with the theorem. Thus, they were making statements about the physical world, statements they believed were not only true but also unique. Space, they believed, conforms to Euclid's axioms, and moreover, it could be no other way. Or so they thought.

Plato and Aristotle went further, adding a particularly esthetic element to the laws of astronomy. To them the circle was the perfect figure. Every point being equally distant from the center, the circle has perfect symmetry; no other figure is as symmetric. So Plato, Aristotle, and their later followers believed that no other figure could control and regulate the motions of planets. They believed the heavens were composed of a set of elegant crystal shells, perfectly transparent, perfectly round, and moving with perfect clockwork precision. They could not conceive of it being any other way.

The Greeks had an equally elegant theory of terrestrial phenomena, which paralleled modern-day physicists' hopes for a unified theory. They believed that all earthly matter was made of four elements: earth, air, water, and fire. Each had its appropriate place and tended to migrate toward that place. Fire was the lightest and, therefore, tended to rise. Earth, being the heaviest, tended to sink to the lowest elevation. Water and air were in between. Four elements and one dynamical principle: you would be surprised at how much they could explain. The only thing missing was uniqueness. I don't see why there can't be additional elements: earth, air, fire, water, red wine, cheese, and garlic.

In any case, astronomers, alchemists, and chemists threw a wrench into the Greek schemes. Johannes Kepler knocked the circle from its high pedestal and replaced it with the more varied and less symmetrical elliptical planetary orbits. But Kepler, too, believed in Pythagorean mathematical harmony. At the time there were only five known planets: Venus, Mars, Jupiter, Saturn, and of course, Earth. Kepler was deeply impressed by the fact that there are exactly five regular polyhedra, five Platonic solids: the tetrahedron, octahedron, icosahedron, cube, and dodecahedron.[4] Kepler was seduced by what he saw as an inevitable and beautiful correspondence between the planets and the Platonic solids. And so he built a mathematical model of the universe as a set of nested polyhedra, with the goal of explaining the radii of their orbits. I'm not sure we would find it elegant, but he did, and that's the important thing. Elegant or not, the five Platonic solids are unique. The theory, of course, was utter nonsense.

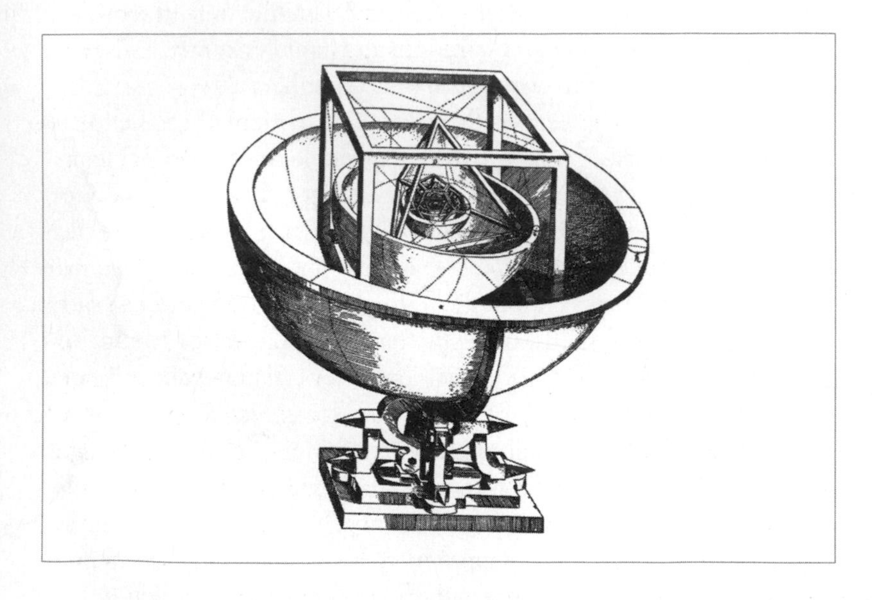

4. The five Platonic solids are the only polyhedra (three-dimensional analogs of polygons) whose faces are all identical polygons. The tetrahedron, octahedron, and icosahedron are all fashioned from triangles. The cube and dodecahedron are made of squares and pentagons, respectively.

At the same time alchemists could not make progress without recognizing the need for more than four elements. By the end of the nineteenth century, they had identified almost one hundred elements, and nature was losing some of its simplicity. The periodic table put some order into chemistry, but it was a far cry from the simplicity and uniqueness that the Greeks had demanded.

But then, in the early twentieth century, Bohr, Heisenberg, and Schrödinger discovered the principles of quantum mechanics and atomic physics, thus providing a precise foundation for chemistry. The number of elements swung back to only four: not the Greek four, but the photon, electron, proton, and neutron. All of chemistry can be (in principle only) unambiguously derived from the quantum mechanical theory of these four elementary particles. Simplicity, elegance, and uniqueness were again gaining the upper hand. The basic principles of relativity, quantum mechanics, and the existence of the four elements would give rise to all chemical behaviors if only we had the power to solve the equations. This was very close to the physicist's ideal.

But, alas, it was not to be. Elementary particles were discovered wholesale: neutrinos, muons, so-called strange particles, mesons, and hyperons, none of them having a place in the simple scheme of things. They played no important role in the description of matter, yet there they were, muddying the waters. Elementary-particle physics in the 1960s was a hopeless muddle of Greek and Latin letters denoting hundreds of supposed elementary particles. As a young physicist hoping to find beauty and elegance in the laws of nature, I found the whole mess depressing.

But then the 1970s improved the outlook. Quarks replaced the protons, neutrons, and mesons as the constituents of the nucleus, and — as discussed in chapter 1 — the single quantum field theory called Quantum Chromodynamics (QCD) could explain everything about protons, neutrons, mesons, and nuclei and the less familiar, so-called strange particles (particles containing strange-quarks). The number of elements had decreased somewhat. At the same time electrons and neutrinos were understood to be twins related by a deep, underlying symmetry. The tug of war was again shifting to the advantage of simplicity. Finally, the Standard Model came into full existence in the mid-1970s, providing a complete description of all known phenomena (or so it is sometimes claimed), but in terms of about thirty arbitrary parameters. The struggle between elegance and clumsiness is ongoing and shows no signs of a final victorious resolution.

String Theory and Unraveling the Myth

Now to String Theory — is it beautiful, as string theorists will tell you, or is it the overcomplicated monstrosity that critics of the theory claim? But before discussing the esthetics, let me fill you in about why we need String Theory at all. If, as I said, the Standard Model describes all known phenomena, why are theoretical physicists driven to search for a deeper mathematical structure? The answer is that the Standard Model does not describe all known phenomena. There is at least one obvious exception: gravity. Gravitation is the most familiar force in daily life, and probably the most fundamental, but it is not found anywhere in the Standard Model. The graviton (the quantum of the gravitational field) is not on the list of Standard Model particles. Perhaps the most interesting objects of all, black holes, have no place in the theory. Although Einstein's classical theory of gravity may be the most beautiful of all theories, it just doesn't seem to fit into the quantum world.

For most purposes gravity is completely unimportant for elementary-particle physics. As we will see in later chapters, the gravitational force between particles — for example, quarks in a proton — is many orders of magnitude weaker than the other forces of nature. Gravity is far too feeble to play a role in any experiment involving elementary particles, at least for the foreseeable future. For this reason traditional elementary-particle physicists have been content to completely ignore the effects of gravity.

But there are two practical reasons for wanting a deeper understanding of the connections between gravity and the microscopic quantum world. The first has to do with the structure of elementary particles. Although the gravitational force is negligible for electrons in an atom (or quarks in a proton), as the distance between particles gets smaller, gravity begins to assert itself. All forces become stronger as the separation decreases, but the gravitational force increases more rapidly than any other. In fact by the time two particles get within a Planck distance of each other, the force of gravity is a good deal stronger than electric forces or even the forces that bind quarks together. If the "Russian nesting doll" paradigm (things made of ever smaller things) continues to hold sway, the ordinary elementary particles may turn out to be made of even tinier objects that are in some sense held together by gravity.

The second practical reason for understanding the links between gravity and quantum theory involves cosmology. In the next chapter we will see that gravity is the force that governs the growth of the universe. When the universe was very young and expanding at a stupendous rate, gravity and quantum mechanics were important in equal measure. A lack of understanding of the connection between these two great theories will ultimately frustrate our efforts to get to the bottom of the Big Bang.

But there is a third reason that physicists are driven to combine quantum theory with general relativity: an esthetic reason. For a physicist, unlike a poet, the biggest crime against esthetics is the crime of inconsistency. Even worse than an ugly theory, incompatible principles are an attack on the basic values we hold dear. And for most of the twentieth century, gravity and quantum mechanics have been incompatible.

That's where String Theory comes in. We won't get to the details of String Theory until chapter 7, but for now let us just say that it is a mathematical theory that consistently unifies gravity with quantum mechanics. Many theoretical physicists (I count myself among them) have a strong feeling that String Theory is our best hope for eventually reconciling these two great but clashing pillars of modern science. What is it about String Theory that makes us feel that way? We have tried lots of other approaches, but they typically flop at the outset. An example is the attempt to build a quantum field theory based on general relativity. The mathematics rapidly becomes inconsistent. But even if the equations made sense, there would be an esthetic disappointment. In every such attempt gravity is an optional "add-on." By that I mean that gravity is just added to some preexisting theory such as Quantum Electrodynamics. There is nothing at all inevitable about it. But String Theory is different. The existence of both gravity and quantum mechanics is absolutely essential for mathematical consistency. String Theory makes sense *only* as a theory of "quantum gravity." That is not a small thing, given the way these two giants — gravity and quantum mechanics — have been at war with each other for most of the twentieth century. I would say that this inevitability is beautiful.

In addition to its close connection with gravity, String Theory looks as if it has connections with ordinary elementary-particle physics. By no means have we understood exactly how to incorporate the Standard Model into String Theory, but it does have all the elements that go into modern particle theory. It has particles — fermions and bosons — that resemble electrons, quarks, photons, gluons, and all of the others. In addition to gravitational forces, at work are forces similar to electric and magnetic forces between charged particles and even forces similar to those that bind quarks into protons and neutrons. None of these things is put in by hand. Like gravity, they also are inevitable mathematical consequences of the theory.

Excitingly, all of String Theory's consequences have unfolded in a mathematically consistent way. String Theory is a very complex mathematical theory with very many possibilities for failure. By failure I mean internal inconsistency. It is like a huge high-precision machine, with thousands of parts. Unless they all fit perfectly together in exactly the right way, the whole thing will come to a screeching halt. But they do fit together, sometimes as a consequence of mathematical miracles. String Theory is not only a physical theory about nature. It is also a very sophisticated mathematical structure that has provided a great deal of inspiration for pure mathematicians.

But is String Theory beautiful? Does String Theory live up to the standards of elegance and uniqueness that physicists demand? Are its equations few and simple? And, most important, are the Laws of Physics implied by String Theory unique?

Elegance requires that the number of defining equations be small. Five is better than ten, and one is better than five. On this score, one might facetiously say that String Theory is the ultimate epitome of elegance. With all the years that String Theory has been studied, no one has ever found even a single defining equation! The number at present count is zero. We know neither what the fundamental equations of the theory are nor even if it has any. Well then, what is the theory, if not a collection of defining equations? We really don't know.

As for the second question — are the Laws of Physics defined by String Theory unique? — here we can be more definite. Although no one can identify the defining equations, the methodology of the theory is very rigorous. It could easily have failed any of a large number of mathematical consistency tests. It didn't, but it was thought that the very tight mathematical constraints would lead to either a completely unique theory or, at most, a very small number of possibilities. There was a great sense of euphoria in the mid-1980s, when string theorists thought they were zeroing in on the final answer, a single, unique theory that would explain why the world is the way it is. It was also believed that the deep and often miraculous mathematical properties of the theory would guarantee that the cosmological constant was exactly zero.

The superintellectual, rarefied atmosphere of the Institute for Advanced Study, in Princeton — once the home of both Albert Einstein and J. Robert Oppenheimer — was the center of this excitement. And at the center of the center were some of the greatest mathematical physicists in the world. Edward Witten and the people around him seemed to be making rapid strides toward a unique answer. That was then.

Today we know that the success "just around the corner" was a mirage. As we learned more about the theory, three unfortunate things began to happen. Number one was that new possibilities kept turning up, new mathematically consistent versions of what was supposed to be a unique theory. During the 1990s the number of possibilities grew exponentially. String theorists watched with horror as a stupendous Landscape opened up with so many valleys that almost anything can be found somewhere in it.

The theory also exhibited a nasty tendency to produce Rube Goldberg machines. In searching the Landscape for the Standard Model, the constructions became unpleasantly complicated. More and more "moving parts" had to be introduced to account for all the requirements, and by now it seems that no realistic model would pass muster with the American Society of Engineers — not for elegance in any case.

Finally, adding insult to injury, the potential candidates for a vacuum like the one we live in all have a nonzero cosmological constant. The hope that some elegant mathematical magic of String Theory will guarantee a zero value for the cosmological constant is rapidly fading.

Judged by the ordinary criteria of uniqueness and elegance, String Theory has gone from being Beauty to being the Beast. And yet the more I think about this unfortunate history, the more reason I think there is to believe that String Theory is the answer.

Is Nature Elegant?

"The great tragedy of science — the slaying of a beautiful hypothesis by an ugly fact."

— THOMAS HENRY HUXLEY

String Theory has no lack of enemies who will tell you that it is a monstrous perversion. Among them are condensed-matter theorists who think the right theory is emergent. Condensed-matter physics is the study of the properties of ordinary matter in solid, liquid, or gaseous form. According to this school, space and time emerge from some unspecified microscopic objects in the same way that crystal lattices and superconductors emerge from the collective behavior of large numbers of atoms. In many cases emergent behavior hardly depends on the particular microscopic details. In the view of condensed-matter physicists, the world may emerge from such a wide variety of microscopic starting points that there is no point in trying to identify the microscopic details. Instead, it is argued, physicists should be trying to understand the rules and mechanisms of emergence itself. In other words, they should study condensed-matter physics.

The trouble with this view is that no ordinary condensed-matter system can ever behave anything like a universe regulated by quantum mechanics together with Einstein's laws of gravity. Later, when we meet the *Holographic Principle*, in chapter 10, we will see that there are profound reasons for this. The idea that there are many microscopic starting points that can lead to a world with gravity may be true, but none is anything like the ordinary materials that condensed-matter physicists study.

Another source of criticism is from some (certainly not all) high-energy experimental physicists who are annoyed that the new phenomena implied by String Theory are too remote from experiment, as if that were the theorists' fault. These physicists are troubled because they can't see how their experiments can ever address the questions that string theorists are trying to answer. They suggest that theorists keep to problems that directly address the near-term future experimental agenda. This is an extremely myopic view. In the present age, high-energy physics experiments have become so large and complicated that they take decades to complete. Brilliant young theoretical physicists are like restless explorers. They want to go where their curiosity about the world takes them. And if it's out into the great sea of the unknown, so be it.

Most really good experimental physicists don't pay too much attention to what theorists think. They build the machines they can build and do the experiments they can do. Most really good theoretical physicists don't pay much attention to what experimenters think. They build their theories based on their own instincts and go where intuition leads them. Everyone hopes that at some point the two paths will cross, but exactly when and how is anybody's guess.

Finally, there are proponents of other theories. That's as it should be. Other avenues need to be explored, but as far as I can tell, none of these theories is very well developed. At present they have very little to say.

What I have never heard is criticism based on the unfortunate inelegance or the lack of uniqueness of String Theory.[5] Either of these tendencies might be thrown back at the string theorists as evidence that their own hopes for the theory are misguided. Perhaps part of the reason that the enemies of String Theory haven't pounced is that string theorists have kept their Achilles heel under wraps until fairly recently. I suspect that now that it is becoming more public, partly through my own writings and lectures, the kibitzers on the sidelines will be grinning and loudly announcing, "Ha ha, we knew it all along. String Theory is dead."

My own guess is that the inelegance and lack of uniqueness will eventually be seen as strengths of the theory. A good, honest look at the real world does not suggest a pattern of mathematical minimality. Below is a list of the masses of the elementary particles of the Standard Model, expressed in terms of the electron mass. The numbers are approximate.

Particle	Mass
photon	0
gluon	0
neutrino	less than 10^{-8} but not zero
electron	1

5. This remark was written in the spring of 1994, but by the time I completed writing *The Cosmic Landscape* a year later, the vultures had descended in force.

up-quark	8
down-quark	16
strange-quark	293
muon	207
tau lepton	3447
charmed-quark	2900
bottom-quark	9200
W-boson	157,000
Z-boson	178,000
top-quark	344,000

There is very little pattern here other than the obvious increase as we go down the list.

The numbers don't seem to have any simple connection to special mathematical quantities like π or the square root of two. The only reason any pattern exists at all is that I purposely listed the particles in order of increasing mass.

These dozen numbers are just the tip of an iceberg. We know with certainty that in the Standard Model at least twenty additional independent coupling constants governing a wide range of different forces belie claims of simplicity. Even that list is probably far from exhaustive. There is more to the world than just the Standard Model of particle physics. Gravitation and cosmology introduce many new constants, such as the mass of dark-matter particles.[6] The consensus among particle physicists, especially those who expect supersymmetry to be a feature of nature, is that well over one hundred separate constants of nature are in no known way related. Far from being the simple, elegant structure sometimes suggested by physicists, the current most fundamental description of nature seems like something Rube Goldberg himself might have designed. A Rube Goldberg theory, then, may be fitting.

While the Standard Model is a huge advance in describing elementary particles, it doesn't explain itself. It is rather complicated, far from unique, and certainly incomplete. What, then, is special about our beloved Standard Model? Absolutely nothing — there are 10^{500} others,

6. See chapter 5 for an explanation of dark matter.

just as consistent. Nothing, that is, except that it permits — maybe even encourages — the existence of life.

Cosmologists are not usually as infected by the elegance-uniqueness bug as string theorists — probably because they are more likely to take a good hard look at nature rather than at mathematics. What some of them see is a bunch of remarkable coincidences:

- The universe is a fine-tuned thing. It grew big by expanding at an ideal rate. If the expansion had been too rapid, all of the material in the universe would have spread out and separated before it ever had a chance to condense into galaxies, stars, and planets. On the other hand, if the initial expansion had not had a sufficient initial thrust, the universe would have turned right around and collapsed in a big crunch much like a punctured balloon.

- The early universe was not too lumpy and not too smooth. Like the baby bear's porridge, it was just right. If the universe had started out much lumpier than it did, instead of the hydrogen and helium condensing into galaxies, it would have clumped into black holes. All matter would have fallen into these black holes and been crushed under the tremendously powerful forces deep in the black hole interiors. On the other hand, if the early universe had been too smooth, it wouldn't have clumped at all. A world of galaxies, stars, and planets is not the generic product of the physical processes in the early universe; it is the rare and, for us, very fortunate, exception.

- Gravity is strong enough to hold us down to the earth's surface, yet not so strong that the extra pressure in the interior of stars would have caused them to burn out in a few million years instead of the billions of years needed for Darwinian evolution to create intelligent life.

- The microscopic Laws of Physics just happen to allow the existence of nuclei and atoms that eventually assemble themselves into the large "Tinkertoy" molecules of life. Moreover, the laws are just right, so that the carbon, oxygen, and other necessary elements can be "cooked" in first-generation stars and dispersed in supernovae.

The basic setup looks almost too good to be true. Rather than following a pattern of mathematical simplicity or elegance, the laws of nature seem specially tailored to our own existence. As I have repeatedly said, physicists hate this idea. But as we will see, String Theory seems to be an ideal setup to explain why the world is this way.

Let us return now to hard science issues. In the next chapter I will explain the surprising — amazing may not be too strong a word — cosmological developments that have been pushing physics and cosmology toward a new paradigm. Most significantly I will explain what we have learned about the early prehistory of our universe — how it arrived at its present precarious condition — and the shocking facts concerning the 120th decimal place of the cosmological constant.

CHAPTER FIVE

Thunderbolt from Heaven

> "I'm astounded by people who want to 'know' the universe when it's hard enough to find your way around Chinatown."
>
> — WOODY ALLEN

Alexander Friedmann's Universe

Mention of the year 1929 brings shudders to anyone old enough to remember it: bank runs, Wall Street suicides, mortgage foreclosures, unemployment. It was the year that brought on the Great Depression. But it wasn't all bad. On Wall Street the stock market did collapse like a popped balloon, but out in sunny California Edwin Hubble discovered the Big Bang, an explosion out of which the entire known universe was born. As previously noted, contrary to what Einstein had thought back in 1917, the universe changes and grows with time. According to Hubble's observations, the distant galaxies are all rushing away from us, as if they had been shot out of a gigantic cannon, a cannon that could shoot in all directions, and from every location, simultaneously. Hubble not only discovered that the universe is changing: he discovered that it is growing like an expanding balloon!

Hubble's technique for measuring the motion of a galaxy was a familiar one. The light from a galaxy was sent through a spectroscope that breaks it up into its component wavelengths. Isaac Newton, way back in the seventeenth century, did exactly that when he sent white sunlight through a triangular prism. The prism, a simple spectroscope, broke the sunlight into all the colors of the rainbow. Newton concluded, rightly, that sunlight is a composite of red, orange, yellow, green, blue, and violet light. Today, we know that each color of the spectrum corresponds to a wave of a particular (wave)length.

If one looks very carefully at the spectrum of starlight, some extremely narrow dark *spectral lines* can be seen superimposed on the rainbow of colors.

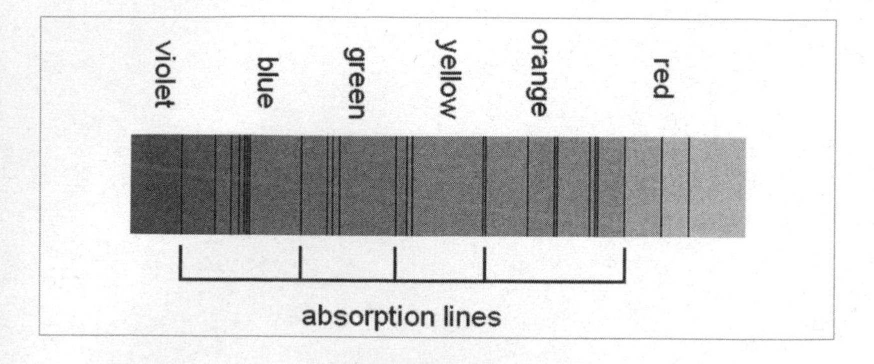

These mysterious lines of missing light are called *absorption lines*. They indicate that something along the line of sight has absorbed certain discrete wavelengths (colors) without disturbing the rest of the spectrum. What causes this curious phenomenon? The quantum mechanical behavior of electrons.

According to Bohr's original quantum theory of the atom, electrons in atoms move in quantized orbits. Newtonian mechanics would allow the electron to orbit at any distance from the nucleus. But quantum mechanics constrains them to move like motor vehicles that are required, by law, to stay in definite lanes. Being between lanes violates the traffic laws: being between quantized orbits violates the laws of quantum mechanics. Each orbit has its own energy, and for an electron to jump from one orbit to another, the energy of the electron has to change. If an electron jumps from an outer to an inner orbit, it must radiate a photon to carry off the excess energy. Conversely, an inner electron can jump to a more distant orbit only if it gains some energy, possibly by absorbing a photon.

Normally an electron moves in the innermost available orbit that is not blocked by other electrons (remember that the Pauli exclusion principle prevents any two electrons from occupying the same quantum state). But if the atom is struck by another object, an electron can absorb some energy and make a *quantum jump* to a new orbit, one farther from the nucleus. The atom is temporarily *excited,* but eventually the electron will emit a photon and drop back to its original orbit. The light radiated in this way has definite wavelengths that are characteristic of the type of atom. Each individual chemical element has its own signature, a set of spectral lines that corresponds to these quantum jumps.

If a photon of the right color falls on an unexcited atom, the reverse process can happen: the photon can be absorbed while the electron jumps to a more energetic orbit. This has an interesting effect on starlight passing through hydrogen gas in the atmosphere surrounding a star. The hydrogen will deplete the starlight of precisely those colors that characterize the hydrogen spectrum. If helium or carbon or any other element is present, it will also leave its distinguishing mark on the starlight. Studying the spectra of starlight is the way we know about the chemical makeup of stars. But our interest right now is not the chemical composition but rather the velocity of the star. The point is that the exact details of the absorption spectrum, as seen from earth, depend on the relative velocity between us and the star. The key is the Doppler effect.

If you have heard the siren of a police car as it speeds past, then you've experienced the Doppler effect. The high-pitched whine, "eeee," of the approach gives way to the lower sound, "oooo," as the siren recedes. During the approach the sound waves coming toward you are bunched up, and conversely, while the car moves off, they are stretched out. Because wavelength and frequency are closely related, you hear "eeeeooooo." You could amuse yourself by trying to estimate how fast the police car is moving by the magnitude of the frequency shift.

But the Doppler effect is not just an amusement for pedestrians. For astronomers it is nothing less than the key to the structure and history of the universe. The Doppler effect happens to all kinds of waves: sound waves, vibrational waves in crystals, even water waves. Try wiggling your finger in the water while hanging off the side of a slowly moving boat.[1] The ripples spreading out along the direction of motion are bunched up. The ones going in the reverse direction are stretched.

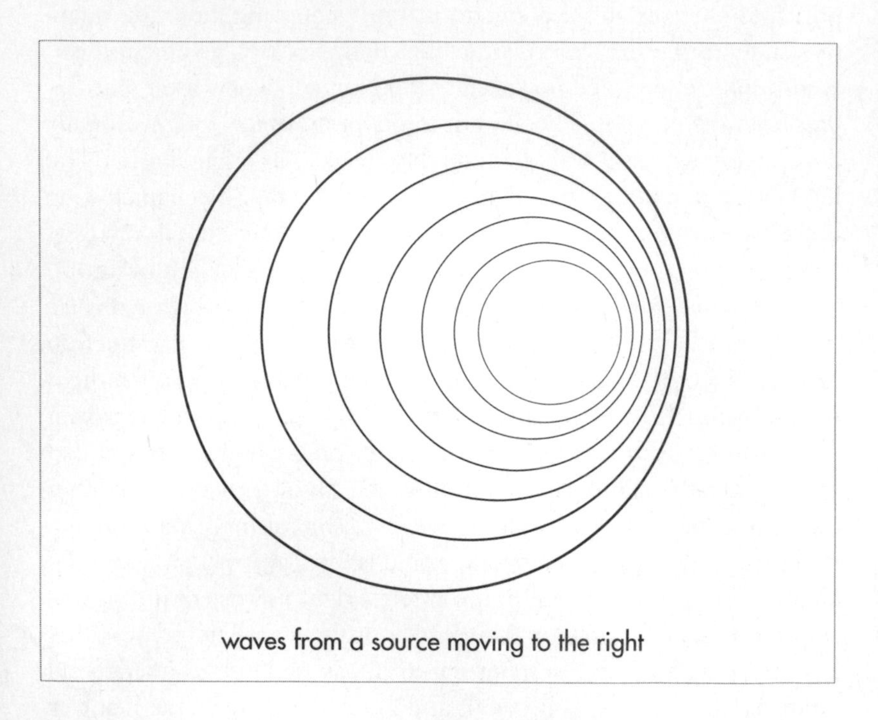

waves from a source moving to the right

Luckily for astronomers light emitted by a moving object does the same thing. A rocket-propelled lemon moving away from you might have the color of an orange or even a tomato if it were going fast enough.[2] While it's moving toward you, you might mistake it for a lime or even a giant blueberry. This is because light from sources moving away from the observer is *redshifted* and light from approaching sources is *blueshifted*. This applies just as well to the light from galaxies as from lemons. Moreover, the amount of shift is a measure of the velocity of the galaxy, relative to the earth.

Hubble used this phenomenon to determine the velocity of a large number of galaxies. He took very accurate spectra of the light coming

1. The boat has to be moving slower than the velocity of the surface waves.
2. To observably change color the lemon would have to move with a significant fraction of the speed of light.

from each galaxy and compared the spectral lines with similar spectra taken in the laboratory. If the universe were static as Einstein originally thought, the galactic and laboratory spectra would have been identical. What Hubble found surprised him and everyone else. The light from every distant galaxy was distinctly on the red side. No doubt about it, Hubble knew that they were all moving away from us. Some galaxies were moving slowly, and some were racing away, but except for a few very nearby galaxies, they were all outward bound. This must have puzzled Hubble. It meant that in the future the galaxies would spread out to ever-farther distances. Even more bizarre, it meant that in the past the galaxies were closer to us, at some point maybe even on top of us!

Hubble was also able to make a rough determination of the distances to the various galaxies, and he found a pattern: the more distant the galaxy, the faster its recessional velocity. The closer galaxies were hardly moving, but the most distant ones were speeding away with tremendous velocities. On a piece of graph paper, Hubble made two axes: on the horizontal axis he plotted the distance to each galaxy; on the vertical axis, its velocity. Each galaxy was plotted as a single point on the graph. What he found was extraordinary; most of the points fell on or near a straight line.

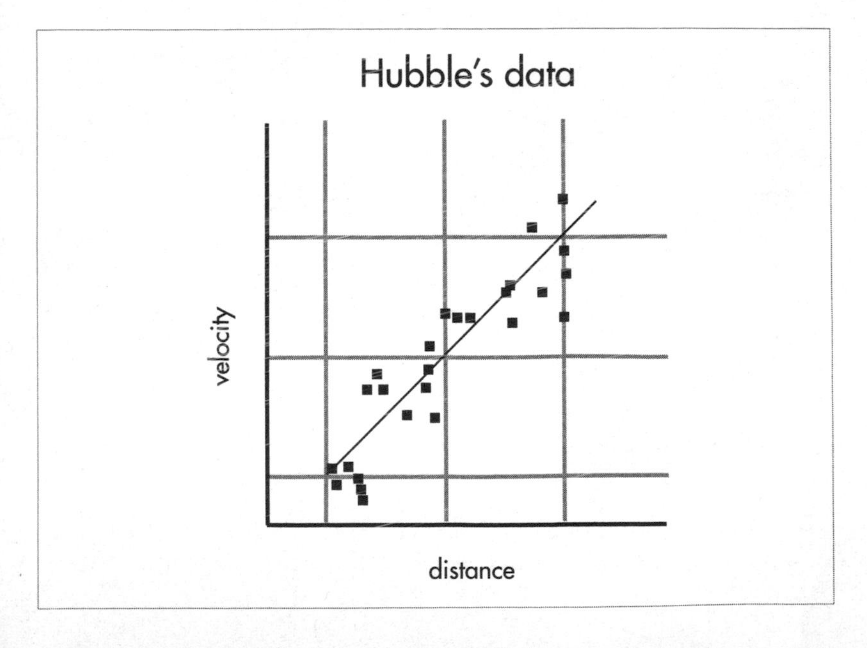

This meant that the recessional velocity not only increased with distance but was directly proportional to the distance. One galaxy, twice as far as another, appeared to recede twice as fast. This was a new, totally unexpected, regularity in the universe: a new cosmological law, Hubble's Law. *Galaxies are receding away from us with velocity proportional to their distance.* An even more precise formulation is: *galaxies are receding away from us with velocity equal to the product of their distance and a numerical parameter called the Hubble constant.*[3]

Well, actually it wasn't completely unexpected. Alexander Friedmann was a Russian mathematician who had studied Einstein's theory of the universe, and in 1922 he published a paper claiming that Einstein might have been wrong in his 1917 paper. He argued that if the universe were not static, if it were changing as time evolved, then the cosmological constant would be superfluous. The Friedmann universe was, like Einstein's, a closed-and-bounded 3-sphere. But unlike Einstein's, Friedmann's universe grew with time. If Einstein's universe was like a static balloon, Friedmann's was like the surface of an expanding balloon. Get yourself a balloon and mark its surface with dots to represent the galaxies. Sprinkle them more or less uniformly. Then slowly blow it up. As the balloon expands, the dots get farther apart, every dot receding from every other dot. No dot is special, but each sees all the others moving away. This was the essence of Friedmann's mathematical universe.

What you may also notice if you are able to watch the dots is that the farther away they are from one another, the faster they separate. In fact the dots would do exactly what Hubble's galaxies did. Hubble's Law is the law of dots on the surface of an expanding balloon. Unfortunately Friedmann died in 1925, before he could learn of Hubble's discovery or the fact that his — Friedmann's — work had laid the foundation for all future cosmology.

Let's review some of that cosmology.

3. The term *Hubble constant* is somewhat misleading, since it changes with time. Long in the past the Hubble constant was much larger than it is today.

The Cosmological Principle and the Three Geometries

> "Only two things are infinite, the universe and human stupidity, and I'm not sure about the former."
>
> — ALBERT EINSTEIN

A couple of years ago I had the good fortune to be invited to South Africa in order to give some lectures in one of the universities. While I was there my wife and I made a trip to the Krueger National Park. The park is an enormous expanse of African veldt and home to all the large mammals of the continent. It was a fabulous experience. In the mornings and evenings we would go out in a Land Rover to view and photograph wildlife. We saw hippos, a rhino, Cape buffalo, a pride of lions devouring antelope, and most impressive of all, an angry rogue male elephant. But for me the most powerful sight of all was the southern sky on a dark, moonless night. The southern sky is much richer than the northern sky that I'm used to, and Krueger is almost completely free of light pollution. The sight of the Milky Way stretched clear across the sky is truly awe-inspiring. But the humbling sense of immensity is deceptive. The entire Milky Way together with all the visible stars is an infinitesimal corner of a much vaster space, homogeneously filled with a hundred billion galaxies, which can be seen only through a large telescope. And even that is a tiny portion of a much bigger cosmos.

According to my dictionary the word *homogeneous* means "uniform in structure or composition throughout." When applied to oatmeal it means nice and smooth, i.e., without lumps. Of course if you look at the oatmeal with a magnifying glass, it hardly looks homogeneous. The point is that when you say something is homogeneous, you must qualify the statement by adding "on scales larger than some specified size." Well-stirred oatmeal is homogeneous on scales larger than an eighth of an inch. Farmer Brown's wheat field in the middle of Kansas is homogeneous on scales larger than a few feet.

Well, not quite. The oatmeal is only homogeneous on scales from an eighth of an inch to the size of the bowl. Farmer Brown's field is homogeneous on scales bigger than ten feet but smaller than a mile. On scales of a mile or more, the countryside looks like a crazy quilt of rectangular fields. The right thing to say is that Farmer Brown's field is homogeneous on scales between ten feet and a fraction of a mile.

Through the naked eye the African night sky is very *inhomogeneous.* The Milky Way is a bright, narrow band of light that divides a much darker background. But a look through a large telescope reveals billions of galaxies that are, on the whole, homogeneously distributed through the observable universe. According to astronomers the universe appears to be homogeneous and *isotropic* on scales larger than a hundred million light-years up to at least fifteen billion light-years. The fifteen billion light-year limit is certainly an underestimate that just represents our inability to see farther.

Returning to my dictionary, I found the following definition for the term *isotropic:* "identical in all directions; invariant with respect to direction." Isotropic is not the same as homogeneous. Here's an example. Once while I was diving near a coral reef in the Red Sea, I saw a huge school of closely spaced, small, thin fish that filled quite a large volume homogeneously. For some strange reason, until I got too close, they all faced the same direction. The school appeared homogeneous over some range of scales, but definitely not isotropic. Every place within the school was like every other place, but every direction was not at all like every other direction. The direction the fish faced was special.

Cosmologists and astronomers almost always assume that the universe is homogeneous and isotropic; no matter where you are in the universe and which direction you are facing, you see the same thing. I don't mean the nearby details but the overall, large-scale features of the universe. Cosmologists call this assumption the cosmological principle. Of course calling it a principle does not make it right. Originally it was just a guess, but gradually better and better observations of several kinds have convinced astronomers and cosmologists that the universe is indeed homogeneous and isotropic over scales ranging from a few hundred million light-years to at least a few tens of billions of light-years. Beyond that we don't know for sure because there is a limit to our observations. It doesn't matter how big our telescope, objects farther than fourteen billion light-years are impossible to observe. The reason is the simple fact that the universe is only about fourteen billion years old. In that time light could not have traveled more than fourteen billion light-years; light from more distant places just hasn't reached us yet. In fact it's a pretty safe bet that the universe is homogeneous and isotropic out to distance scales much larger than the observable part of the universe. But like Farmer Brown's field, the universe may become a crazy quilt at a large enough distance: a patchwork of pocket universes.

For now, let's adopt the very conventional point of view that the cosmological principle is correct out to the largest scales. This raises an interesting question: what kind of overall spatial geometry is compatible with the cosmological principle? By spatial geometry I mean the shape of space. Let's begin with two-dimensional examples. A 2-sphere is a particular geometry. So are ellipsoids, pear shapes, and banana shapes.[4]

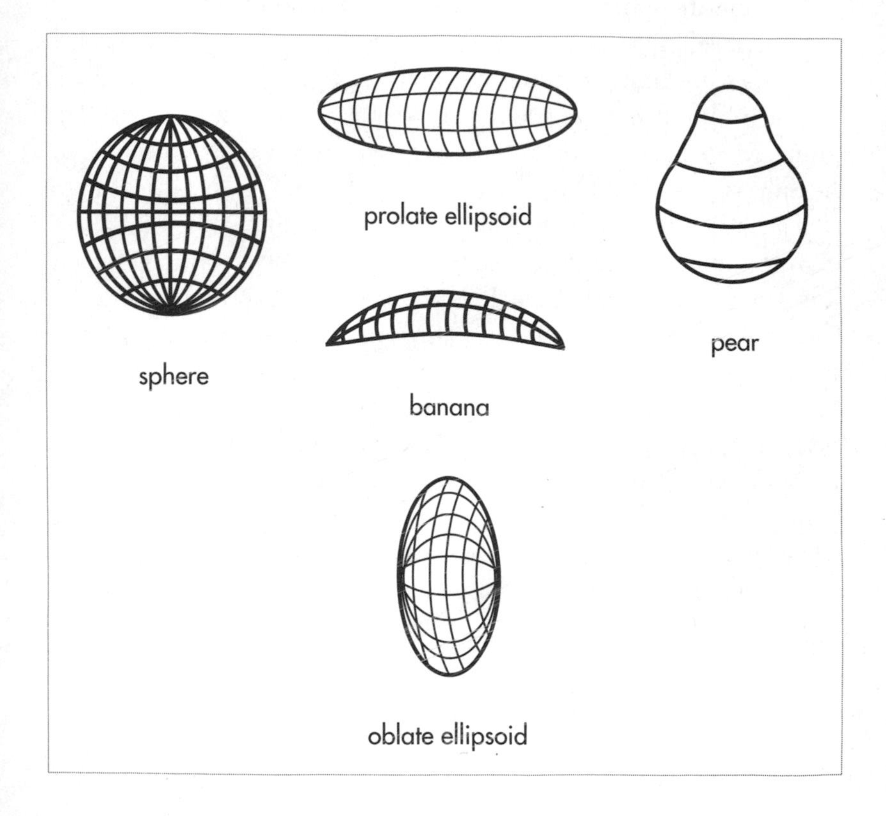

Among this list, only a sphere is homogeneous and isotropic. Like a circle, a sphere has perfect symmetry; every point is exactly like every

4. The shape, of course, refers to the surface of the figure.

other point. An ellipsoid, while not as symmetric as a sphere, still has a good deal of symmetry. For example, its mirror image looks just like itself. But not every place on the ellipsoid is the same as every other. The pear and banana are even less symmetric.

One way to describe the properties of a surface is by its curvature. The curvature of the sphere is absolutely uniform. Mathematically speaking it is a space of uniform positive curvature. The ellipsoid is also a positively curved space, but it is more curved in some places than others. For example, the prolate ellipsoid, which is shaped somewhat like a submarine, is curved more near its ends than at its waist. Of all these examples only the sphere is uniformly curved and homogeneous.

Spheres, ellipsoids, and the surfaces of fruits are closed-and-bounded geometries, meaning that they are finite in extent but without edges. But the truth is that no one knows if the universe is finite in extent. No cosmic Magellan has ever circumnavigated it. It's entirely possible that the universe goes on forever, in which case it is unbounded, or infinite.

If we allow the possibility that the universe is infinite, then there are two more homogeneous, isotropic geometries. The first is obvious: the infinite flat plane. Think of it as a sheet of paper that goes on and on forever. There are no landmarks on the infinite plane to tell you where you are or which way you are facing. And unlike the surface of the sphere, the plane is not curved: mathematically it has zero curvature. Positive curvature for the sphere, zero curvature for the plane, and finally, the last homogeneous isotropic geometry, the negatively curved "hyperbolic geometry." To help visualize this, think of a piece of duct pipe bent to a right angle. At the outer "elbow," the sheet metal is positively curved like the sphere. The inner curved surface is the place where the curvature is negative.

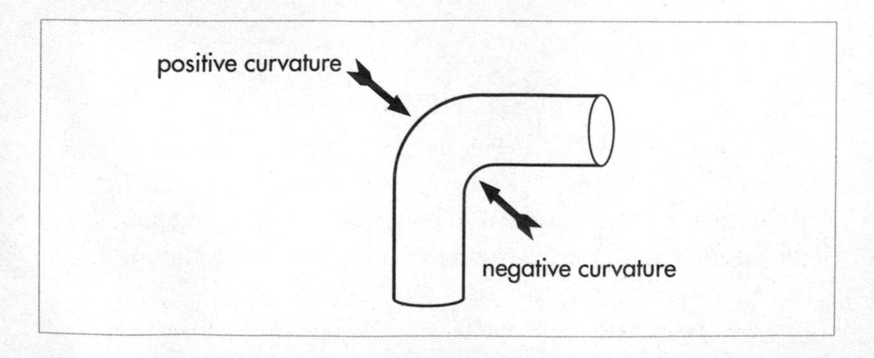

But of course the duct-pipe elbow is not homogeneous. The inner curved region is not at all the same as the outer positive curvature region. A better example is the surface of a saddle. Imagine continuing the saddle shape forever to form an unbounded negatively curved surface. It's not easy to visualize, but it's perfectly possible.

All three of these surfaces — sphere, plane, and hyperbolic geometry — are homogeneous. Moreover, all three of them have three-dimensional analogs: the 3-sphere, ordinary Euclidean three-dimensional space, and the most difficult to visualize, the hyperbolic three-dimensional space.

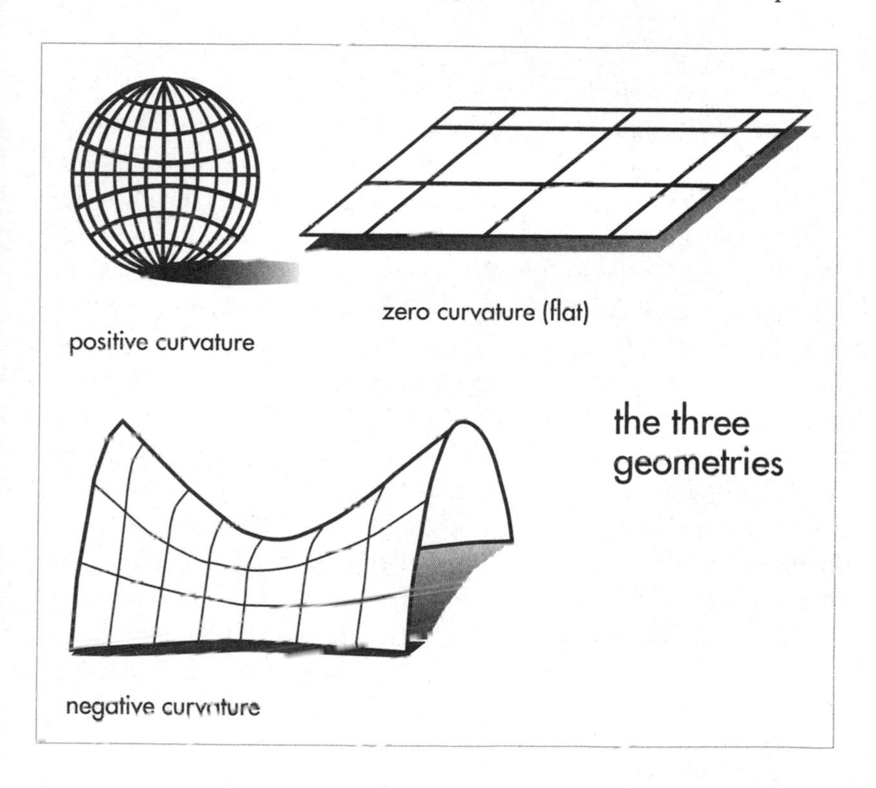

the three geometries

To envision the three standard types of cosmologies, think of each surface as a rubber sheet (or balloon in the case of the sphere) and fill the surface with dots to represent galaxies. Then start stretching the surface so that the dots begin to separate and the distance between any two grows with time. That's all there is to it. You now have a rough idea of the three homogeneous, isotropic cosmologies. Cosmologists refer to these three cases as $k = 1$, $k = 0$, and $k = -1$. It's just shorthand for positive curvature (the sphere), zero curvature (flat space or the plane), and negative curvature (hyperbolic space).

Is the universe finite and bounded as Einstein thought, or is it unbounded, filled with an endless infinity of stars and galaxies? The question fascinated cosmologists throughout the twentieth century and since, but the answer has proved elusive. In the rest of this chapter, I will tell you what has been discovered in the recent past and how it bears on the answer.

The Three Fates

About a month ago I was home working on this book when I was disturbed by a knock on my front door. When I answered it, three very neatly dressed young people handed me a leaflet. I don't usually bother arguing with proselytizers, but when I saw the title of the booklet — *Are You Prepared for the End of the Universe?* — I couldn't resist asking them a few questions. When I asked them how they knew anything about the end of the universe, they told me that modern-day scientists had confirmed the biblical account of Armageddon and that the end of the universe was a scientific certainty.

They were probably right. Modern-day scientists do predict that the universe — at least the universe *as we know it* — will come to an end. Every reasonable cosmological theory says so. When and how it will happen varies from one set of assumptions to another, but all agree that it won't happen for some tens of billions of years.

Broadly speaking there are two "end-of-the-world" scenarios. To understand them, think of a stone being thrown vertically into the air. Actually, I want to forget the air. Let's throw the stone from an airless asteroid. One of two things can happen. The gravitational pull of the asteroid may be sufficient to pull the stone back, or it may not be. In the first case the stone will reverse its outward motion and come crashing back down, but in the second case it will overcome the gravitational attraction and fly off forever. It all depends on whether the initial velocity of the stone is faster than the *escape velocity*. The escape velocity depends on the mass of the asteroid: the bigger the mass, the larger the escape velocity.

According to the General Theory of Relativity, the fate of the universe is much like the fate of that stone.[5] The galaxies (and other material in the universe) have been shot out of the Big Bang explosion and are now flying away from each other. Meanwhile, gravity is working to pull them back. To put it another way, the balloonlike universe is growing, but gravitating matter is slowing down the expansion. Will the expansion keep going, or will gravity reverse it and eventually cause the universe to start shrinking? The answer is quite similar to the case of the asteroid and the stone. If there is enough mass in the universe, it will reverse direction and eventually collapse in a dreadful, big, superheated crunch. On the other hand, if there is not enough mass, the universe will keep expanding indefinitely. In this case the end may be gentler, but eventually the universe will become so thinned out that it will die a cold death.

For both the stone and the universe there is a third possibility. The stone might be precisely at the escape velocity. This would require a perfect balance between gravitational attraction and outward velocity. If you do the math in this case, you will find that the stone keeps going but at an ever-decreasing velocity. The same is true of the universe. If there is an exact balance between mass density and outward expansion, the universe will eternally expand but at an ever-decreasing rate.

Geometry Is Fate

Three possible geometries and three possible fates; is there a connection? Indeed there is. Einstein's theory of gravity (without a cosmological constant) relates geometry to the presence of mass; mass affects geometry. The Newtonian dictum that "mass is the source of the gravitational field" is replaced by "mass warps and bends space." That's the link that relates the three geometries to the three fates. The details are in the difficult mathematics (tensor calculus and Riemannian geometry) of general relativity, but the result (with no cosmological constant) is easy to understand:

5. For the moment I am completely ignoring the possibility that there is a cosmological constant. As we will see, a cosmological constant can significantly alter the conclusions.

1. If the mass density in the universe is large enough to reverse the outgoing expansion, it will distort space into a sphere, a 3-sphere that is. This is the case of a closed-and-bounded universe. And its fate is a final crunch or, in the technical jargon, a *singularity.* This case is called the *closed universe,* or the $k = 1$ universe.

2. If the mass density is less than the minimum amount needed to close the universe, then it is also insufficient to reverse the motion. In this case it distorts space into a hyperbolic geometry. The hyperbolic universe expands forever. It's called the *open universe,* or the $k = -1$ case.

3. If the universe is right on the knife-edge, between open and closed, then the geometry of space is flat, uncurved, Euclidean space, but the universe endlessly expands, albeit at an ever-diminishing rate. This is called the flat universe, and it is labeled $k = 0$.

So, which is it?

Some say the world will end in fire,
Some say in ice.
From what I've tasted of desire
I hold with those who favor fire.
But if it had to perish twice,
I think I know enough of hate
To know that for destruction ice
Is also great
And would suffice.

— *Robert Frost,* "Fire and Ice"

When I asked the three young missionaries if it would be the hot death or the cold death, they said that it all depended on me. Very probably it would be the hot death unless I changed my ways.

Physicists and cosmologists are less certain of the final reckoning. For decades they have tried to determine which of the three fates will rule the last days. The first way to find out is very direct: use telescopes to look out into the distant reaches of space and count all the mass that can be seen — stars, galaxies, giant clouds of dust, and everything else that can be seen or deduced. Is the gravitational pull of all that material enough to turn the expansion around?

We know how fast the universe is expanding today. Hubble determined that the velocity of a distant galaxy is proportional to its distance — the factor of proportionality being the Hubble constant. This quantity is the best measure of the expansion rate: the larger the Hubble constant, the faster every galaxy is moving away from us. The units of the Hubble constant are velocity per unit distance. Astronomers usually quote it as "kilometers per second per megaparsec." Everyone will recognize kilometers per second as a unit of velocity. One kilometer per second is about three times the speed of sound, i.e., Mach 3. The megaparsec is less familiar. It's a unit of length, convenient for the study of cosmology. One megaparsec is about three million light-years, or thirty million trillion kilometers, a little more than the distance to our neighboring galaxy, Andromeda.

The value of the Hubble constant has been repeatedly measured over the years and has been the subject of a lively debate. Astronomers agreed it was somewhere between fifty and one hundred kilometers per second per megaparsec, but only in the recent past has the answer been resolved as about seventy-five in these units. The implication is that at a distance of one megaparsec, the galaxies are receding with a velocity of 75 km/sec. At two megaparsecs their velocity is 150 km/sec.

Now 75 km/sec sounds awfully fast by terrestrial standards. At that rate it would take about ten minutes to circumnavigate the earth. But it's not at all fast from the viewpoint of a physicist or astronomer. For example, the pinwheel motion of the Milky Way imparts a velocity to the earth that's ten times faster. And by comparison with the speed of light, it's a snail's pace.

In fact, according to Hubble's Law, the Andromeda Galaxy should be receding from us at about 50 km/sec — but in reality it is moving toward us. It is so close that the Hubble expansion is counteracted by the gravitational pull of our galaxy. However, Hubble's Law was never intended to be exact for a galaxy as close as Andromeda. When we consider galaxies that are far enough apart to escape each other's gravity, the law works very well.

Nevertheless, the expansion is slow, and it would take very little mass density to turn it around.

Knowing the expansion rate, it is a straightforward application of Einstein's equations to compute how large a mass density would be required to prevent the universe from eternally growing. The answer? Just 10^{-25} kilograms per cubic meter would be the knife-edge value: just barely enough to eventually reverse the outward flow of the galaxies. That's not much. It's roughly the mass of fifty protons in a cubic meter. A tiny bit more would be enough to curve the universe into a 3-sphere and turn the Big Bang into a disastrous big crunch. If the density were exactly this critical value, the universe would be flat (i.e., $k = 0$).

Astronomers search the sky for matter in the form of stars, gas, and dust clouds: all the matter in the universe that emits or scatters light. Assuming that the universe is homogeneous, we can count up all the glowing mass in the general vicinity of our galaxy and measure the average cosmic mass density. The number is remarkably small, only one proton mass per cubic meter: too small by a factor of fifty to close the universe. The obvious implication is that we are living in an infinite open ($k = -1$) universe with negative curvature and that it will go on expanding forever.

But astronomers and cosmologists have always been wary of jumping to this conclusion. Unlike physics, where being wrong by a factor of fifty is a disgrace, astronomy, until very recently, was a crude science. Estimates could easily be off in either direction by factors of ten or one hundred. Given that the mass density might have had any value, the fact that it came out so close to the critical density filled cosmologists with suspicion. And they were right to be suspicious.

There is another way to determine the mass of a galaxy besides just measuring the light from it, a much more direct and reliable way. And that's to use Newton's laws. Let's return to the asteroid and the stone. Now, instead of moving vertically, the stone is moving in a circular orbit around the asteroid. The gravity of the asteroid keeps the stone in orbit. The key observation, which goes back to Newton, is that by measuring the velocity of the stone and the radius of its orbit, you can determine the mass of the asteroid. In a similar fashion, by measuring the velocity of stars in the outermost parts of a rotating galaxy, astronomers can measure the galaxy's mass. And what do they find?

The galaxies are all heavier than the astronomers had thought. Roughly speaking, every galaxy is about ten times more massive than all the visible stars and interstellar gas that it contains. The remaining nine-tenths of the mass is a mystery. It is almost certainly not made of the things that comprise ordinary matter: protons, neutrons, and electrons. Cosmologists call it dark matter: dark because it gives off no light.[6] Nor does this ghostly matter scatter light or allow itself to be visible in any form, except through its gravity. So strange is modern science. For all these years — since the time of John Dalton — all matter was thought to be the usual stuff of chemistry. But now it seems that 90 percent of all the matter in the universe is something we know nothing about.

As astronomers were in the slow process of convincing themselves that dark matter is really there, theoretical physicists were busy postulating all sorts of new elementary particles for all kinds of reasons. Neutrinos were a very early example, superpartners were another, but they certainly don't exhaust the imaginative list of hypothetical particles that were postulated for one reason or another. No one knows for sure what dark matter is, but the most likely solution is that there are new, heavy elementary particles that we haven't discovered yet. Perhaps they are the nonidentical superpartner twins of ordinary particles — the bosonic partners of neutrinos or even the fermion partner of the photon. Perhaps they are a totally unsuspected class of elementary particles that no theorist has dreamed up yet. Whatever they are, they are heavy — they have mass and gravitate — but they have no electric charge to scatter or emit light. That's all we really know. They must be all around us, constantly passing through the earth and even our bodies, but we can't see them, feel them, or smell them. Without electric charge, they have no direct way of interacting with our senses. Very sensitive particle detectors are being built so that we may learn more about these mysterious objects, but for now it's enough to know that they make galaxies ten times heavier than we thought.

The question of whether the universe is open and infinite or closed and finite has haunted astronomy for as long as there have been as-

6. Don't confuse dark matter with dark energy. *Dark energy* is another term for vacuum energy.

tronomers. A closed universe with a finite number of galaxies, stars, and planets is intuitively understandable, but an unbounded universe is almost incomprehensible. We have gotten closer to having enough matter to close the universe — tantalizingly close. Originally we were shy of the critical density by a factor of fifty. Now it is only a factor of five, but we are much more confident that we know the amount of mass that is out there. Could it be that the Hubble constant was not measured accurately? If it were smaller by a factor of two or three, then the mass density would be very close to closing the universe. So much hinges on getting this right that we want to close any possible loophole in the reasoning.

Astronomers have been closing in on the value of the Hubble constant for almost eighty years with ever more sophisticated instruments. It now seems very unlikely that it can be small enough to allow the universe to be closed. If this were the end of the story, then we would have to conclude that the cosmic mass density was insufficient to close the universe — but we're not done yet.

The other way to determine if the universe is open, closed, or flat is very direct. Imagine a very large triangle in space, a triangle of cosmic proportions. To ensure that the sides are straight, we might take them to be the paths of light rays. A cosmic surveyor might measure the angles of the triangle, and if she were a student of Euclidean geometry, she might conclude that the sum of the angles should add up to 180 degrees — two right angles. The ancient Greeks were sure of it; they couldn't conceive of space being any other way.

But modern geometers know that the answer depends on the geometry of space. If space is flat like Euclid thought, the sum of the three angles would add up to 180 degrees. On the other hand, if space is a sphere, the angles would add up to a total greater than 180 degrees. Less easy to visualize, the angles of a triangle in a negatively curved space will always sum to something less than 180 degrees.

Sending a team of cosmic surveyors billions of light-years to the corners of an immense triangle is not feasible, and even if it were, it would take billions of years to get there and billions of years more to get the result back to earth. But the ingenuity of astrophysicists is unbounded, and believe it or not, they devised a way to do the job without ever leaving the earth. I will return to how they did it after I explain the *cosmic*

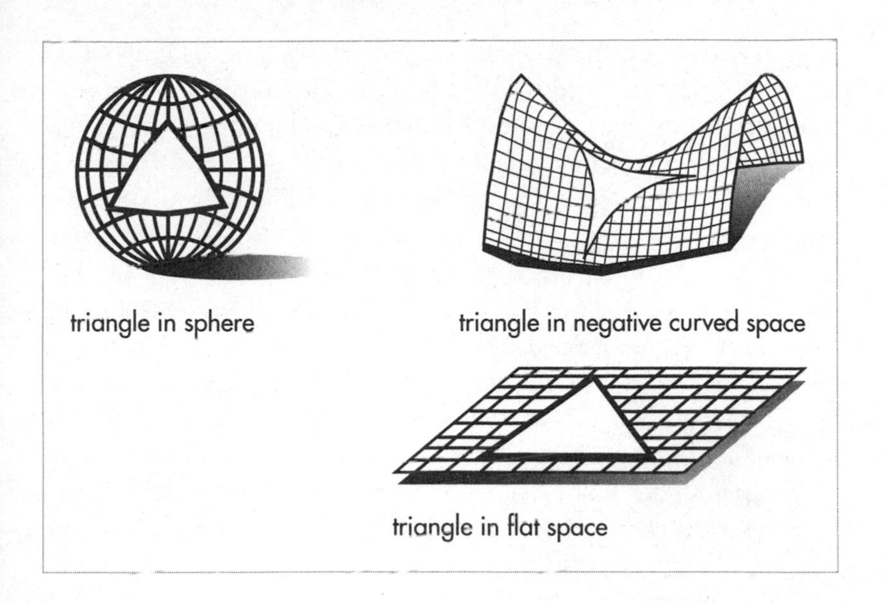

microwave background, or CMB. But the result is easy to state: space appears to be flat! The angles add up just as Euclid assumed. Or at least they add up to 180 degrees to within the accuracy of the experiment.

By now, dear reader, you must realize something is terribly wrong. We have two ways to determine if the universe is open, closed, or flat and two incompatible answers. The amount of mass in the universe appears to be five times too small to either close the universe or even to make it flat. But surveying cosmic triangles seems to leave little doubt that the geometry of the universe *is* flat.

The Age of the Universe and the Oldest Stars

Imagine a cosmic movie, a biography that follows the universe from its birth in fire to its present old age. But instead of viewing the movie in the ordinary way — from birth to old age — we run it backward, on rewind, so to speak. Instead of expanding, we see it contracting. The galaxies appear to move according to a reverse version of the Hubble Law — their velocity being proportional to their distance but approaching us instead of receding. Let's follow one of those distant galaxies as it approaches us. Using Hubble's Law (run backward), we can determine its velocity. Let's say the galaxy is one megaparsec away. Hubble's Law tells us that it is approaching with a velocity of seventy-five kilometers a second. Knowing how far away it is and how fast it is moving, it's an easy exercise to determine how long it will be until the galaxy is on top of us. I will do it for you. The answer is about fifteen billion years. That's the answer if we assume that the galaxy moves with a steady constant speed.

What if we started with a galaxy two megaparsecs away instead of one? Hubble's Law tells us that it is moving twice as fast as the previous galaxy: twice as far but twice as fast. It too will arrive in our lap in fifteen billion years. In fact the same is true for any distant galaxy. According to this reckoning all the galaxies will merge into an undifferentiated mass in about fifteen billion years in the reverse movie.

But galaxies don't move with uniform speed as they approach. In the forward version of the movie, gravity slows them as they recede. Thus, in the backward version, gravity speeds them up as they fall toward one another. This means that it should take less time for them to collide. When cosmologists carry out the correct calculation (in the forward version), they find that the galaxies were crowded together in a dense mass about ten billion years ago. This would mean that it has been only ten billion years since the hydrogen and helium gases began to differentiate themselves into the clumps that eventually became galaxies. To say it concisely, the universe, according to this reckoning, is ten billion years old.

Determining the age of the universe has been a bumpy ride. Originally Hubble underestimated the distances to the galaxies by about a factor of ten. This led him to conclude that the universe started its expansion a mere one billion years ago. But by Hubble's time, rocks two billion years old had already been dated by their radioactivity. Obviously there was an error, and it was soon found. But a modern version of the problem still exists. Astronomers and astrophysicists, who study the detailed properties of stars in our galaxy, find that the oldest stars are older than the universe. They are about thirteen billion years old. The child is older than the parent!

In short, three big problems affect our thinking about the universe. First, there is the contradictory evidence concerning the geometry of space, whether it is open, closed, or flat. Second, is it really younger than the oldest stars? And third, the mother of all problems: is there a cosmological constant as Einstein originally believed, and if not, why not? Are these problems connected? Of course they are.

The Solution

Perhaps the resolution is that our theory of gravity — the General Theory of Relativity — is just plain wrong. In fact some physicists have jumped to this conclusion. These physicists usually try to make modifications in the theory that will only affect the gravitational force at very great distances. Personally, I don't find much merit in these schemes. They are usually very contrived, often violate fundamental principles, and in my opinion, are quite unnecessary.

Another possible way out is to suppose that astronomers are taking the precision of their data too seriously. You can make a good living betting against experimental data that contradict the prevailing expectations. Such data are almost always wrong, and further experimentation usually proves it. In this case I would have bet against the astronomical data, not the theory. But it seems I would have lost my bet. As the data have improved over the last few years, they reinforce the fact that observation and theory are at odds with each other. There really is something wrong.

But one possibility lurks just beneath the surface that cannot be easily dismissed. What if there is a small cosmological constant after all? What if Einstein's greatest blunder was really one of his greatest discoveries? Could that resolve the conflicts?

When we considered whether the observable mass in the universe would be enough to render it flat or closed, we completely ignored the possibility of vacuum energy. That would be a mistake in a world with a cosmological constant. Einstein's equations say that *all* forms of energy affect the curvature of space. Energy and mass are the same thing, so vacuum energy must be counted as part of the mass density of the universe. The ordinary and dark matter together add up to about 30 percent of the mass needed to flatten or close the universe. The obvious way out of the dilemma is to make up the missing 70 percent in the form of a cosmological constant. This would mean that the vacuum-energy density was a little more than twice the mass of ordinary and dark matter combined, about thirty proton masses per cubic meter.

Because the cosmological constant represents a repulsive force, it would have an effect on the way that the universe expands. The early phase of the expansion would not be much affected, but as the distance grows between galaxies, so too does the repulsive force. Eventually the cosmological constant can accelerate the outward motion of the galaxies, causing the Hubble expansion to pick up speed.

Let's run it backward. The galaxies are falling inward, but now the extra repulsion slows them down. The initial estimate of their inward velocity (the one we make today) overestimates how fast they will be moving as they grow closer. Failure to account for the vacuum energy will lead us to underestimate the length of time until the galaxies all merge. In other words, if there were a cosmological constant but we didn't know it, we would find the universe appearing younger than it really is. Indeed, if we include the effects of a vacuum energy equal to about thirty proton masses per cubic meter, the ten-billion-year lifetime of the universe gets stretched to about fourteen billion years. That's perfect because it makes the universe just a little older than the oldest stars.

These conclusions concerning the existence of a cosmological constant are so important that I want to repeat them. The existence of a small cosmological constant, representing 70 percent of the energy in the universe, solves the two biggest puzzles of cosmology. First, the additional energy is just enough to make the universe flat. This fact removes the awkward discrepancy between the observed flatness of space and the fact that the mass in the universe was insufficient to render it flat.

The second paradox that is eliminated by the cosmological constant is the equally awkward discrepancy that the oldest stars appear older than the universe. In fact, the same vacuum energy — 70 percent of the total — remarkably, is exactly what is needed to make the universe a little older than these ancient stars.

Type I Supernovae

Over the last decade the historical accuracy of the universe's biography has been greatly improved. We now know the history of the expansion in much greater detail. The trick involves a class of distant events called *Type I supernovae.* A supernova is a cataclysmic event in which a dying star collapses under its own weight and becomes a neutron star. The supernova is so unimaginably violent that when it occurs in a galaxy it can outshine the billions of stars that galaxy comprises. Supernovae are easy to spot even in very distant galaxies.

All supernovae are interesting, but something is very special about Type I supernovae. They originate from double star systems in which an ordinary star and a white dwarf are orbiting each other at a relatively close distance. The white dwarf star is a dead star that didn't have quite enough mass to collapse to a neutron star.

As the two stars revolve around each other, the gravity of the white dwarf gradually sucks matter away from the ordinary star and, in this way, slowly increases its own mass. At some very precise point, when the mass is just right, the white dwarf can no longer support its own weight, and it implodes, creating a Type I supernova. The behavior of the final collapse doesn't depend on the original mass of the white dwarf, or for that matter, its companion. In fact these events are believed to occur in a unique way and always give the same amount of light. An astronomer would say they all have the same luminosity.[7] Astronomers can tell, with a good degree of certainty, how far away they are by how bright they appear.

The velocity of the galaxy in which the supernova is embedded can also be easily determined using the Doppler method. And once we know both the distance and the velocity of the distant galaxy, the Hubble constant is easy to determine. But the special thing about very distant galaxies is that their light was given off long in the past. A galaxy five billion light-years away radiated the light that we now see five billion years ago. When we measure the Hubble parameter on earth today, we are really measuring the value that it had five billion years ago.

By concentrating on galaxies at a variety of different distances, we effectively measure the history of the Hubble parameter. In other words, Type I supernovae allow us to know a great deal about the history of the universe during the various stages of its evolution. And most important,

7. Luminosity is a measure of the rate at which an object gives off energy in the form of light. The luminosity of an electric lightbulb is measured in watts. If two objects have the same luminosity, the closer of them will appear brighter. By measuring the apparent brightness of the photographic images of Type I supernovae, their distance from us can be determined.

they allow us to compare our real universe with mathematical models, with and without cosmological constants. The results are unambiguous. The expansion of the universe is accelerating under the influence of a cosmological constant, or something very much like it. To theoretical physicists like myself, this is a stunning reversal of fortune that cannot help but change our entire outlook. For so long we were trying to explain why the vacuum energy is exactly zero. Well, it seems that it is not zero. The first 119 decimal places of the cosmological constant cancel, but then, in the 120th, incredibly, a nonzero value results. To make matters even more interesting, its value is just about what Weinberg predicted it would be based on the Anthropic Principle!

Light from Creation

Because light travels with a finite velocity, great telescopes that look to tremendous distances are also looking far back into the past. We see the sun as it was eight minutes ago, the nearest star as it was four years ago. Early humans were first beginning to stand straight when the light started its two-million-year journey from the nearest galaxy, Andromeda.

Oldest of all is the light that has been traveling to us for about fourteen billion years. This light started before the earth or even the oldest stars were formed. Indeed, the hydrogen and helium had not yet begun the process of differentiation into galaxies. So hot and dense were these gases that the atoms were all ionized. It was as close to creation as nature will ever allow us to see, at least if the messenger is electromagnetic radiation.

Think of the universe as a series of concentric shells with us at the center. There are, of course, no real shells out there, but nothing prevents us from dividing space up in that way. Each successive shell is farther away than the last. Each shell also represents an earlier (time) epoch than the previous. By looking deeper and deeper, we are, in effect, running the movie of the universe backward.

The deeper we look, the more densely populated the universe appears. In the reverse movie of the universe, the matter gets progressively denser as if some giant piston were squeezing it ever tighter. That piston is, of course, gravity. Moreover, it is a property of matter that as it is compressed, it grows hotter as well as denser. Today, the average temperature of the universe is only about 3 degrees above absolute zero, or –270 degrees centigrade. But as we follow the universe into the past, the temperature rises, first to room temperature, then to the boiling point, and eventually to the temperature on the surface of the sun.

The sun is so hot that the atoms that it is composed of have been torn apart by their violent thermal motion. The nuclei are intact, but the more loosely attached electrons have been torn free and can roam throughout the sun's hot gases, which are now electrically conducting *plasma*.[8]

Electrical conductors are generally the least transparent of materials. The freely moving electrons easily absorb and scatter light. This scattering of light makes the sun opaque. But as we move outward to the sun's surface, the temperature and density decrease to the point where it becomes transparent. That is where we see the sun's surface.

Now let us travel backward in time and outward in space to the last visible shell, where the conditions are similar to the sun's surface. Again the light comes to us from a surface like the sun's — a giant shell of hot plasma surrounding us on every side. Astronomers call it the *surface of last scattering*. Sadly, looking through the conducting plasma to an even earlier and more distant shell is no more possible than looking through the sun.

Immediately after the Big Bang, the light from the surface of last scattering was every bit as bright as the sun's surface. That raises an interesting question: why, when we look at the sky around us, don't we see the bright glare of ionized hot primordial plasma? To ask it another way, why isn't the sky uniformly illuminated with the same brightness that we would see if we were to look straight into the sun? Fortunately, the Doppler shift rescues us from that awful prospect. Because of the Hubble expansion, the plasma that originally emitted the primordial light is receding away from us with a large velocity. In fact, using the Hubble Law, we can calculate the velocity of this recession, and the result is only slightly less than the speed of light. This means that the emitted radiation was Doppler redshifted way past the visible and infrared, all the

8. *Plasma* is just another word for gas that has had its atoms ionized. In other words, some of the electrons have been torn free of the nucleus and are free to wander through the gas, unattached to atoms.

way to the microwave spectrum. Here, one of the earliest discoveries of quantum mechanics plays an important role: the energy of a photon depends on wavelength in such a way that a microwave photon has about one thousand times less energy than a photon of visible light. For this reason, the photons that eventually reach us from the surface of last scattering are not very potent. They have no more effect on our retinas than the radio waves that continually surround us.

There is another way to understand the diminished potency of the cosmic radiation by the time it reaches us. The photons from the surface of last scattering were very hot, about as hot as the sun's surface. They filled space, forming a kind of photon gas, and like all gases, when they expand, they cool. The expansion of the universe, since the time of the Big Bang, cooled the photon gas to the point where it lost most of its energy. Today, the CMB (cosmic microwave background) radiation is very cold: fewer than 3 degrees above absolute zero. The two explanations of the CMB's loss of power are mathematically completely equivalent.

George Gamow was the first to have the idea of a Big Bang. Soon after, two of his younger colleagues, Ralph Alpher and Robert Herman, got the idea for CMB as a kind of leftover afterglow. They even estimated the temperature of the radiation today, and got 5 degrees: within two degrees of the right answer. But physicists at that time believed that such weak radiation could never be detected. They were wrong, but it took until 1964 for the CMB to be accidentally discovered.

At that time, the Princeton cosmologist Robert Dicke wanted to test the idea of CMB by measuring the radiation left over from the hot Big Bang. While he was in the process of building a detector, two young Bell Laboratory scientists were doing precisely the kind of experiment that Dicke was aiming for. Arno Penzias and Robert Wilson were scanning the sky for microwave signals, not for the purpose of discovering the birth of the universe, but for communications technology. They couldn't identify a strange background static that was getting in the way of their real goal. Legend has it that they thought it was bird droppings on the detector.

Princeton University and Bell Labs are close neighbors in central New Jersey. As fate would have it, Dicke found out about the Penzias-Wilson "noise" and realized that it was the CMB from the Big Bang!

Dicke got in touch with the Bell Labs scientists and told them what he thought was going on. Subsequently, Penzias and Wilson got the Nobel Prize for the discovery. It is one of those twists of fate that had Princeton and Bell Labs been farther apart, Dicke might have finished his experiment and been the first to make the discovery.

The Penzias-Wilson detector was a crude affair mounted on the roof of Bell Labs. By contrast, the most modern CMB detectors are extremely sophisticated and are mounted in space, high above the atmosphere. The detectors can be pointed in different directions to measure the CMB from each point in the sky. The results are presented as a kind of map of the sky.

One of the most striking features of the CMB is how dull these maps are. To a very high degree of precision, the microwave-sky is a featureless, homogeneous expanse. It seems that, in early times, the universe was almost perfectly homogeneous and isotropic. The microwave radiation coming from the surface of last scattering is almost identical in all directions of the sky. This extraordinary degree of homogeneity is somewhat puzzling and needs an explanation.

As smooth as the universe was at that early time, it could not have been perfectly smooth. Some small, primordial lumpiness had to be there to seed the formation of galaxies. If the seeds were too weak, galaxies would not have formed; if too strong, the lumps would have grown too rapidly and collapsed to black holes. Cosmologists strongly suspected that under this boring homogeneous background, the seeds of future galaxies were there to see. Even better, theoretical cosmologists had a pretty good idea how strong the density contrasts had to be in order to create the galaxies as we see them now. The difference between the microwave intensity in different directions would have to be about 100,000 times smaller than the average intensity.

How on earth is it possible to detect such incredibly small density contrasts? The answer is that you don't do it on earth. You've got to get high above the polluting environment of the planet. The first experiments to see small variations in the microwave radiation were done by detectors suspended from balloons flying above the South Pole. The South Pole is good for a number of reasons, not the least of which is that a balloon doesn't wander very far from its launch site. The prevailing winds would carry a balloon clear around the world, but around the world is not very far when you're at the South Pole. The experiment was named Boomerang!

High above the South Pole, microwave detectors compared the intensity at pairs of locations and automatically determined the difference between them. Theorists had their expectation, but no one knew for sure that anything of interest would be seen. Perhaps the sky would just continue to be a featureless, gray background. Then they would have to go back to the drawing board and redesign the theories of galaxy formation. Everyone who had any interest in cosmology waited for the jury's verdict with nervous anticipation. The verdict that came back was everything a defense attorney could hope for. The theorists told the truth. Lumps in the cosmic oatmeal were there and at exactly the right strength — 10^{-5} — one part in 100,000.

Outer space is an even better place from which to measure the cosmic microwaves. The data from the orbiting Wilkinson Microwave Anisotropy Probe, known by the nickname WMAP (pronounced "double-u map"), are so incredibly precise that they not only measured that 10^{-5} lumpiness but also detected the roiling, oscillating motions of huge blobs of hot plasma that radiated the CMB.

The large blobs of coherently moving plasma were not at all unexpected. Theoretical cosmologists had predicted that the expansion of the universe would start the lumps in the plasma oscillating like ringing bells. At first the smaller lumps should begin contracting and expanding. Later, with a lower frequency, larger blobs would join in: a perfectly predictable symphony. The detailed calculations indicated that at any time the largest visible oscillating blobs would have a certain definite size. Thus, when WMAP saw such oscillating blobs, cosmologists already knew a great deal about the size of the largest ones.

Knowing the size of the largest oscillating blobs had an incredible serendipitous payoff: it now became possible to survey cosmic triangles and measure the curvature of space. Here's how it was done.

Suppose you know the size of an object and also just how far away it is. This will enable you to predict how big it will look in the sky. Consider the moon. The moon is about 2,000 miles in diameter and about 240,000 miles away. Just from that information I can predict that it will occupy an angle of half a degree in the sky. By pure coincidence the sun is four hundred times bigger than the moon but also four hundred times farther away. The result is that the sun and the moon look the same size in the sky, namely half a degree. If we were on the moon looking at the 8,000-mile-diameter earth, it would look four times bigger than the moon as seen from the earth, i.e., two degrees.

Actually, in making these claims I made a tacit assumption, namely, that space is flat. Think of the diameter of the moon as the third side of a triangle. The two other sides are straight lines from our point on the earth to two diametrically opposite points on the moon.

If space is flat between the moon and the earth, my claims are correct. But if space is appreciably curved, the situation is different. For example, if space is positively curved, the moon will look bigger than a half degree. The opposite is true if the curvature is negative.

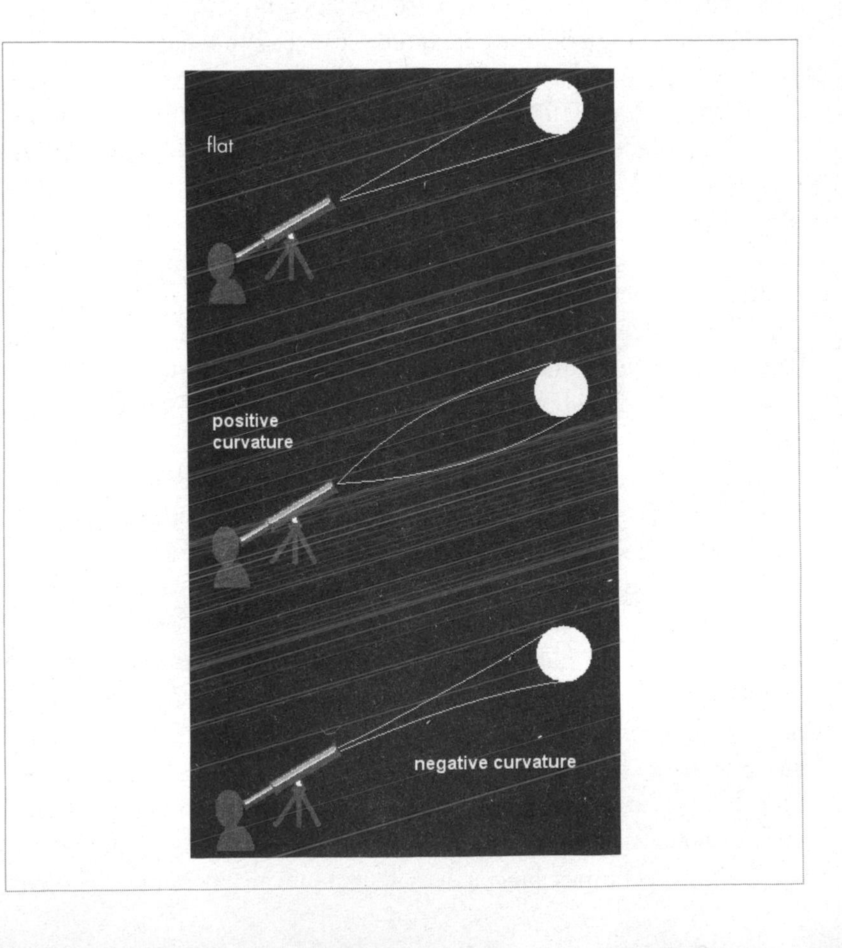

Now suppose we had independent confirmation that the diameter of the moon is 2,000 miles and that it was 240,000 miles away. We can use the apparent size of the moon to tell us the curvature of space. To a very high degree of accuracy, space is flat between us and the moon.

Let's get back to surveying the cosmos. Here is what we know: the largest oscillating blobs that were active at the time the CMB was emitted were about 200,000 light-years in diameter. Bigger blobs than that had not yet begun to ring.

Today, the source of the CMB is about ten billion light-years away, but at the time the CMB started its journey, our distance from the surface of last scattering was a thousand times smaller, i.e., ten million light-years away. That's enough to compute how big the largest CMB blobs should look to WMAP if space is flat, namely, about two degrees: as big as the earth seen from the moon. If space is not flat, the apparent size of the blobs would tell us how curved it is.

What did WMAP find? It found that Euclid was right! Space is flat.

Let me qualify that a bit. By measuring triangles on the surface of the earth, it is possible to tell that the earth is a curved sphere. But in practice, unless we can measure very big triangles, we would find that they behave as if the earth were flat. Obviously, Columbus could not convince the king of Spain that the earth was round by drawing a few triangles near the king's palace. He would have had to measure triangles at least a few hundred miles on a side, and even then he would have had to do it with great accuracy. All Columbus could say by surveying small triangles was that the earth is very big.

The same is true for cosmic surveying: all we can really conclude is that the universe is flat on scales of ten or twenty billion light-years. If the universe is finite, it is a lot bigger than the portion we can see.

So here is what we know with good confidence. First, the ordinary mass in the universe, stars, gas clouds, and dust, is not sufficient to make the universe flat. By past standards it's not so far off, only by a factor of fifty. But cosmology is no longer a qualitative science. By today's standards it's not close at all. Without other hidden sources of matter, the universe would be ruled open and negatively curved. But there is more matter in the universe, about ten times more, that we know about only by its gravitational effects. It may be made up of new elementary particles that hardly interact with the usual kind. These dark-matter particles, if that's what they are, would fill the galaxy, passing right through the sun, the earth, and even us. But they are still not enough to make the universe flat or closed. If the universe is flat, another kind of mass or energy must be pervading space.

Second, the age of the universe appears to be too young unless the history of its expansion is different than expected. The only conventional explanation is that there is a cosmological constant which accelerates the expansion. Although completely unexpected, it is confirmed by Type I supernova data that provide a kind of reverse film clip of the evolution. The best explanation of the age problem is that a cosmological constant exists at just about the level predicted by Weinberg's anthropic argument.

Third, cosmic microwave data directly show that the universe was extremely homogeneous in early times. Moreover, it is also very large, large enough to appear flat to cosmic surveyors. The bottom line is that the universe is many times bigger than the portion that we can see, and its expansion is accelerating under the influence of a small cosmological constant.

Inflation

It used to be a joke in the United States how Soviet Communist ideologues would claim that everything had been invented first in Russia. This included the radio, television, lightbulb, airplane, abstract painting, and baseball. In my own field of physics, there was sometimes truth to the joke. Soviet physicists were so badly isolated that a number of extremely important discoveries went unnoticed in the West. One of them was a remarkable guess about how the universe began. More than a quarter century ago, the young cosmologist Alexei Starobinsky had the idea that the universe started out with a brief period of prodigious exponential expansion. I'm not sure exactly what his motivation was, but in any case, only a few other isolated Russians appreciated Starobinsky's thought until some time later, when a young physicist in my own university rediscovered the idea. Alan Guth was a young postdoc working on high-energy theoretical physics at the Stanford Linear Accelerator Center (SLAC).

When I first met him, in 1980, I assumed that he was working on the ordinary problems of particle physics. At that time very few elementary-particle physicists knew much about cosmology. I was an exception because two years earlier Savas Dimopoulos and I had worked on the problem of why nature made so many more particles than antiparticles. My friend Bob Wagoner, one of the early pioneers of cosmology, had asked me if particle physics provided any explanation for the overwhelming preponderance of matter over antimatter. Dimopoulos and I had the right idea, but we were so ignorant of basic cosmology that we had confused the horizon size with the scale factor. That's like an auto mechanic not knowing his steering wheel from a hole in his muffler. But under Bob's wing we learned fast and eventually wrote the first paper outside the USSR about a subject to be called baryosynthesis. Ironically, barosynthesis was another subject that had first been invented in the USSR, this time by the great Andrei Sakharov, twelve years earlier.

Anyway, despite the fact that I was interested in the subject, I don't think I knew that Guth was interested in cosmology, that is, I didn't know until he gave a seminar on something that he called inflationary cosmology. I imagine I was one of the two or three people in the room who knew enough to be impressed.

Alan was after big game: the biggest. Why was the universe so big, flat, and so extremely homogeneous? To see why this is such a puzzle, let's go back to the CMB and focus on two separated points on the sky. At the time when the CMB was produced by the hot plasma, those two points were a certain distance apart. In fact, if they were more than a few degrees apart, the distance at that time would have been large enough that no light or other signal could possibly have gotten from one point to the other. The universe was only about half a million years old, so if the points were separated by more than half a million light-years, they could never have been in contact. If they had never been in contact, what made those two places so similar? In other words, how did the universe become so homogeneous that the CMB looked exactly the same in every direction?

To make this point clearer, let's return to the balloon theory of the universe. Imagine that the balloon began in a deflated state, badly shriveled up, with lots of wrinkles like a dried prune. As the balloon expanded, the wrinkles would have begun to smooth out. At first, small wrinkles;

then later, the bigger wrinkles would be ironed out. There is a rule about how wrinkles smooth out: a wrinkle of a given size can get smoothed only if there is enough time for a wave to propagate across the wrinkle. In the case of the universe, that means enough time for a light wave to cover the distance.

If there were insufficient time for large wrinkles to smooth out when the CMB originated, we should see them imprinted on the sky map. But we don't see such wrinkles. Why was the universe so smooth? Might there have been a long prehistory, hidden from view by the opaque early plasma, during which the wrinkles were stretched out? That's what Inflation theory is all about — a prehistory during which wrinkles were removed.

Alan immediately seized on the possibility that Starobinsky's exponential expansion might well be the key to this puzzle. The universe, according to Guth, had inflated like a balloon, but an extraspecial balloon. A real balloon will inflate only so far, and then it will burst. Alan's universe grew *exponentially*, and in a short time it became enormous. You can think of the Inflation as taking place before usual cosmology began. By the time the *conventional* Big Bang started, the universe had already grown to immense proportions. And in growing, all the wrinkles and inhomogeneities got stretched out so that the universe became exceedingly smooth.

I knew the idea was very good, but I didn't know how good. My guess is that even Alan didn't know how good it was. Certainly no one could guess that within twenty-five years Inflation would be the centerpiece for a new Standard Model of cosmology.

To understand the mechanism behind Inflation, we have to understand how a universe with a positive cosmological constant behaves. Remember that a positive cosmological constant gives rise to a universal repulsive force proportional to distance. The effect is to force the distance between galaxies to grow. This can happen only if the balloon that they are drawn on — space itself — expands.

Vacuum energy or mass has an unusual property. Ordinary mass density, like that due to the galaxies, dilutes when the universe grows. The mass density in the form of ordinary matter is about one proton per cubic meter. Suppose the radius of the universe doubled over some billions of years but the number of protons in the universe remained fixed.

Then the mass density would obviously decrease. In fact it would decrease by a factor of eight. Double the radius again, and the number of protons per cubic meter decreases to one sixty-fourth of its present value. The same is true of the dark-matter component.

But vacuum energy is very different. It's a property of empty space. When empty space expands it's still just empty space, and the energy density is exactly what it was originally. No matter how many times you double the size of the universe, the vacuum energy density stays the same, and its repulsive effect never diminishes!

By contrast, ordinary matter thins out and eventually becomes ineffective at slowing down the expansion. After a sufficient amount of expansion, all forms of energy will be diluted away except for vacuum energy. Once this happens there is nothing to counteract the repulsive effects of the vacuum energy, and the universe expands exponentially. If the cosmological constant were large enough to double the size of the universe in one second (it's not), then it would become four times as big in two seconds, eight times as big in three seconds, sixteen times, thirty-two times, and so on. Things that are close to us now would soon be rocketing away faster than the speed of light.

The real universe is in the early stages of this kind of exponential expansion. It won't bother us very much since the cosmological constant is only strong enough to double the size of the universe over a period of tens of billions of years. But imagine that for some unknown reason, in the very early universe, the cosmological constant were much bigger, perhaps a hundred orders of magnitude bigger. This may sound like a strange thought experiment, but remember that the hard thing to understand is why today's cosmological constant is so ridiculously small. Make it one hundred orders of magnitude larger, and it becomes ordinary, at least from the theoretical physicists' point of view.

If the cosmological constant were that big early on, it would cause the universe to double in a tiny fraction of a second. In one second the universe would grow from the size of a proton to something vastly larger than the known universe. This is real Inflation of the kind envisioned by Starobinsky and Guth.

The reader may wonder what kind of double-talk allows me to speak of different cosmological constants in the early and late universe, i.e., during Inflation and now. After all, aren't constants constant? Stop now and think *Landscape.* The cosmological constant in a given region of the Landscape is nothing but the local altitude. A picture of a bit of Landscape is worth a thousand words.[9] The picture below is a very simplified version of a Landscape that might resemble our neighborhood. The little ball represents the universe, rolling along, seeking a valley where the vacuum energy is minimum.

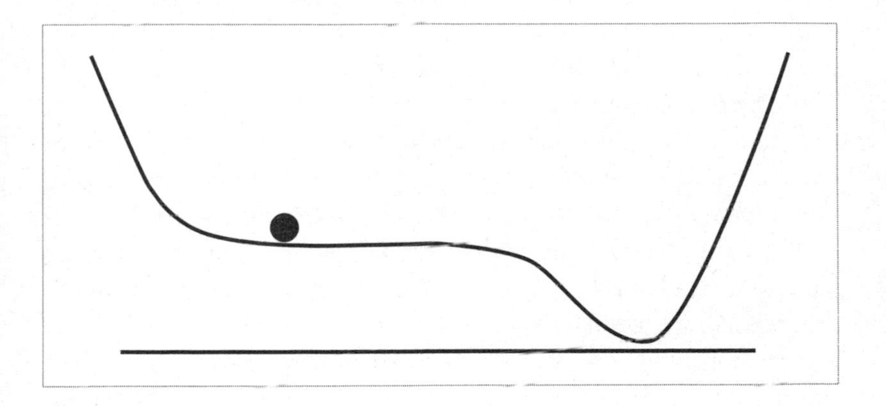

Some unknown history of the universe placed it on a relatively broad, high shelf overlooking a deep valley of almost zero altitude (here is where Guth's Inflation begins). How the universe arrived on the shelf is a question for another day. Because the shelf is so flat, the universe rolled very slowly at first. While it was on the shelf, the vacuum energy (altitude) was practically unchanging. To state it differently, the altitude of the plateau served as a cosmological constant while the universe was resting on the shelf.

And, as I'm sure you guessed, as it slowly rolled, it inflated because the vacuum energy was large and positive. If the shelf was flat enough and the rolling slow enough, the universe would double many times before coming to the steep descent down to the valley. This was the inflationary era, although in a more modern form than Starobinsky and

9. The picture shows a one-dimensional landscape, but the real thing is multidimensional. Think of the illustration as a one-dimensional slice through a much more complicated multidimensional Landscape. An analogy would be a one-dimensional road passing over hills and valleys in a two-dimensional Landscape.

Guth first proposed. If the universe doubled one hundred times or more during this period, it would have grown to such large proportions that it would be as flat and homogeneous as the CMB requires.

Eventually the rolling brought the universe to the edge of the shelf and then down into the valley, where it came to rest. If the altitude at that point is not quite zero, then the long-term future of the universe will have a small cosmological constant. If by chance the cosmological constant is small enough in the valley, and other conditions are right, galaxies, stars, planets, and life could form. If not, that particular pocket would be sterile. All of known cosmology took place during a roll from one value of the cosmological constant to a much smaller one. Can anyone seriously doubt that there was more to the history and geography of the universe than this brief episode and this tiny pocket?

But wait! Something is wrong with this picture. If the universe inflates to such a large degree, it can be expected to be incredibly homogeneous. All the wrinkles would have been ironed out so completely that there would be no variations at all in the CMB. But we know that without some small wrinkles to seed the galaxies, the universe would have remained smooth indefinitely. We seem to have overdone the homogenizing.

The solution to this puzzle involves an idea so radical and surprising that at first you might be tempted to dismiss it as pie-in-the-sky speculation. But it has withstood the test of time and is currently one of the cornerstones of modern cosmology. Once again its initial discovery took place in Russia, by a young cosmologist named Slava Mukhanov, who was studying Starobinsky's work. History repeats itself: Mukhanov's work was unknown outside the USSR until several groups working in the United States independently rediscovered it.

Quantum mechanics and its jittery consequences are normally thought to apply to the world of the very small, not galaxies and other cosmic-scale phenomena. But it now appears all but certain that galaxies and other large-scale structures are remnants of original minute quantum fluctuations that were expanded and enhanced by the unrelenting effect of gravity.

The idea that the universe is at an exact point in the Landscape is a little too simple. Like everything else, quantum fields such as the Higgs field have the jitters. Quantum mechanics is enough to ensure that the fields do fluctuate from point to point in space. No amount of Inflation can iron out the random quantum fluctuations that every field must have. This is true in our vacuum today, and it was true during the rapid exponential expansion of Inflation. But rapid Inflation does something to these fluctuations that doesn't happen to any appreciable degree in our very slowly expanding universe. It stretches out the old wrinkles but keeps replacing them with new ones. New wrinkles on top of old wrinkles, all expanding as the universe expands. By the time Inflation ended and the universe tipped over the edge of the ledge, the accumulated quantum wrinkles had built up and formed the minute density contrasts that eventually grew to become galaxies.

These frozen-in quantum wrinkles also imprinted themselves on the surface of last scattering, and we can see them as the tiny variations of brightness in the cosmic microwave vacuum. The connection between the quantum theory of the microscopic world and the large-scale structure of the astronomical and cosmological world is one of the greatest achievements of cosmology.

Let me finish this chapter by summarizing the two most important things we have learned from cosmological observations during the last decade. First, we have found a real shocker: there really is a cosmological constant. The first 119 decimal places cancel, but astonishingly, in the 120th the result is not zero!

The second point of enormous interest is that the theory of Inflation has strong support from the study of cosmic background radiation. The universe apparently grew exponentially for some period of time. It is all but certain that the entire universe is many, many orders of magnitude bigger than the part we can see.

These are both great discoveries, but they are also disturbing. If we reached into a bag of random numbers and pulled out generic values for the constants of nature, neither a small cosmological constant nor a suitable period of Inflation would be likely outcomes. Both require an enormous degree of fine-tuning. As we've seen before, the universe appears to have been specially designed. More about this specialness in the next chapter.

CHAPTER SIX

On Frozen Fish and Boiled Fish

For explaining physics to an audience of nonphysicists, analogies and metaphors are obviously invaluable. But for me they are also tools for thought, my own idiosyncratic tools. Often I convince myself of the truth of some difficult point by inventing an analogy that applies similar questions to a more ordinary context.

The Anthropic Principle has created more confusion and irrelevant philosophical claptrap than anything that has come out of science for quite some time. Incessant argument occurs over its meaning, how it should be used to explain and predict, when it is legal, when it is not, when it is sensible, and when it is nonsense. The surest guide for me is to build an analogy about the more familiar world, where good old common sense can clear the air. More than a decade ago, I made up a parable to convince myself that the Anthropic Principle can make some sense.

A Birthday Present for Tini

It's an old tradition for well-known physicists to celebrate their sixtieth birthdays with parties, but these birthday parties usually consist of a couple of long days of continuous physics seminars — without music. I had to give a lecture at one such party for an old friend, Martinus Veltman. Tini — a bristly, bearded, colorful ogre of a Dutchman — looked like a cross between Orson Welles playing Macbeth and Saddam Hussein when he came out of his spider hole. Tini recently won the Nobel Prize for his work with Gerard 't Hooft that developed the mathematics of the Standard Model.

Because Tini was one of the first people to recognize the problem of vacuum energy, I thought I would give a birthday talk called "Tini and the Cosmological Constant." What I wanted to speak about was the Anthropic Principle and Steve Weinberg's calculation of galaxy formation. But I also wanted to explain how the Anthropic Principle could make good scientific sense. So as usual I made up an analogy.

Instead of asking why the cosmological constant is so precisely fine-tuned, I substituted a similar question: why is the temperature of the earth finely tuned to be in the narrow range in which liquid water can exist? Both questions ask how it happens that we live in a very unlikely environment that seems perfectly tailored to our own existence. To answer my question I proposed the following parable about intelligent fish.[1]

A Fish Story

Once upon a time, on a planet completely covered by water, lived a race of big-brained fish. These fish could survive only at a certain depth, and none had ever seen either the surface above or the bottom below. But their big brains made them very smart and also very curious. In time their questions about the nature of water and other things became very sophisticated. The most brilliant among them were called fyshicists. The fyshicists were wonderfully clever, and in a few generations they came to understand a great deal about natural phenomena, including fluid dynamics, chemistry, atomic physics, and even the nuclei of atoms.

Eventually some of the fyshicists began to question why the laws of nature are what they are. Their sophisticated technology allowed them to study water in all its forms, especially ice, steam, and of course, the liquid state. But with all their efforts still one thing stumped them. With all the possible values from zero to infinity, how could they account for the fact that the background temperature, T, was fine-tuned to be in the very narrow range that allowed H_2O to exist in its liquid form? They tried

1. The story was previously published in *The New Scientist* (November 1, 2003).

many things, including symmetries of various kinds, dynamical relaxation mechanisms, and many other ideas, but nothing could explain it.

Closely allied with the fyshicists were another group, the codmologists, who were also studying their watery world. The codmologists were less interested in the ordinary depths, where the big-brained fish lived, than they were in discovering if an upper boundary to their water-world existed. The codmologists were well aware that much of the water-world was not habitable, the pressure being wrong for their big brains. Journeying by fin to the upper reaches was by no means possible. Their big brains would explode if exposed to the very low water pressure in these regions. So instead, they speculated.

It happened that one school of thought among the codmologists held a very radical (some said ridiculous) idea about the fine-tuning of T. And they had a name for the idea — the Ickthropic Principle. The I.P. maintained that the temperature was in the liquid water range because only in this case could fish exist to observe it!

"Garbage!" said the fyshicists. "That's not science. It's religion. It's just giving up. And besides, if we agree with you, everyone will laugh at us and take away our funding."

Now not all of the codmologists meant the same thing by the Ickthropic Principle. In fact it was hard to find any two who agreed. One thought that it meant that the Head Angel Fish had made the world just for the purpose of accommodating big-brained fish. Another thought that the quantum-wave function of the waterverse was a superposition of all values of T and only by observing it did some ancestral fish "collapse the wave function."

A small number of codmologists, led by Andrei-the-Very-Big-Brained and Alexander Who-Swims Deep, held a very extraordinary idea. They believed that a stupendously big space existed beyond the upper water boundary. In this very big space, many other bodies similar in some ways to their water-world but different in other ways might exist. Some worlds would be unimaginably hot, so hot that the hydrogen nuclei might even fuse to form helium and then perhaps grow even hotter. Other worlds would be so cold that frozen methane would exist. Only a tiny fraction of the bodies would be at temperatures conducive to the formation of fish. Then it would be no mystery why T was fine-tuned.

As every angler knows, most places are fishless, but here and there conditions are just right. And that's where the fish are.

But the fyshicists sighed and said, "Oh Lord, there they go again with their fishy ideas. Just ignore them." The end.

The story was a complete flop. Loud sighs and moans from the audience were audible during the seminar. Afterward people avoided me. Tini himself was less than impressed. The Anthropic Principle affects most theoretical physicists the same way that a truckload of tourists in the African bush affects an angry bull elephant.

Anthropic Landscapes

No one, knowing what we do about astronomy, would doubt that the codmologists got it right. The story suggests that there are situations where an anthropic (or "ickthropic") explanation makes sense. But what are the rules? When is anthropic reasoning appropriate? When is it inappropriate? We need some guiding principles.

First, there is the obvious: an anthropic explanation of proposition X can make sense only if there is a strong reason to believe that the existence of intelligent life would be impossible unless X is true. For the big-brained fish, it's clear: too hot, and we get fish soup; too cold, and we get frozen fish. In the case of the cosmological constant, Weinberg provided the reasoning.

When you start to think about what it takes for life to be possible, the Landscape becomes a nightmarish minefield. I've already explained how a large cosmological constant would have been fatal, but there are many other dangers. The requirements for a universe fall into three main categories: the Laws of Physics must lead to organic chemistry; the essential chemicals must exist in sufficient abundance; and finally, the universe must evolve to create a large, smooth, long-lived, gentle environment.

Life is of course a chemical process. Something about the way atoms are constructed makes them stick together in the most bizarre combinations: the giant crazy Tinkertoy molecules of life — DNA, RNA, hundreds of proteins, and all the rest. Although chemistry is usually regarded as a separate branch of science — it has its own university departments and its own journals — it is really a branch of physics: that branch which deals with the outermost electrons in the atom. These *valence* electrons, hopping back and forth or being shared between atoms, give the atoms their amazing abilities to combine into a diverse array of molecules.

How is it that the Laws of Physics allow marvelously intricate structures like DNA that hold themselves together without collapsing, flying apart, or destructing in some other way? To some degree it is luck.

As we saw in chapter 1, the Laws of Physics begin with a list of elementary particles like electrons, quarks, photons, neutrinos, and more, each with special properties such as mass and electric charge. No one knows why the list is what it is or why the properties are exactly what they are. An infinite number of other lists is possible. But a universe filled with life is by no means what one would expect from a random choice of the list. Eliminating any of these particles (electrons, quarks, and photons) or even changing their properties a modest amount would destroy conventional chemistry. This is obviously so for electrons and quarks, which make up the atom and its nucleus, but it may be less obvious for the photon. Photons are of course the little "bullets" that make up light. True enough, without them we couldn't see, but we could still hear, feel, and smell, so maybe the photon isn't so important. Thinking that, however, is a big mistake: the photon happens to be the glue that holds the atom together.

What keeps the valence electrons in orbit around the central core of the atom? Why don't they just fly off and say adios to the protons and neutrons? The answer is the electrical attraction between the oppositely charged electrons and atomic nucleus. Electrical attraction is different from the attraction between a fly and a strip of flypaper. The flypaper may be very sticky and hold on fiercely, but once you separate the fly even a little, the flypaper immediately lets go. The fly flies off, and unless it is stupid enough to come back, it is completely free. In physics jargon the flypaper force is strong but *short range* — it doesn't reach out over large distances.

A short-range force of the flypaper type would be useless in binding electrons to the nucleus. The atom is a miniature solar system, and the all-important valence electrons are the most distant planets: the Plutos and Neptunes of the atom. Only a force that reaches out to large distances can keep them from flying off into the "outer space" beyond the boundaries of the atom.

Long-range forces, the kind that grab from a distance, are uncommon. Of all the many different types of forces in nature, only two are long range. Both are familiar, the most familiar being gravity. When we jump off the earth, gravity pulls us back. It reaches out hundreds of millions of miles to hold the planets in their orbits around the sun and tens of thousands of light-years to keep the stars confined within galaxies. That's the kind of force that is needed to tie the outer electrons to the central nucleus. Of course it's not gravity that holds the atom together: it's far too weak.

Another familiar long-range force acts between a magnet and an iron paper clip. The magnet doesn't have to be in direct contact with the paper clip to attract it. A strong magnet tugs on the clip even from a remote distance. But more relevant for the atom, the electric cousin of magnetic force is a long-range force that acts between electrically charged particles. Much like the gravitational force, except vastly stronger, the electric force binds the valence electrons the same way that gravity ties Pluto to the sun.

As I explained in chapter 1, electric forces between charged particles are caused by photons that are exchanged between the charges.[2] The ultralight photons (remember that they have no mass) are capable of jumping long distances to create long-range forces that bind the distant valence electrons to the nucleus. Remove the photon from the list, and there would be nothing to hold the atom together.

The photon is very exceptional. It is the only elementary particle, other than the graviton, that has no mass. What if it were less exceptional and had mass? Feynman's theory tells us how to compute the force when a hypothetical massive photon jumps between nucleus and electron. What one finds is that the heavier the photon, the less far it is able to jump. Were the photon mass even a tiny fraction of the electron mass, instead of being a long-range force, electric interactions would become short-range "flypaper forces," totally incapable of holding on to

2. This theme will be taken up again in chapter 7 in the context of String Theory.

the distant valence electrons. Atoms, molecules, and life are entirely dependent on the curious fact that the photon has no mass.

The range of the electric force is not the only feature that is essential for atoms to work properly. The strength of the force (how hard it pulls on the electrons) is critical. The force holding the electron to the nucleus is not very large by the standards of ordinary human experience. It's measured in billionths of a pound. What is it that determines the strength of electric forces between charged particles? Again, Feynman's theory tells us. The other ingredients in a Feynman diagram, besides particles, are vertex diagrams. Remember that every vertex diagram has a numerical value — the coupling constant — and for photon emission the coupling constant is the fine structure constant α, a number about equal to the fraction $1/137$. The smallness of α is the ultimate mathematical reason why electric forces are much weaker than their nuclear counterparts.

What if the fine structure constant were bigger, say about one? This would create several disasters, one of which would endanger the nucleus. The nuclear force that holds the nucleons (protons and neutrons) together is a flypaper force: short range and strong. The nucleus itself is like a ball of sticky flies. Each nucleon is stuck to its nearest neighbors, but if it can just separate from the others by a tiny bit, it's free to fly off.

There is something working against the nuclear force, competing with it, to repel the protons from one another. The protons are, of course, electrically charged. They attract the negative electrons because they have the opposite charge (opposite charges attract; like charges repel). The neutrons are electrically neutral and, therefore, don't play a role in the balance of electric forces. Protons, on the other hand, are positively charged and electrically repel one another. In fact if a nucleus has more than about one hundred protons, the repulsive long-range electric forces are enough to blow it apart.

What would happen if the electric force were as strong as the nuclear force? Then all complex nuclei would be unstable. In fact the electric force could be a good deal weaker than the nuclear force and still endanger nuclei like carbon and oxygen. Why is the fine structure constant small? No one knows, but if it were big, there would be no one to ask the question.

Protons and neutrons are no longer considered to be elementary particles. Each is composed of three quarks. As discussed in chapter 1, there are several different species of quarks labeled up, down, strange, charmed, bottom, and top. While the names are quite meaningless, the differences between the types of quarks are important. A quick look at the list of particle masses in chapter 3 reveals that the quark masses vary over a huge range from roughly 10 electron masses for the up- and down-quarks to 344,000 electron masses for the top-quark. Physicists puzzled for some time about why the top-quark is so heavy, but recently we have come to understand that it's not the top-quark that is abnormal: it's the up- and down-quarks that are absurdly light. The fact that they are roughly twenty thousand times lighter than particles like the Z-boson and the W-boson is what needs an explanation. The Standard Model has not provided one.

Thus, we can ask what the world would be like if the up- and down-quarks were much heavier than they are. Once again — disaster! Protons and neutrons are made of up- and down-quarks. (Particles made of strange-, charmed-, bottom-, and top-quarks play no role in ordinary physics and chemistry. They are of interest mainly to high-energy physicists.) According to the quark theory of protons and neutrons, the nuclear force (force between nucleons) can be traced to quarks hopping back and forth between these nucleons.[3] If the quarks were much heavier, it would be much more difficult to exchange them, and the nuclear force would practically disappear. With no sticky flypaper force holding the nucleus together, there could be no chemistry. Luck is with us again.

Remember that in terms of the Landscape, our universe rests in a valley where all the fortunate coincidences are true. But in generic regions of the Landscape, things can be very different. The fine structure constant could easily be larger, the photon massive, quarks heavier, or even worse, electrons, photons, or quarks might not be on the list. Any one of these would be enough to eliminate our presence.

Even if all the standard particles existed with the right mass and the right forces, chemistry could still fail. One thing more is needed: elec-

3. A more technically correct description is that pions — particles composed of quarks and antiquarks — transmit the force between nucleons. If the masses of the up- and down-quarks were larger, the mass of pions would also increase. This would have the effect that the nuclear force would be drastically modified.

trons must be fermions. The fact that fermions are so exclusive — you can't put more than one in a quantum state — is essential to chemistry. Without the Pauli exclusion principle, all electrons in an atom would sink down to the lowest atomic orbits, where they would be much more difficult to dislodge. Ordinary chemistry is completely dependent on the Pauli principle. If electrons suddenly turned into the more sociable bosons, life based on carbon chemistry would go poof. So you see that a world with ordinary chemistry is far from generic.

Physicists often use words differently from the way they are ordinarily used. When you say that something exists, you probably mean that it can be found somewhere in the universe. For example, if I were to tell you that black holes exist, you might ask me where you can find one. Black holes do exist in the ordinary sense: they are actual astronomical objects found, for example, at the centers of galaxies. But suppose I told you tiny black holes no bigger than a speck of dust exist. Again you might ask where they are found. This time I would answer that there is none; it takes a huge amount of mass to squeeze itself into a black hole. No doubt you would get annoyed and say, "Stop pulling my leg. You told me they exist!"

What physicists (especially of the theoretical variety) mean by the term *exist* is that the object in question can exist *theoretically*. In other words, the object exists as a solution to the equations of the theory. By that criterion perfectly cut diamonds a hundred miles in diameter exist. So do planets made of pure gold. They may or may not actually be found somewhere, but they are possible objects consistent with the Laws of Physics.

Long-range weak forces and short-range strong forces, acting among fermions, lead to the *existence* of complex atoms like carbon, oxygen, and iron. That's nice, but I mean it in the theoretical sense. "What more," you may ask, "is needed to make sure that complex atoms exist in *my* ordinary sense? What is required to actually produce those atoms and make them abundant in the universe?" The answer is not so simple. Complex atomic nuclei are not likely to result from random collisions of particles, even in the early hot universe.

In the first minutes after the Big Bang, there were no atoms or even nuclei. Hot plasma composed of protons, neutrons, and electrons filled all space. The high temperature prevented nucleons from sticking together to form nuclei. As the universe cooled, protons and neutrons stuck together and formed the primordial elements.[4] But apart from tiny traces of other elements, only the very simplest of nuclei formed: hydrogen and helium.

Moreover, as medieval alchemists discovered, it's not easy to transmute one element into another. So where, then, did all the carbon, oxygen, nitrogen, silicon, sulfur, iron, and other familiar chemical elements come from? The answer is that the intensely hot nuclear furnace of a star can do what no alchemists ever could — transform the elements, one into another. The cooking process is nuclear fusion, the same kind of fusion that powers nuclear weapons. Fusion combines the hydrogen nuclei in all sorts of permutations and combinations. The results of these nuclear reactions were the familiar elements.

The chain of nuclear reactions in stars that starts with the lightest elements and leads to iron is complicated. A couple of examples will illustrate the point. The most familiar example is the fusion reaction that begins with hydrogen and produces helium. Here is where the weak interactions (diagrams with W- and Z-bosons) come in. The first step is the collision of two protons.[5] Many things can happen when two protons collide, but if you know the Feynman diagrams for the Standard Model, you can find one that ends up with a proton, a neutron, a positron, and a neutrino.

The positron finds a wandering electron in the star, and together they self-destruct into photons that eventually become the star's thermal energy (heat). The neutrino just zips away and disappears with almost the speed of light. That leaves one sticky proton and one sticky neutron that stick together to form an isotope of hydrogen called deuterium.

Next, a third proton strikes the deuterium nucleus and sticks to it. The nucleus with two protons and a neutron is a form of helium called helium-three (^{3}He), but it's not the stable kind of helium that we use to fill balloons. That stuff is called helium-four (^{4}He).

The story continues: two ^{3}He nuclei collide. All together that means four protons and two neutrons. But they don't all stick together. Two of

4. When I say elements I am usually referring to the nucleus. The binding of electrons to the nucleus took a lot longer.

5. Hydrogen is the simplest of the elements. Its nucleus is a single proton.

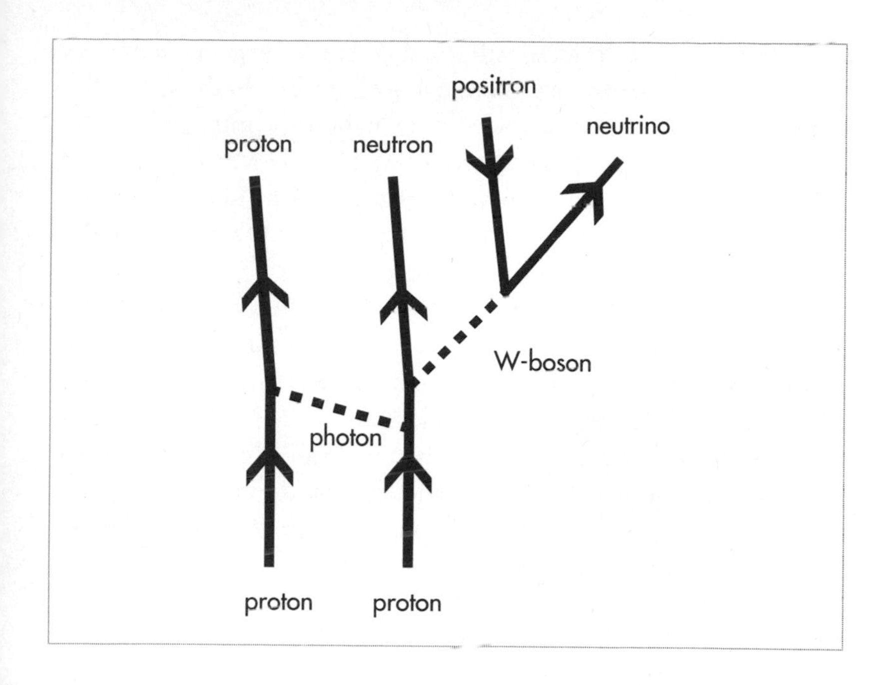

the protons fly off and leave a nucleus with two protons and two neutrons. That's an ordinary ^{4}He nucleus. You don't need to remember all that. Very few physicists do.

Most of the nuclear reactions that take place in stars consist of a single proton colliding with an already present nucleus and increasing its atomic weight by one unit. Sometimes the proton turns into a neutron by giving off a positron and a neutrino. Sometimes a neutron will become a proton, electron, and antineutrino. In any case, inside the star, step-by-step, the original hydrogen and helium nuclei turn into heavier elements.

But what good are the complex elements locked up inside stars? Science-fiction stories might posit strange forms of life made of swirling hot plasma that thrive at millions of degrees, but real life needs a cooler environment. Sadly, the carbon and oxygen remained imprisoned in the star's interior throughout the entire lifetime of the star.

But stars don't live forever.

Eventually all stars, our sun included, will run out of fuel. At that point a star collapses under its own weight. Before the fuel runs out, stars are kept in equilibrium by the heat and pressure generated by nuclear reactions. There are two competing tendencies in the star. Like a nuclear bomb, it wants to explode, while at the same time gravity is trying to crush it under its own enormous weight. These two tendencies, exploding and imploding, are kept in balance as long as there is fuel to burn. But once the fuel runs out, there is nothing to resist the pull of gravity, and the star implodes.

There are three possible endpoints to the implosion. A star like our sun is relatively light, and it will collapse only until it forms a white dwarf. A white dwarf is made of more or less ordinary material — protons, neutrons, and electrons — but the electrons are squeezed up against one another to a far greater degree than in ordinary materials. It's the Pauli exclusion principle that keeps the electrons from collapsing even further. If all stars ended up as white dwarfs, the freshly cooked elements would remain imprisoned inside them.

On the other hand, if the star is many times heavier than the sun, the force of gravity will be irresistible. The inevitable disastrous collapse will end in the most violent process imaginable — the formation of a black hole. Elements trapped in black holes would be even less available than those in white dwarfs.

But there is a middle ground. Stars within a certain range of masses collapse past the white dwarf stage but not all the way to a black hole. In these stars the electrons, in a sense, get squeezed out, while the protons turn into neutrons, and the end result is a solid ball of incredibly dense neutron matter: a neutron star. Surprisingly, the weak interactions play an indispensable role. Each proton, as it becomes a neutron, gives off two particles, a positron and a neutrino. The positrons quickly combine with the electrons in the star and disappear.

Such an event, called a supernova, is not a gentle one. A supernova can outshine an entire galaxy with a hundred billion stars!

In everyday physics and chemistry, neutrinos are of no importance at all. They can pass through light-years of lead without disturbing it one bit. Neutrinos from the sun are continually passing through the earth, through our food and drink, and through our bodies with no effect at all. But our existence is totally dependent on them. The neutrinos flying out of the supernova implosion are so numerous that, despite their feebleness, they create an enormous pressure, pushing matter in

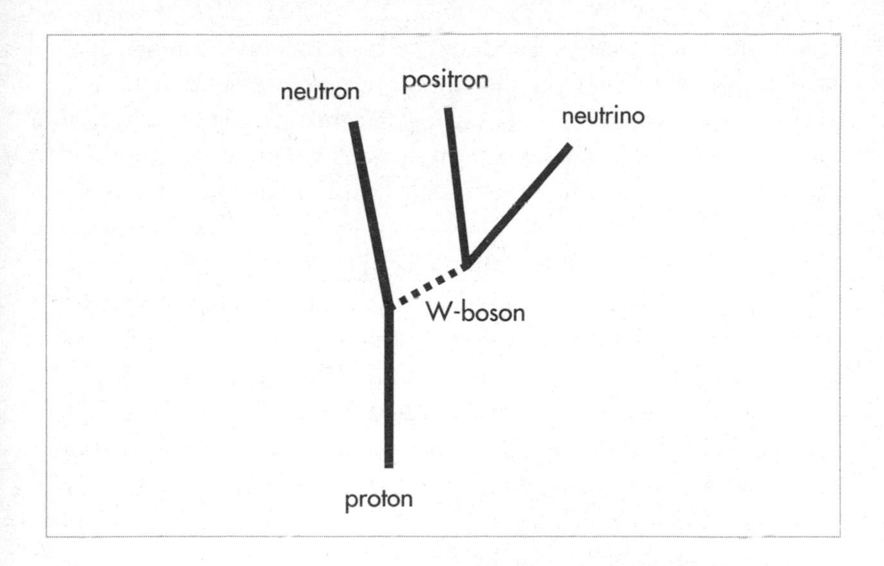

front of them. The pressure exerted by the neutrinos blows off the outer layers of the collapsing star and, in the process, sprays out the complex nuclei that were cooked before the star collapsed. So as its final act, the star in its death throes donates its complex nuclei to fill the universe with matter.

The Crab Nebula is the remnant of a supernova explosion. The explosion was seen on Earth in the year 1054.

Our sun is a youngster. The universe is about fourteen billion years old, but the sun was born late in its history, only five billion years ago. By that time generations of stars had formed and died and there were already enough heavy elements to form the solar system. We are fortunate indeed that the ghostly neutrino exists — in the ordinary sense of the word.

There are multiple ways that things could go wrong with the nuclear cooking. If there were no weak interactions or if neutrinos were too heavy, protons could not turn into neutrons during the cooking. The cooking of carbon is sensitive to the details of the carbon nucleus. One of the great scientific events of the twentieth century occurred when the cosmologist Fred Hoyle was able to predict one of these nuclear details just from the fact that we are here. In the early 1950s Hoyle argued that there is a "bottleneck" in the cooking of elements in stars like the sun. There appeared to be no way for the cooking to proceed past atomic number 4 — helium. Nuclear cooking usually goes forward one proton at a time to form a heavier element, but there is no stable nucleus with atomic number 5, so there is no easy way to get past helium.

There is one way out. Two helium nuclei can collide and stick together to form a nucleus with atomic number 8. That nucleus would be the isotope beryllium 8. Later, another helium nucleus could collide with the beryllium and form a nucleus with atomic number 12 — good old carbon 12, the stuff of organic chemistry. But there is a fly in this ointment.

Beryllium 8 is a very unstable isotope. It decays (falls apart) so rapidly that there is not enough time for the third helium nucleus to collide before the beryllium disappears — unless an unlikely coincidence occurs. If by accident there were an excited state — a so-called resonance — of the carbon nucleus with exactly the right properties, the probability for the beryllium to capture a helium nucleus would be much higher than expected. The likelihood of such a coincidence is very small, but when Hoyle suggested that such a coincidence might solve the problem of cooking the heavy elements, experimental nuclear physicists went right to work. And BINGO, the excited state was discovered with exactly the properties that Hoyle guessed. Just a small increase or decrease in the energy of the excited carbon nucleus, and all the work of making galaxies and stars would have been in vain; but as it is, carbon atoms — and thus, life — can exist.

The properties of Hoyle's carbon resonance are sensitive to a number of constants of nature, including the all-important fine structure constant. Just a few percent change in its value, and there would have been no carbon and no life.[6] This is what Hoyle meant when he said that "it looks as if a super-intellect has monkeyed with physics as well as with chemistry and biology."

But again, it would do no good for the nuclear physics to be "just right" if the universe had no stars. Remember that a perfectly homogeneous universe would never give birth to these objects. Stars, galaxies, and planets are all the result of the slight lumpiness at the beginning. Early on, the density contrast was about 10^{-5} in magnitude, but what if it had been a little bigger or a little smaller? If the lumpiness had been much less, let's say, 10^{-6}, in the early universe, galaxies would be small and the stars, very sparse. They would not have had sufficient gravity to hang on to the complex atoms that were spewed out by supernovae; these atoms would have been unavailable for the next generation of stars. Make the density contrast a little less than that, and no galaxies or stars would form at all.

What would happen if the lumpiness were larger than 10^{-5}? A factor of one hundred larger, and the universe would be full of violent, ravenous monsters that would swallow and digest galaxies before they were even finished forming. Don't worry; I haven't lost my mind. The "megamonsters" are huge black holes. Remember that gravity is the agent that works on the regions with slight excess mass density and pulls them together to form galaxies. But if the overdensities were too strong, gravity would work too quickly. The gravitational collapse of these regions would go right past the galaxy stage and evolve into black holes. All matter would quickly be gobbled up and destroyed at the infinitely violent central singularity of the black hole. Even density contrasts a factor of

6. There is debate over just how sensitive the existence of carbon is to the various constants. Some people would put it in the range of a couple of percent. Others, Steven Weinberg among them, would put the number at roughly 10 or 15 percent. But all would agree that some fine-tuning is needed to ensure an abundant supply of carbon.

ten stronger could endanger life by creating too many collisions between celestial objects in the solar system.

A lumpiness of about 10^{-5} is essential for life to get a start. But is it easy to arrange for this amount of density contrast? The answer is most decidedly no! The various parameters governing the inflating universe must be chosen with great care in order to get the desired result. More of Hoyle's monkeying?

There is a lot more. The laws of particle physics include the requirement that every particle has an antiparticle. How then did the universe get to have such a large preponderance of matter over antimatter? Here is what we think happened:

When the universe was very young and hot, it was filled with plasma that contained almost exactly equal amounts of matter and antimatter. The imbalance was extremely small. For every 100,000,000 antiprotons, there were 100,000,001 protons. Then, as the universe cooled, particles and antiparticles combined in pairs and annihilated into photons. One hundred million antiprotons found 100,000,000 partners and, together, they committed suicide, leaving 200,000,000 photons and just 1 leftover proton. These leftovers are the stuff we are made of. Today, if you take a cubic meter of intergalactic space, it will contain about 1 proton and 200,000,000 photons. Without the slight initial imbalance, I would not be here to tell you (who would not be here to read) these things.

Another essential requirement for life is that gravity be extremely weak. In ordinary life gravity hardly seems weak. Indeed, as we age, the daily prospect of fighting gravity gets more and more daunting. I can still hear my grandmother saying, "Oy vey, I feel like a thousand pounds." But I don't ever recall hearing her complain about electric forces or nuclear forces. Nonetheless, if you compare the electric force between the nucleus and an atomic electron with the gravitational force, you would find the electric force is about 10^{41} times larger. Where did such a huge ratio come from? Physicists have some ideas, but the truth is that we really don't know the origin of this humongous discrepancy between electricity and gravity despite the fact that it is so central to our existence.[7] But we can ask what would have happened if gravity had been a

7. It is called the gauge hierarchy problem, and no universal agreement on its solution has been reached.

little stronger than it is. The answer again is that we would not be here to talk about it. The increased pressure due to stronger gravity would cause stars to burn much too fast — so fast that life would have no chance to evolve. Even worse, black holes would have consumed everything, dooming life long before it began. The large gravitational pull might even have aborted the Hubble expansion and caused a big crunch very shortly after the Big Bang.

Just how seriously should we take this collection of lucky coincidences? Do they really make a strong case for some kind of Anthropic Principle? My own feeling is that they are very impressive, but not so impressive that they would have pushed me past the tipping point to embrace an anthropic explanation. None of these accidental pieces of good luck, with the exception of the remarkable weakness of gravity, involves extraordinarily high-precision (precision to many decimal places) fine-tuning. And even the feebleness of gravity has a possible explanation involving the magic of supersymmetry. Taken together, these coincidences do seem like an unlikely bunch of accidents; but accidents, after all, do happen.

However, the smallness of the cosmological constant is another matter. To make the first 119 decimal places of the vacuum energy zero is most certainly no accident. But it was not just that the cosmological constant was very small. Had it been even smaller than it is, had it continued to be zero to the current level of accuracy, one could have gone on believing that some unknown mathematical principle would make it exactly zero. The event that hit us like the proverbial ton of bricks was the fact that in the 120th place the answer was not zero. No missing mathematical magic is going to explain that.

But even the cosmological constant would not have been enough to tip the balance for me. For me the tipping point came with the discovery of the huge Landscape that String Theory appears to be forcing on us.

When Do Anthropic Explanations Make Sense?

Suppose that you and I were partners in the business of creating life-friendly universes. Your job is to think of all the necessary ingredients and to create a design. My job is just to search the Landscape for a location that satisfies your requirements. You would come up with a design. Then I would go off and search the Landscape. If the Landscape had only a handful of valleys, I almost certainly would not find what you were looking for. I would tell you that you are on a fool's errand because the thing you're looking for is incredibly improbable.

But if you knew a bit about String Theory, you might question my judgment: "Are you sure you looked everywhere: at every nook and cranny, in every valley? There are 10^{500} of them, you know. Surely with that number, it must be possible to find what we are looking for. Oh, and don't bother looking at the average valleys. Look for exceptional ones."

This suggests a second criterion for an acceptable anthropic explanation. The number of mathematically consistent possibilities must be so large that even very unlikely requirements will be met in at least a few valleys.

This second requirement has real force only in the context of a precise theory of the Landscape. To give an example, the codmologists could point to Newton's theory of gravity and argue that the equations permit circular planetary orbits at any distance from a star. The very distant orbits have frozen planets, where water and even methane freeze to ice. The orbits lying close to the star have hot planets, where water instantly boils. But somewhere in between a point must exist where the temperature is right for liquid H_2O. The theory has so many solutions that among them there must be some that are just right.

Strictly speaking a planet cannot orbit at any distance. Solar systems are a lot like atoms, the sun and planets replacing the atomic nucleus and the electrons. As Niels Bohr first understood, electrons can orbit only in definite *quantized* orbits. The same reasoning applies to planets. But fortunately, the possible orbits are so numerous and densely spaced that for practical purposes any distance is possible.

It was not enough for the codmologists to know that the requirements for life are mathematically consistent. They also needed a universe that is so big and diverse that it actually does contain almost everything that can exist. The known universe has 10^{11} galaxies, each with 10^{11} planets, for a grand total of 10^{22} opportunities to satisfy the special requirement for liquid water. With that many planets there is near certainty that many will be habitable.

Following, then, are the requirements:

To explain proposition X anthropically, we should first of all have reason to believe that not-X would be fatal to the existence of our kind of life. In the case of the cosmological constant, this is exactly what Weinberg found.

Even if X seems wildly unlikely, a rich enough Landscape with enough valleys may make up for it. This is where the properties of String Theory are beginning to have impact. At a few universities in the United States and Europe, the exploration of the Landscape has begun. As we will see, all signs point to an unimaginable diversity of valleys: perhaps more than 10^{500} of them.

And last but certainly not least, the cosmology implied by the theory should naturally lead to a supermegaverse, so large that all the regions of the Landscape will be represented in at least one pocket universe. Once again, String Theory, when combined with the idea of Inflation, fills the bill. But that's for later chapters.

The Anthropic Principle is the bête noire of theoretical physics. Many physicists express an almost violent reaction to it. The reason is not hard to imagine. It threatens their paradigm, the paradigm that says that everything about nature can be explained by mathematics alone. Are their arguments justified? Do they even make sense?

Let's look at some of the objections from the viewpoint of the big-brained fish. The objection that the Anthropic Principle is religion, not science, is clearly off the mark. In the view of Andrei and Alexander, there is no need for the hand of God to fine-tune the world for the benefit of her children. If anything, most of the world is a very inhospitable place, far more deadly than the fyshicists ever imagined. In fact the Ickthropic Principle, in the form proposed by Andrei and Alexander, completely removes the mysterious from the fyshicists' mystery.

A more relevant objection is that physics loses its predictive power. To a large extent that is true if what we want to predict is the temperature of our planet, the amount of sunlight it receives, the precise length of the yearly cycle, the height of the tides, the amount of salt in the ocean, and other environmental facts. But to reject the ickthropic explanation of some of the parameters of the environment on the basis that predictivity would be lost is clearly irrational. Requiring complete predictivity has an emotional basis that has nothing to do with hard facts of planetary science.

The complaint that the big-brained fish are giving up the traditional quest for scientific explanation is also expressing a psychological disappointment but obviously has no scientific merit. At some point fyshicists' hopes turn into dogmatic religion.

Of all the criticisms of the Anthropic Principle that I have heard, only one strikes me as serious science. It was leveled by two of my close friends, Tom Banks and Mike Dine, who don't like my ideas.[8] Here's how it goes:

Suppose that there is a fine-tuning in nature that has no anthropic value. I'll give you an example. The earth and the moon are the same apparent size in the sky. In fact the disk of the moon is so close in size to the disk of the sun that during a solar eclipse the moon blocks out the disk of the sun almost exactly. That is very lucky for solar astronomers: it allows them to make observations that they could not otherwise make. For example, they can study the corona of the sun during the eclipse. They can also measure the precise amount that light rays are bent by the gravity of the sun. But this unusually precise fine-tuning has no particular value for making life possible on the earth. Moreover, it is likely that the majority of habitable planets don't have moons that match their suns so precisely. The probability of selecting a planet with such solar-lunar fine-tuning if we arbitrarily selected a random habitable planet is very small. So unless we believe in unlikely coincidences, the explanation for our world must be something other than random choice subject only to the anthropic constraint.

8. By the time I finished writing *The Cosmic Landscape*, Dine became one of the chief supporters of the view that some features of nature are likely to be environmental and can be understood only from an anthropic perspective. Banks remains skeptical.

The moon-sun coincidence is not really much of a problem. The precision with which the moon matches the sun is not phenomenal. The difference is about 1 percent. One-percent coincidences happen about 1 percent of the time. It is nothing more than a lucky accident. But what if the moon and sun matched to one part in a trillion trillion trillion? That seems so unlikely that it would require an explanation. Something in addition to the Anthropic Principle would have to be at work. It might cast doubt on the idea that the unexplainable specialness of the universe has anything to do with the success of life.

There is at least one very unusual feature of the Laws of Physics that seems very finely tuned with no anthropic explanation in sight. It has to do with the proton, but let's first review the properties of its almost identical twin, the neutron. The neutron is an example of an unstable particle. Neutrons, not bound inside a nucleus, will last only about twelve minutes before disappearing. Of course the neutron has mass, or equivalently energy, which cannot just disappear. Energy is a quantity that physicists say is *conserved*. That means the total amount of it can never change. Electric charge is another exactly conserved quantity. When the neutron disappears, something with the same total energy and charge must replace it. In fact the neutron decays into a proton, an electron, and an antineutrino. The initial and final energy and electric charge are the same.

Why does the neutron decay? If it didn't, the real question would be, why doesn't it decay? As Murray Gell-Mann once quoted T. H. White, "Everything which is not forbidden is compulsory." Gell-Mann was expressing a fact about quantum mechanics: quantum fluctuations — the quantum jitters — will eventually make anything happen unless some special law of nature expressly forbids it.

What about protons — can they decay, and if so, what do they become? One simple possibility is that the proton disintegrates into a photon and a positron. The photon has no charge, and the proton and positron have exactly the same charge. It ought to be possible for protons to disintegrate into photons and positrons. No principle of physics prevents it. Most physicists expect that given enough time, the proton will decay.

But if the proton can decay, it means that all atomic nuclei can disintegrate. We know that atomic nuclei of atoms like hydrogen are very stable. The lifetime of a proton must be many times the age of the universe.

There must be a reason why the proton lives so long. Can that reason be anthropic? Certainly our existence places limitations on the lifetime of the proton. It obviously cannot be too small. Let's suppose the proton lives one million years. Then I would not have to worry very much about my protons disappearing during my life. But since the universe is about ten billion years old, if the proton lived only a million years, they all would have disappeared long before I was born. So the anthropic requirement for the proton lifetime is a lot longer than a human lifetime. The proton must last at least fourteen billion years.

Anthropically, the lifetime of the proton may have to be a good deal longer than the age of the universe. To see why, let's suppose that the proton lifetime were twenty billion years. The decay of an unstable particle is an unpredictable event that can happen any time. When we say that the proton lifetime is twenty billion years, we mean that, statistically, the average proton will last that long. Some will decay in one year, and some in forty billion years.

Your body has about 10^{28} protons. If the proton lifetime were twenty billion years, about 10^{18} of those protons would decay every year.[9] This is a negligible fraction of your protons, so you don't have to worry about disappearing. But each proton that decays in your body shoots out energetic particles: photons and positrons and pions. These particles moving through your body have the same effects as exposure to radioactivity: cell damage and cancer. If 10^{18} protons decay in your body, they will kill you. So the anthropic constraints on proton decay may be stronger than what you naively think. As far as we know, a lifetime of a million times the age of the universe — 10^{16} years — is long enough not to jeopardize life. On anthropic grounds we can rule out all valleys of the Landscape where the average proton lifetime is less than this.

But we know that the proton lives vastly longer than 10^{16} years. In a tank of water with roughly 10^{33} protons, we would expect to see one pro-

9. The difference between twenty billion and ten billion years is not important for our discussion. If the proton lifetime were ten billion (10^{10}) years, that would mean that one proton out of ten billion would decay every year. If we then multiply this by the 10^{28} protons in your body, the number that would decay in one year is $10^{28}/10^{10} = 10^{18}$.

ton decay each year if the lifetime were 10^{33} years. Physicists, hoping to witness a few protons decaying, have constructed huge underground chambers filled with water and photoelectric detectors. Sophisticated modern detectors can detect the light from just a single decay. But so far, no cigar; not a single proton has ever been seen to disintegrate. Evidently the lifetime of the proton is even longer than 10^{33} years, but the reason is unknown.

To compound the problem, we also don't know any reason why the String Theory Landscape should not have valleys in which the Laws of Physics are life-friendly but where protons live for only 10^{16} or 10^{17} years. Potentially the number of such valleys could vastly outweigh those with much greater lifetimes.

This is a serious concern but probably not a showstopper. Unfortunately we don't have nearly enough information about the Landscape to know what percentage of its habitable valleys have such very long proton lifetimes. But there is some reason for optimism. The Standard Model with no modification does not permit the proton to decay at all! This has nothing to do with the Anthropic Principle; it is simply a mathematical property of the Standard Model that the proton cannot disintegrate. If the typical habitable environment requires something fairly similar to the Standard Model, then proton stability may go along for the ride.

But we know that the Standard Model is not the full story. It does not contain gravity. Even though the Standard Model may be a very good description of ordinary physics, it nonetheless must break down. This could happen many ways. Theories called Grand Unified Theories (GUTs) are, despite their awful name, very attractive. The simplest generalization of the Standard Model to a GUT brings the proton lifetime to just around 10^{33} or 10^{34} years.

Other extensions of the Standard Model are not so safe. One of them, based on supersymmetry, can lead to significantly shorter proton lifetimes unless it is appropriately adjusted. We need more information before we can draw far-reaching conclusions. Fortunately, particle physics experiments in the near future may bear on the validity of the Standard Model and also on the reasons for the unusual stability of the proton. Stay tuned for a few years.

Philosophical Objections

In the abstract of a paper titled "Scientific Alternatives to the Anthropic Principle," the physicist Lee Smolin writes, "It is explained in detail why the Anthropic Principle cannot yield any falsifiable predictions, and therefore cannot be a part of science."[10]

Smolin's paper goes on in the introduction to say:

> I have chosen a deliberatively provocative title, in order to communicate a sense of frustration I've felt for many years about how otherwise sensible people, some of whom are among the scientists I most respect and admire, espouse an approach to cosmological problems that is easily seen to be unscientific. I am referring of course to the anthropic principle. By calling it unscientific I mean something very specific, which is that it fails to have a necessary property to be considered a scientific hypothesis. This is that it be falsifiable. According to [the philosopher] Popper, a theory is falsifiable if one can derive from it unambiguous predictions for doable experiments such that, were contrary results seen, at least one premise of the theory would have been proven not to apply to nature.

Richard Feynman once remarked, "Philosophers say a great deal about what is absolutely necessary for science, and it is always, so far as one can see, rather naive, and probably wrong." Feynman was referring to Popper among others. Most physicists, like Feynman, don't usually think much about philosophy: not unless they are trying to use it to prove someone else's theory is unscientific.

Frankly, I would have preferred to avoid the kind of philosophical discourse that the Anthropic Principle excites. But the pontification, by the "Popperazzi," about what is and is not science has become so furious in news reports and Internet blogs that I feel I have to address it. My opinion about the value of rigid philosophical rules in science is the

10. Lest anyone get the idea that Smolin and I are enemies, this is not at all the case. In fact Smolin is a good friend for whom I have a great deal of admiration. Nevertheless, our opinions on this particular topic are strongly at odds.

same as Feynman's. Let me quote from a debate that appeared on the Internet site edge.org. The quote is from a short essay that I wrote in response to Smolin's paper. Smolin's arguments can also be found there. They are thoughtful and interesting.

> Throughout my long experience as a scientist I have heard unfalsifiability hurled at so many important ideas that I am inclined to think that no idea can have great merit unless it has drawn this criticism. I'll give some examples:
>
> From psychology: You would think that everybody would agree that humans have a hidden emotional life. B. F. Skinner didn't. He was the guru of a scientific movement called behaviorism that dismissed anything that couldn't be directly observed as unscientific. The only valid subject for psychology according to the behaviorist is external behavior. Statements about the emotions or the state of mind of a patient were dismissed as unfalsifiable and unscientific. Most of us, today, would say that this is a foolish extreme. Psychologists today are deeply interested in emotions and how they evolved.
>
> From physics: In the early days of the quark theory, its many opponents dismissed it as un-falsifiable. Quarks are permanently bound together into protons, neutrons and mesons. They can never be separated and examined individually. They are, so to speak, hidden behind a different kind of veil. Most of the physicists who made these claims had their own agendas, and quarks just didn't fit in. But by now, although no single isolated quark has ever been detected, there is no one who seriously questions the correctness of the quark theory. It is part of the bedrock foundation of modern physics.
>
> Another example is Alan Guth's inflationary theory. In 1980 it seemed impossible to look back to the inflationary era and see direct evidence for the phenomenon. Another impenetrable veil called the "surface of last scattering" prevented any observation of the inflationary process. A lot of us did worry that there might be no good way to test inflation. Some — usually people with competing ideas — claimed that inflation was un-falsifiable and therefore not scientific.

I can imagine the partisans of Lamarck criticizing Darwin, "Your theory is un-falsifiable, Charles. You can't go backward in time, through the millions of years over which natural selection acted. All you will ever have is circumstantial evidence and an un-falsifiable hypothesis. By contrast, our Lamarckian theory is *scientific* because it is falsifiable. All we have to do is create a population that lifts weights in the gym every day for a few hours. After a few generations, their children's muscles will bulge at birth." The Lamarckists were right. The theory is easily falsified — too easily. But that didn't make it better than Darwin's theory.

There are people who argue that the world was created 6000 years ago with all the geological formations, isotope abundances, dinosaur bones, in place. Almost all scientists will point the accusing finger and say "Not falsifiable!" And I would agree. But so is the opposite — that the universe was not created this way — un-falsifiable. In fact that is exactly what creationists do say. By the rigid criterion of falsifiability "creation-science" and science-science are equally unscientific. The absurdity of this position will, I hope, not be lost on the reader.

Good scientific methodology is not an abstract set of rules dictated by philosophers. It is conditioned by, and determined by, the science itself and the scientists who create the science. What may have constituted scientific proof for a particle physicist of the 1960's — namely the detection of an isolated particle — is inappropriate for a modern quark physicist who can never hope to remove and isolate a quark. Let's not put the cart before the horse. Science is the horse that pulls the cart of philosophy.

In each case that I described — quarks, inflation, Darwinian evolution — the accusers were making the mistake of underestimating human ingenuity. It only took a few years to indirectly test the quark theory with great precision. It took 20 years to do the experiments that confirmed inflation. And it took 100 years or more to decisively test Darwin (some would even say that it has yet to be tested). The powerful methods that biologists would discover a century later were unimaginable to Darwin and his contemporaries. Will it be possible to test eternal inflation and the Landscape? I certainly think so although it may be, as in the case of quarks, that the tests will be less direct, and involve more theory, than some would like.

After this material was written, I thought of a couple of additional examples of overzealous Popperism. An obvious one is the S-matrix theory[11] of the 1960s, which said that since elementary particles are so small, any theory that attempts to discuss their internal structure is unfalsifiable and, therefore, not science. Again, no one takes that seriously today.

A famous example from the late nineteenth century involves one of Einstein's heroes, Ernst Mach. Mach was both a physicist and a philosopher. He was an inspiration for Wittgenstein and the logical positivists. At the time when he was active, the hypothesis that matter was made of atoms was still an unproved conjecture, and it remained that way until Einstein's famous 1905 paper on Brownian Motion unequivocally demonstrated that matter has an atomic structure.

Even though Boltzmann had shown that the properties of gases could be explained by the atomic hypothesis, Mach insisted that it was not possible to prove the reality of atoms. He allowed that they might be a useful mnemonic, but he argued strenuously that the impossibility of falsifying them undermined their status as real science.

Falsification, in my opinion, is a red herring, but confirmation is another story. (Perhaps this is what Smolin really meant.) By confirmation I mean direct positive evidence for a hypothesis rather than absence of negative evidence. It is true that the theory of Eternal Inflation described in chapter 9 and the existence of multiple pocket universes cannot be confirmed in the same way that the big-brained fish could confirm their version of the Ickthropic Principle. Without violating any laws of nature, the codmologists could construct a pressurized, water-filled submarine to take them to the surface and observe the existence of planets, stars, and galaxies. They could even visit these astronomical bodies and confirm for themselves the enormous diversity of environments. Unfortunately there are insurmountable (see, however, chapter 12) reasons why the analogous option is not available to us. The key concept is the existence of cosmic horizons that separate us from other

11. See chapter 7.

pocket universes. In chapters 11 and 12, I discuss horizons and the question of whether they are really ultimate barriers to collecting information. But certainly the critics are correct that *in practice*, for the foreseeable future, we are stuck in our own pocket with no possibility of directly observing other ones. Like quark theory, the confirmation will not be direct and will rely on a great deal of theory.

As for rigid philosophical rules, it would be the height of stupidity to dismiss a possibility just because it breaks some philosopher's dictum about falsifiability. What if it happens to be the right answer? I think the only thing to be said is that we do our best to find explanations of the regularities we see in the world. Time will shake out the good ideas from the bad, and they will become part of science. The bad get added to the junk heap. As Weinberg emphasized, we have no explanation for the cosmological constant other than some kind of anthropic reasoning. Will it be one of the good ideas that become science or one that winds up in the junk? No rigid rules of philosophers, or even scientists, can help very much. Just as generals are always fighting the last war, philosophers are always parsing the last scientific revolution.

Before concluding this chapter I want to discuss one more favorite objection to the Anthropic Principle. This argument is that the Anthropic Principle isn't wrong, it's just a silly tautology: Of course the world has to be such that it supports life. Life is an observed fact. And of course it's true that if there were no life, there would be no one to observe the universe and ask the questions we are asking. But so what? The principle says nothing beyond the fact that life formed.

This is a kind of willful missing of the point. As usual I find it helpful to rely on an analogy. I call it the Cerebrothropic Principle. The Cerebrothropic Principle is intended to answer the question, "How did it happen that we developed such a big, powerful brain?" This is what the principle says:

"The laws of biology require the existence of a creature with an extraordinarily unusual brain of about fourteen hundred cubic centimeters because without such a brain there would be no one to even ask what the laws of biology are."

That is extremely silly even though true. But the Cerebrothropic Principle is really shorthand for a longer, much more interesting, story. In fact two stories are possible. The first is creationist: God made man with some purpose that involved man's ability to appreciate and worship God. Let's forget that story. The whole point of science is to avoid such stories. The other story is far more complex and, I think, interesting. It involves several features. First of all it says that the Laws of Physics and chemistry allow for the possible existence of computer-like systems of neurons that can exhibit intelligence. In other words the Landscape of biological designs includes a small number of very special designs that have what we call intelligence. That's not trivial.

But the story requires more — a mechanism to turn these blueprint designs into actual working models. That's where Darwin comes in. Random copying errors together with natural selection have a tendency to create a tree or bush of life whose branches fill every niche, including a niche for creatures that survive by their brainpower. Once all that is understood, the question, "Why did I wake up this morning with a big brain?" is exactly answered by the Cerebrothropic Principle. Only a big brain can ask the question.

The Anthropic Principle can also be silly. "The Laws of Physics have to be such that they allow life because if they weren't, there wouldn't be anyone to ask about the Laws of Physics." The critics are quite right — by itself, it's silly. It simply states the obvious — we are here, so the laws of nature must permit our existence — without providing any mechanism for how our existence influenced the choice of laws. But taken as shorthand for the existence of a fantastically rich Landscape and a mechanism for populating the Landscape (chapter 11) with pocket universes, it is not at all trivial. In the next few chapters, we will see evidence that our best mathematical theory provides us with such a Landscape.

CHAPTER SEVEN

A Rubber Band–Powered World

The large number of lucky accidents I've described so far, including the incredible fine-tuning of the cosmological constant, make a strong case for at least keeping an open mind to anthropic arguments. But these accidents alone would not have persuaded me to take a strong position on the issue. The success of Inflation (Inflation implies an enormous universe) and the discovery of a bit of vacuum energy made the Anthropic Principle appealing, but in my own mind, the "straw that broke the camel's back" was the realization that String Theory was moving in what seemed to be a perverse direction. Instead of zeroing in on a single, unique system of physical laws, it was yielding an ever-expanding collection of Rube Goldberg concoctions. I felt that the goal of a single unique string world was an ever-receding mirage and that the theorists looking for such a unique world were on a doomed mission.

But I also sensed an extraordinary opportunity in the approaching train wreck: String Theory might just provide the kind of technical framework in which anthropic thinking would make sense. The only problem is that String Theory, while it had a lot of possibilities, didn't seem to have nearly enough. I kept asking my friends, "Are you sure that the number of Calabi Yau manifolds is only a few million?" Without the mathematical jargon, what I was asking them was whether they were quite certain that the number of String Theory vacuums (in other words, valleys in the Landscape) was measured in the millions. A few million possibilities when you are trying to explain the cancellation of 120 decimal places is of no real help.

But all that changed in the year 2000. Raphael Bousso, then a young postdoc at Stanford, together with an old friend, Joe Polchinski, from the University of California at Santa Barbara, wrote a paper explaining how the number of possible vacuums could be so large that there could easily be enough to overcome the unlikelihood of tuning 120 digits. Soon after, my Stanford colleagues Shamit Kachru, Renata Kallosh, and Andrei Linde and the Indian physicist Sandip Trivedi confirmed the conclusion. That was it for me. I concluded that the only rational explanation for the fine-tunings of nature would have to involve both String Theory and some form of anthropic reasoning. I wrote a paper called "The Anthropic Landscape of String Theory," which stirred up a hornet's nest that is still buzzing. This is the first of three chapters (7, 8, and 10) devoted to explaining String Theory.

Hadrons

"Three quarks for muster mark," said James Joyce. "Three quarks for the proton, three quarks for the neutron, and a quark-antiquark pair for the meson," said Murray Gell-Mann. Murray, who has a fetish for words, invented a large fraction of the vocabulary of high-energy physics: *quark*, *strangeness*, *Quantum Chromodynamics*, *current-algebra*, the *eight-fold way*, and more. I'm not sure whether the curious word *hadron* (pronounced hay-dron or ha-dron) was one of Murray's words. Hadrons were originally defined, somewhat imprecisely, as particles that shared certain properties with nucleons (protons and neutrons). Today, we have a very simple and clear definition: hadrons are the particles that are made out of quarks, antiquarks,[1] and gluons. In other words they are the particles that are described by Quantum Chromodynamics (chapter 1).

What does the word *hadron* mean? The prefix *hadr* in Greek means "strong." It's not the particles themselves that are strong — it's a lot easier to break up a proton than an electron — but rather the forces between them. One of the early achievements of elementary-particle physics was to recognize that there are four distinct types of forces between elementary particles. What distinguishes these forces is their strength: how hard

1. Antiquarks are, of course, the antiparticle twins of quarks. They can be thought of either as particles in their own right or as quarks going backward in time.

a pull or push they exert. Weakest of all are the gravitational interactions between particles; then come the so-called weak interactions; somewhat stronger are the familiar electromagnetic forces; and finally are the strongest of all — the nuclear, or strong, interactions. You may find it odd that the most familiar — gravity — is the weakest. But think of it for a moment: it requires the entire mass of the earth to hold us to the surface. The force between an average person standing on the earth's surface and the earth itself is only 150 pounds. Divide that force by the number of atoms in a human body, and it becomes apparent that the force on any atom is minute.

But if the electric forces are so much stronger than gravity, why doesn't the electric interaction either eject us from or crush us to the surface? The gravitational force between any two objects is always attractive (ignoring the effects of a cosmological constant). Every electron and every nucleus in our bodies gravitationally attracts every electron and every nucleus in the earth. That adds up to a lot of attraction, even though the individual forces between the microscopic particles are totally negligible. By contrast, electric forces can be either attractive or repulsive. Opposite charges — an electron and a proton, for example — attract. Two like charges, a pair of electrons or a pair of protons, repel each other. Both our own bodies and the substance of the earth have both kinds of charge — positive nuclei and negative electrons — in equal amounts. The electric forces of attraction and repulsion cancel! But suppose we could temporarily eliminate all the electrons in both ourselves and in the earth. The remaining positive charges would repel with a total force that would be incomparably stronger than the gravitational force. How many times stronger? Roughly one with forty zeros after it, 10^{40}. You would be ejected from the earth with such force that you would be moving with practically the speed of light in no time at all. In truth this could never happen. The positive charges in your own body would repel so strongly that you would be instantaneously blown to smithereens. So would the earth.

Electric forces are neither the strongest nor the weakest of the nongravitational forces. Most of the familiar particles interact through the so-called weak interactions. The neutrino is a good example because it has only weak forces (ignoring gravity). As I have previously explained, the weak forces are not really so weak, but they are very short range.

Two neutrinos have to be incredibly close, about one one-thousandth of a proton diameter, to exert an appreciable force on each other. If they are that close, the force is about the same as the electric force between electrons, but under ordinary conditions the weak forces are only a tiny fraction of the electric.

Finally we come to the strongest of all forces, those that hold the atomic nucleus together. A nucleus is composed of electrically neutral neutrons and positively charged protons. No negative charges are found in a nucleus. Why doesn't it blow up? Because the individual protons and neutrons attract with a nonelectric force about fifty times stronger than the electrical repulsion. The quarks that make up a single proton have even stronger forces binding them together. How is it that our protons and neutrons aren't attracted to the protons and neutrons of the earth by such powerful forces? The answer is that although the nuclear force is powerful, it is also very short range. It is easily strong enough to overcome the electric repulsion of protons, but only when particles are close together. Once they separate by more than a couple of proton diameters, the force becomes negligible. Underlying the strong interactions are the powerful forces between quarks, the elementary particles that make up hadrons.

I often feel a discomfort, a kind of embarrassment, when I explain elementary-particle physics to laypeople. It all seems so arbitrary — the ridiculous collection of fundamental particles, the lack of pattern to their masses, and especially the four forces, so different from each other, with no apparent rhyme or reason. Is the universe "elegant" as Brian Greene tells us? Not as far as I can tell, not the usual laws of particle physics anyway. But in the context of a megaverse of wild diversity, there is a pattern. All of the forces and most of the elementary particles are absolutely essential. Change any of it more than a bit, and life as we know it becomes impossible.

Origins of String Theory

A peculiar ideology insinuated itself into the high-energy theoretical physics of the 1960s. It paralleled almost exactly a fad that had taken hold in psychology. B. F. Skinner was the guru of the *behaviorists*, who insisted that only the external behavior of a human being was the proper material of mind science. According to Skinner, psychologists had no business inquiring into the inner mental states of their subjects. He even went so far as to declare that no such thing existed. The business of psychology was to watch, measure, and record the external behavior of subjects without ever inquiring about internal feelings, thoughts, or emotions. To the behaviorists a human was a black box that converted sensory input into behavioral output. While it is probably true that Freudians went too far in the other direction, the behaviorists carried their ideology to extremes.

The behaviorism of physics was called *S-matrix theory*. Sometime in the early sixties, while I was a graduate student, some very influential theoretical physicists, centered in Berkeley, decided that physicists had no business trying to explain the inner workings of hadrons. Instead, they should think of the Laws of Physics as a black box — a black box called the Scattering Matrix, or S-matrix for short. Like the behaviorists the S-matrix advocates wanted theoretical physics to stay close to experimental data and not wander off into speculation about unobservable events taking place inside the (what was then considered) absurdly small dimensions characteristic of particles like the proton.

The input to the black box is some specified set of particles coming toward one another, about to collide. They could be protons, neutrons, mesons, or even nuclei of atoms. Each particle has a specified momentum as well as a host of other properties like spin, electric charge, and so on. Into the metaphorical black box they disappear. And what comes out of the black box is also a group of particles — the products of collision, again with specified properties. The Berkeley dogma forbade looking into the box to unravel the underlying mechanisms. The initial and final particles are everything. This is very close to what experimental physicists do with accelerators to produce the incoming particles and with detectors to detect what emerges from the collision.

The S-matrix is basically a table of quantum-mechanical probabilities. You plug in the input, and the S-matrix tells you the probability for a given output. The table of probabilities depends on the direction and energy of both the incoming and outgoing particles, and according to the prevailing ideology of the mid-1960s, the theory of elementary particles should be confined to studying the way the S-matrix depends on these variables. Everything else was forbidden. The ideologues had decided that they knew what constituted good science and became the guardians of scientific purity. S-matrix theory was a healthy reminder that physics is an empirical subject, but like behaviorism, the S-matrix philosophy went too far. For me it turned all of the wonder of the world into the gray sterility of an accountant's actuarial tables. I was a rebel, but a rebel without a theory.

In 1968 Gabriele Veneziano was a young Italian physicist living and working at the Weizmann Institute, in Israel. He was not especially ideological about S-matrix theory, but the mathematical challenge of figuring out the S-matrix appealed to him. The S-matrix was supposed to satisfy certain technical requirements, but no one at that time could point to a specific mathematical expression that satisfied the rules. So Veneziano tried to find one. The attack was brilliant. The result, famous today as the "Veneziano amplitude," was extremely neat. But it was not a picture of what particles were made of or of how the processes of collision could be visualized. The Veneziano amplitude was an elegant mathematical expression — an elegant mathematical table of probabilities.

The discovery of String Theory, which in a sense is still ongoing, was full of twists of fate, reversals of fortune, and serendipity. My own involvement with it began sometime in 1968 or early 1969. I was beginning to tire of the problems of elementary particles, especially hadrons, which seemed to have little to offer in the way of deep, new principles. I found the S-matrix approach boring and was beginning to think about the relation between quantum mechanics and gravity. Putting the General Theory of Relativity together with the principles of quantum mechanics seemed far more exciting, even if all the experimental data were about hadrons. But just at that time, a friend from Israel visited me in New York. The friend, Hector Rubinstein, was extremely excited about Veneziano's work. At first I was not very interested. Hadrons were exactly what I wanted to forget about. Mainly out of politeness I decided to hear Hector out.

Hector became so excited while explaining the Italian's idea that I really couldn't follow the details. As far as I could make out, Veneziano had worked out a formula for describing what happens when two hadrons collide. He finally wrote down Veneziano's formula on the blackboard in my office. It immediately struck a chord. It was extremely simple, and features of the formula looked familiar to me. I recall asking Hector,

"Does this formula represent some kind of simple quantum-mechanical system? It looks like it has something to do with harmonic oscillators." Hector didn't know of a physical picture that went with the formula, so I wrote it down on a sheet of paper to remember.

$$A = g \frac{\Gamma(1-\alpha(s))\,\Gamma(1-\alpha(t))}{\Gamma(2-\alpha(s)-\alpha(t))}$$

I was intrigued enough to postpone thinking about quantum gravity and give hadrons another chance. As it turned out I didn't seriously think about gravity again for more than a decade. I pondered the formula above for several months before I began to see what it really represented.

The term *harmonic oscillator* is physics language for anything that can vibrate or swing back and forth with a periodic (repeating) motion. A child on a playground swing or a weight hanging at the end of a spring are familiar harmonic oscillators. Vibrations of a violin string or even the oscillations of the air when a sound wave passes through it are also good examples. If the vibrating system is small enough — the vibrations of atoms in a molecule are an example — then quantum mechanics becomes important, and energy can be added to the oscillator only in discrete steps. I had mentioned the harmonic oscillator to Hector because certain features of Veneziano's formula reminded me of the mathematical properties of quantum-mechanical harmonic oscillators. I imagined a hadron as two weights connected by a spring, vibrating in periodic oscillation — the weights first approaching and then receding from each other. I was clearly playing with forbidden fruit, trying to picture the internal machinery inside elementary particles, and I knew it.

Being tantalizingly close to the answer but not quite being there is maddening. I tried all sorts of quantum-mechanically oscillating systems, attempting to match them with Veneziano's formula. I was able to produce formulas that looked a lot like Veneziano's from the simple weight and spring model, but they weren't quite right. During that period I spent long hours by myself, working in the attic of my house. I hardly came out, and when I did I was irritable. I barked at my wife and ignored my kids. I couldn't put the formula out of my mind, even long enough to eat dinner. But then for no good reason, one evening in the attic I suddenly had a "eureka moment." I don't know what provoked the thought. One minute I saw a spring, and the next I could visualize an elastic string, stretched between two quarks and vibrating in many different patterns of oscillation. I knew in an instant that replacing the mathematical spring with the continuous material of a vibrating string would do the trick. Actually, the word *string* is not what flashed into my mind. A *rubber band* is the way I thought of it: a rubber band cut open so that it became an elastic string with two ends. At each end I pictured a quark or, more precisely, a quark at one end and an antiquark at the other.

I quickly did a few calculations in my notebook to test the idea, but I already knew that it would work. The simplicity of it was stunning. Veneziano's S-matrix formula precisely described two colliding "rubber bands." I didn't know why I hadn't thought of it earlier.

Nothing is quite like the excitement of a new discovery. It doesn't happen often, even for the greatest physicists. You say to yourself, "Here I am, the only one on the planet who knows this thing. Soon the rest of the world will know, but for the moment I am the *only one*." I was young and unknown but with visions of glory.

But I wasn't the *only one*. At just about the same time, a physicist in Chicago was doing the same calculations. Yoichiro Nambu was a good deal older than I and had long been one of the most eminent theoretical physicists in the world. Born in Japan, he came to the University of Chicago as a young physicist right after World War II. Nambu was a star who had the reputation of seeing things long in advance of anyone else. Later I was to find out that yet another physicist in Denmark, Holger Bech Nielsen, was thinking about very similar ideas. I won't deny that I was disappointed when I found out that I wasn't alone in thinking of the "rubber band theory," but being in the same company as the great Nambu had its own satisfactions.

Today's modern String Theory is all about the elusive unification of quantum mechanics and gravity, over which physicists banged their collective head for much of the twentieth century. That means that it is a theory of what the world is like at that fabulously tiny scale of the Planck length, 10^{-33} centimeters. As I have explained, it started out much more modestly as a theory of hadrons. We will see in the next chapter how it morphed into a much deeper fundamental theory, but let's follow its earlier incarnation.

Hadrons are small objects, typically about 100,000 times smaller than an atom. This makes them 10^{-13} centimeters in diameter. It takes an enormous force to bind quarks at such small separation. Hadronic strings, the rubber bands of my imagination, although microscopically small, are prodigiously strong. If you could find a way to attach one end of a meson (one kind of hadron) to a car and the other end to a crane, you could easily lift the car. Hadronic strings are not particularly small on the scale of today's experiments. Modern accelerators are probing nature at scales from a hundred to a thousand times smaller. Just for comparison let me get ahead of the story and tell you what the strength of a string is in the modern reincarnation. In order to hold particles together at the Planck distance, a string would have to be about 10^{40} times stronger than the hadronic strings; one of them could support a weight equal to the entire mass of our galaxy if we could somehow concentrate the galaxy at the surface of the earth.

All hadrons belong to one of three families: baryons, mesons, and glueballs. Nucleons, the ordinary protons and neutrons of nuclear physics, are the most familiar hadrons. They belong to the first family, called baryons.[2] All baryons are composed of three quarks. The quarks are connected to one another by means of three strings in the manner of a gaucho's bola: three strings joined at the center, with three quarks at their ends. The only thing wrong with the bola picture is that the hadronic strings are elastic, much like ideal stretchable bungee cords.

2. The prefix *bary* means "heavy" in Greek. When the names were first coined, the nucleons and their close relatives were the heaviest known particles. Meson indicates something in the middle. Mesons are lighter than nucleons but a good deal heavier than the electron.

The ordinary proton and neutron are the lowest energy configurations of the bola, with the quarks at rest at the ends of very short, unstretched strings.

The quarks at the ends of the strings can be set into motion in a number of ways. The bola can be spun around its center, the centrifugal force stretching the strings and pushing the quarks out from the center. This spinning motion requires energy (remember $E = mc^2$), and that makes spinning hadrons heavier. As noted earlier, the jargon for a particle with extra energy is that it is excited. The quarks can also be excited without rotating. One way is through oscillating motions, moving toward and away from the center, in and out, in and out. In addition the strings themselves can be bent into curved, vibrating patterns almost as if they were plucked with a guitar pick. All of these motions, or at least indirect evidence of them, are routinely seen in real experiments on nucleons. Baryons really do behave like elastic quantum bolas.

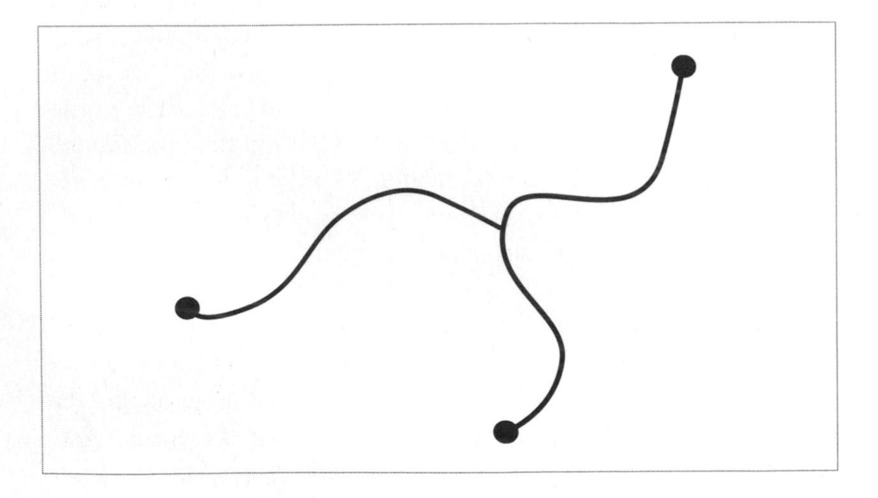

What does it mean that they are quantum bolas? Quantum mechanics implies that the energy (mass) of any vibrating system can be added only in indivisible, discrete steps. In the earliest days of experimental hadron physics, physicists didn't realize that the discretely different quantum states of the vibrating system were really the same object. They gave each energy level a different name and considered them all to be different particles. The proton and the neutron were the baryons with the least energy. The more massive ones had odd names that would mean absolutely nothing to most young physicists today. These particles are nothing more than rotating or vibrating excited states of the proton and neutron. When this was realized, of course, it brought a lot of order and unity to what had been a very messy zoo of particles.

Next come the mesons, the particles that I studied in my attic in 1969. They are simpler than baryons. Each meson is made of a single string with a quark at one end and an antiquark at the other. Mesons, like baryons, can rotate and vibrate in discrete quantum steps. The calculation that I did in the attic represented a fundamental process of interaction between two meson strings.

When two mesons collide they can do a number of things. Because quantum mechanics is a theory of probabilities, it is impossible to predict with certainty how the history of the collision will unfold. One possibility, in fact the most likely one, is that the two mesons will go right past each other, even if it means that the strings pass through each other. But a second, more interesting, possibility is that they can fuse, joining together, to form a single, longer string.

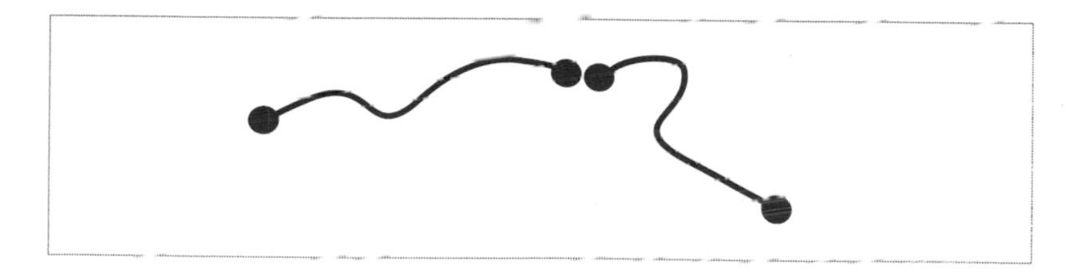

Imagine each string to be a group of dancers holding hands to form a line. At each end the dancers have one hand free (a quark or antiquark), all the others have both hands occupied. Picture the two lines racing toward each other. The only way they are allowed to interact is by a dancer at the end of one line clasping a free hand of the other group. Once they are joined they form a single chain. In this configuration they swing around one another in a complicated dance until somewhere along the chain a dancer releases his neighbor's hand. Then the chain splits into two independent chains, and off they go, separating in some new direction. More precisely but less colorfully, the quark from the end of one string comes together with the antiquark of the other string. They collide and annihilate, as always, when a particle and an antiparticle come together. What they leave over is a single, longer string with a single quark and a single antiquark.

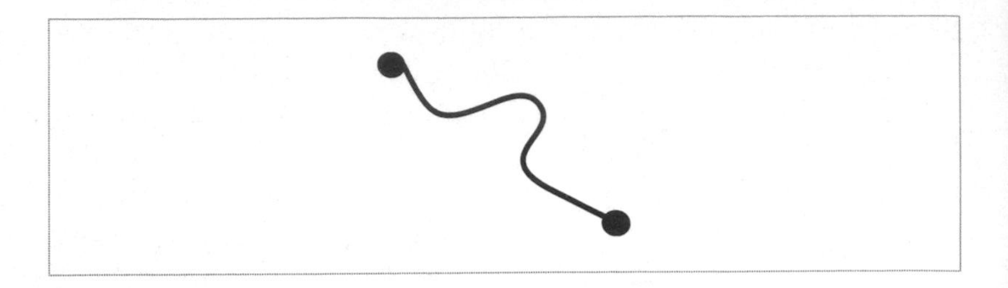

The resulting single string is usually left in an excited state vibrating and rotating. But after a short time, like the chain of dancers, the string can break in two by reversing the process that joined the original strings. The net result is an operation in which a pair of strings comes together, forms a compound string, and then splits back into two strings.

The problem that I had solved in the attic was this: suppose two mesons (strings) were originally moving with a given energy in opposite directions before they collided. What is the quantum-mechanical probability that, after they collide, the resulting pair of strings will be moving along some specified new direction? It sounds like a horribly complicated problem, and it was something of a mathematical miracle that it could be solved.

The mathematical problem of describing an idealized elastic string had already been solved by the early nineteenth century. A vibrating string can be viewed as a collection of harmonic oscillators, one for each separate type of oscillating motion. The harmonic oscillator is one of the few physical systems that can be completely analyzed with very simple high school mathematics.

Adding quantum mechanics to make the string a quantum object was also simple. All that was necessary was to remember that the energy levels of any oscillating system come in discrete units of energy (see chapter 1). This simple observation was sufficient to understand the properties of a single vibrating string; but describing two interacting strings was much more intricate. For that I had to work out my own rules from scratch. What made it possible was that the complexity lasts for only an infinitesimal instant of time when the ends touch and join. Once that happens the two strings become one string, and the simple math of a single string takes over. A little later the single string splits, but again, the complicated event takes only an instant. Thus, with great precision, I was able to follow the two strings as they merged and separated. The results of the mathematical calculation could be compared with Veneziano's formula, and gratifyingly, they agreed exactly.

Baryons are three strings joined at the center, and mesons are a single string with two ends, but what are glueballs? Start with a single chain of dancers. As the dancers move in complicated steps, every so often the two end dancers may run into each other. Not realizing that they belong to the same chain, they clasp hands. The result is a closed circle of dancers with no free end. The same thing can happen to a vibrating meson. In the course of moving, vibrating, and rotating, the two ends come close together. The quark at one end sees the antiquark at the other and doesn't care that it happens to be attached to the same string. It grabs the end like a snake mistaking its tail for a tasty meal. The result is a glueball: a closed loop of string with no ends and no quarks. Many mesons and baryons were known to exist long before String Theory came along, but glueballs were a prediction of String Theory. Today, if you look at a list of the known particles, glueballs and their masses will be listed along with baryons and mesons.

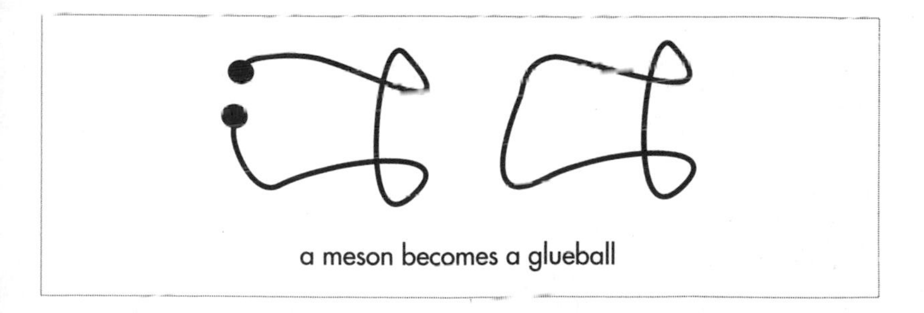

a meson becomes a glueball

Mesons, baryons, and glueballs are complex objects that can vibrate and oscillate in all sorts of patterns. For example, the string connecting the ends of a meson can vibrate like a spring or even a violin string: it can even spin around an axis, with centrifugal force stretching it out to form a whirling, propeller-like hadron. These "excited states" of hadrons are also well-known objects, some of which were discovered in experiments as early as the 1960s.

The connection between the String Theory of hadrons and the Laws of Physics, particularly their expression in terms of Feynman diagrams, is far from obvious. One way of looking at it is that String Theory is a generalization of Feynman diagrams in which a string replaces each point particle. Feynman diagrams are composed from basic units that we discussed in chapter 1: propagators and vertices. Propagators and vertices make a great deal of sense for the infinitely small *point particles* of quantum field. For example, the vertex itself is the point at which the particle trajectories meet. If the particles themselves are not points, it is not at all clear what is meant by the point at which they meet. Here's how the ideas of propagator and vertex make sense for strings. If we start with a point particle and imagine it moving through space-time, it traces out a curve. At each instant it is a point, but as time unfolds, the points trace out a curve. The great Minkowski called such a path through space-time a *world line*, and the terminology stuck.

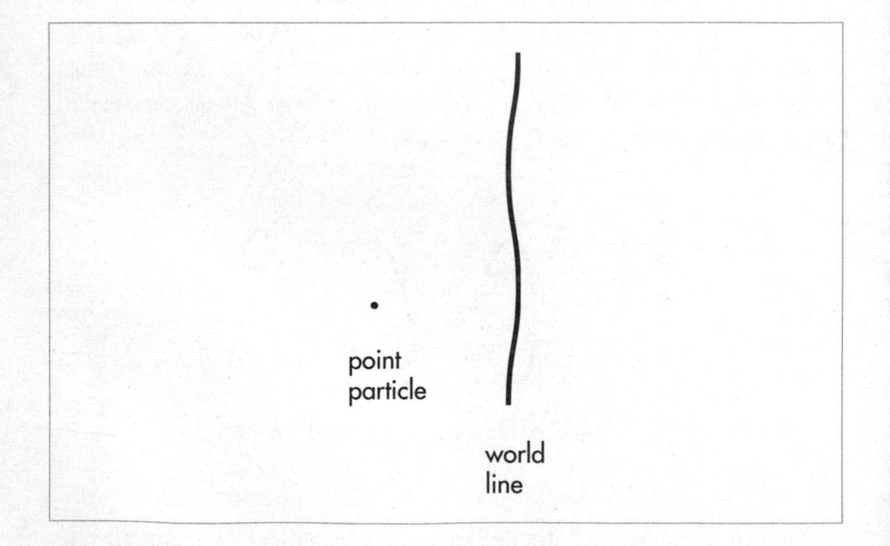

Now let's try to envision the history of a string. Let's take the case of a closed string with no ends. At any instant of time, the string is just a closed curve (loop) through space. Imagine a stroboscope illuminating the string. The first instant we see a loop. The next instant we see the same loop, but now it may have moved to a slightly different location in space. This pattern repeats so that as time unfolds we see a series of loops in space-time, one stacked above the next.

But time really evolves continuously, not in sudden flashes like a stroboscopically lit disco. To represent the history of the string, we need to fill in the spaces between the flashes. The result is a tube through space: a two-dimensional cylinder.

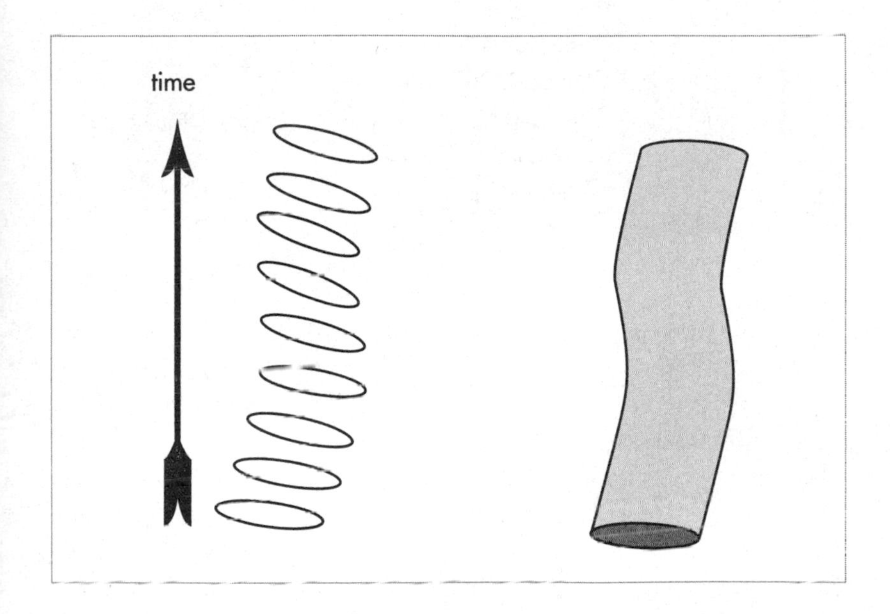

The size of the loop of string might change from one instant to the next. Strings are, after all, like elastic, stretchable rubber bands. They can even get twisted into figure eights or more complex shapes. In that case the neat cylinder will be deformed but still recognizable.

The surfaces swept out in this way could well be called world tubes, in analogy with the world line of the point particle. As it happened that was not the term I originally used to describe them. Instead I used the term *world sheet*, and again, the terminology stuck. But whichever they are called, the cylinder-like world sheet of a string replaces the propagator of the point particle.

A meson with its quark ends may also be described by a world sheet: not a cylinder, but a ribbon with two edges. Again start with the stroboscopic history. This time we see a sequence of open strings, each with a quark and an antiquark at its ends. Filling in the spaces gives the ribbonlike world sheet.

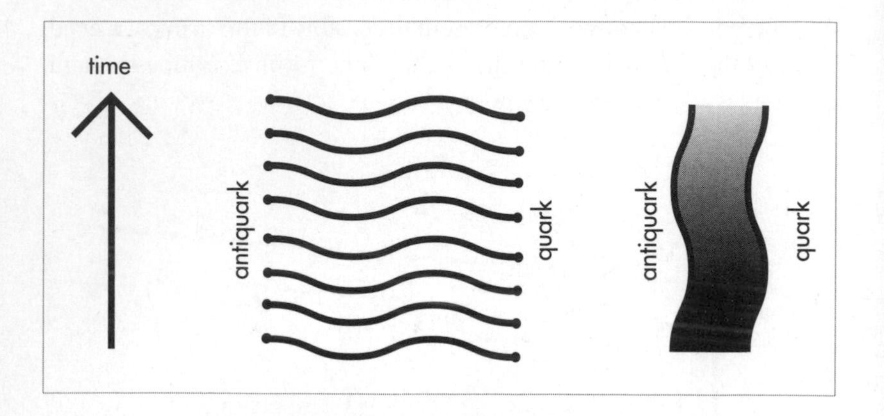

But an interesting theory that can describe all the complexity of interacting colliding particles needs more than just propagators. It also needs vertex diagrams, forks in the road that allow particles to emit and absorb other particles. String Theory is no different.

For an open string the vertex is replaced by the splitting process in which a single string splits in two with a new quark-antiquark pair created at the newly formed ends. Closed strings can also divide by a kind of plumbing diagram, where a single pipe divides; call it a Y-joint.

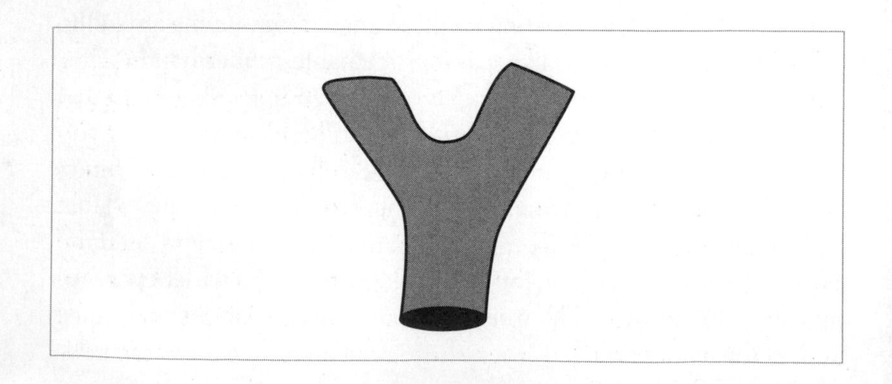

If you follow the action from bottom to top (past to future), you will see a single, closed string splitting and becoming two separate strings that each go off in its own direction. You can also turn the diagram over and see two strings coming together and fusing, to form a single string.

The idea is that plumbing networks, built up from string propagators and Y-joints, replace the usual Feynman diagrams. It was understood very early that the division of a diagram into cylindrical propagators and Y-joints was an artificial one and that the theory was really about world sheets of any shape and topology. The diagrams have openings representing the incoming and outgoing stringlike glueballs, but otherwise they can be of any complexity.

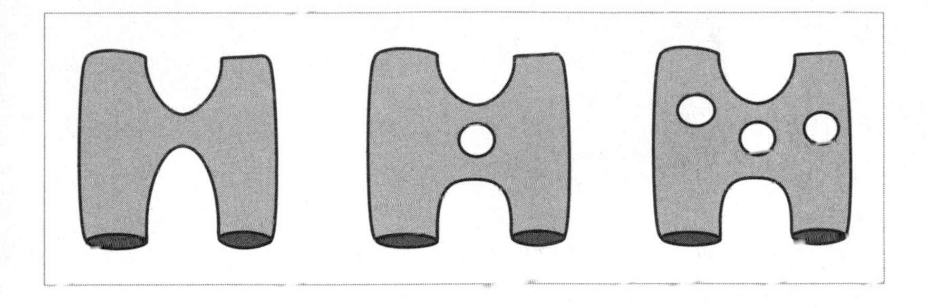

This way of thinking about hadrons is hard to connect with the Standard Model, a theory based on conventional Feynman diagrams (i.e., point particles). The modern Standard Model has what at first sight seems to be an entirely different theory of hadrons — the theory known as Quantum Chromodynamics, or QCD.

According to QCD, hadrons are made of quarks and antiquarks. That much QCD has in common with the String Theory that Nambu and I discovered. But the binding force — the glue — that holds the quarks together has nothing obvious to do with strings. In the same way that an electron can emit a photon, quarks can emit (and absorb) gluons. These gluons, being exchanged between quarks, explain the forces that bind the quarks into hadrons.

Gluons have one feature that makes them more complicated than photons. Charged particles can emit and absorb photons, but photons themselves do not have the capacity to emit photons. Another way to say the same thing is that there is no vertex in which a single photon splits into two photons. Gluons themselves do have this capacity. There is a vertex diagram in which three gluons come together at the vertex. Ultimately this makes gluons and quarks a lot stickier than electrons and positrons.

It sounds like there are two different theories of hadrons — QCD and String Theory. But it was understood, almost from the beginnings of String Theory, that these two kinds of theories might really be two faces of the same theory. In fact the key insight preceded the discovery of QCD by a couple of years.

The bridge between ordinary Feynman diagrams and String Theory became clear when I received a letter in 1970 from Denmark. Holger Bech Nielsen was very enthusiastic about my paper on rubber band theory, and he wanted to share some of his ideas with me. He explained in his letter how he, too, had been thinking about something very similar to elastic strings, but he had a different angle on it.

At about that time, Dick Feynman was arguing that a lot of what was known about hadrons indicated that they were made up of smaller, more fundamental, objects of some sort. He wasn't very specific about what these objects were. He just called them *partons* to indicate that they were the parts that made up hadrons. The idea of combining String Theory with Feynman's parton ideas was something I had been thinking about for quite a while. Nielsen had thought deeply about this and had an extremely interesting vision. He suggested that the smooth, continuous world sheet is really a network or mesh of closely spaced lines and vertices. In other words it is a very complicated, but otherwise ordinary, Feynman diagram composed of a great many propagators and vertices. The mesh becomes finer and finer as more and more propagators and vertices are added. And it becomes better and better approximated by a smooth sheet. The String Theory of hadrons can be pictured in just this way. The world sheets, tubes, and Y-joints are really just very complicated Feynman diagrams involving quarks and a very large number of gluons. When you look at the world sheet from a distance, it appears smooth. But under a microscope it looks like a "fishnet" or "basketball-net" Feynman diagram.[3] The lines in the fishnet represent the propagators of point particles, Feynman's partons or Gell-Mann's quarks and

3. *Fishnet diagram* was the term used by Nielsen.

gluons. But the "fabric" created by these microscopic world lines forms a smooth, almost continuous world sheet.

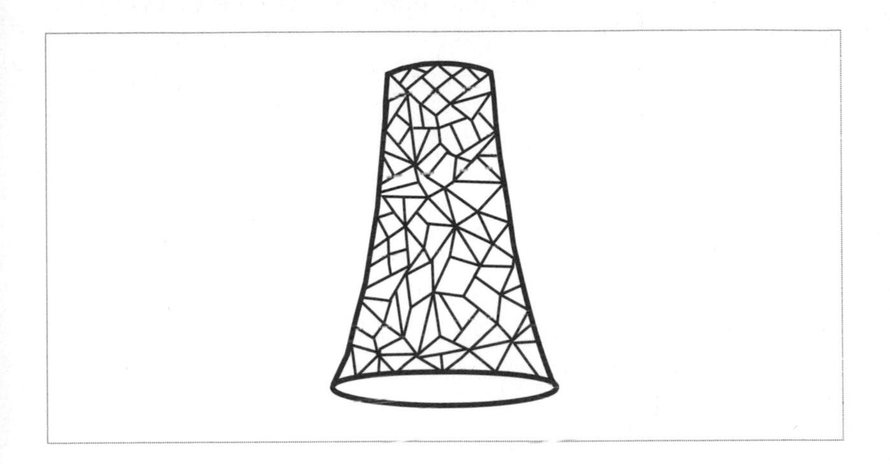

As I said earlier, you can picture a string as a bunch of partons, strung together like a pearl necklace. Feynman's parton theory, Gell-Mann's quark theory, and rubber band theory are all different facets of QCD.

The string or rubber band model of hadrons was not an immediate success. Many theoretical physicists working on hadron physics in the 1960s had a very negative attitude toward any theory that attempted to visualize the phenomenon. As noted, the zealous advocates of the S-matrix theory maintained that the collision is an unknowable black box, a perverse view held with almost messianic fervor. They had only one commandment: "Thou shalt not leave the mass shell." That is, don't look inside the collision to discover the mechanisms taking place. Don't try to understand the composition of particles like the proton. Hostility to the idea that the Veneziano formula represented scattering of two rubber bands persisted, to some extent, until one day when Murray Gell-Mann put his stamp of approval on it.

Murray was the king of physics when I first met him in Coral Gables, Florida, in 1970. At that time the high point of the theoretical physicists' season of conferences was the Coral Gables conference. And the high point of the conference was Murray's lecture. The 1970 event was the first big conference I had ever been invited to — not to lecture, of course, but to be part of the audience. Murray gave his talk on the subject of *spontaneously broken dilatation symmetry*, one of his less successful efforts. I can barely recall the lecture, but I remember very well what transpired afterward. Murray and I got stuck in an elevator together.

I was a totally unknown physicist at the time, and the physics community was in awe of Murray. Needless to say, getting stuck with him excited all my insecurities.

Needing to make conversation, Murray asked what I did. Intimidated, I responded, "I'm working on a theory of hadrons that represents them as some kind of elastic string, like a rubber band." In the unforgettable awful moment that followed, he started to laugh. Not a little chuckle — a big guffaw. I felt like a worm. Then the elevator door opened, and I slunk out with burning ears.

I didn't see Murray again for about two years. Our next meeting was at another conference that took place at Fermilab, a big particle accelerator in Illinois. The Fermilab conference was a really big deal: about one thousand people participated, including the most influential theoretical and experimental high-energy physicists in the world. Once again I was a spectator.

At the very beginning of the conference, before the first lecture, I was standing with a group of friends, and Murray came sauntering over. In front of all these people, he said, "I'm sorry I laughed at you in the elevator that day. I think the work that you are doing is fantastic, and I'm going to spend most of my big lecture talking about it. Let's sit down and talk about it when we get a chance." I went from feeling like a worm to feeling like a prince. The king was going to confer royalty on me!

After that for a couple of days, I chased Murray, asking, "Is this a good time, Murray?" Each time his answer was, "No, I have to talk to someone important."

On the last day of the conference, there was a long waiting line to talk to a travel agent. I needed help changing my air tickets, and I had waited about an hour on the line. Finally I was only two or three people from the travel agent when Murray came over and plucked me out, saying, "Now! Let's talk now. I have fifteen minutes." Okay, I said to myself, this is it. Do it right and you're a prince. Do it wrong, and you're fish bait.

We sat down at an empty table, and I started to explain how the new rubber band theory was related to his and Feynman's ideas. I wanted to explain the fishnet-diagram idea. I remember saying, "I'll begin explaining it in terms of partons."

"Partons? Partons? What the hell is a parton? Put-ons? You're putting me on, right?" I knew I had made a bad mistake, but I didn't know exactly how. I tried to explain, but all I got back was, "Put-on? What is that?" Fourteen of my precious fifteen minutes were gone by the time he said, "These put-ons, do they have charge?" I answered yes. "Do they have SU(3)?" Again I agreed. Then all became clear. He said, slowly, "Oooh, you mean quarks!" I had committed the unpardonable sin of calling the constituents by Feynman's word instead of Murray's. It seems I was the only person in the world who didn't know about the weird rivalry between the two great Caltech physicists.

Anyway, I had one or two minutes to spill out what I was thinking, and then Murray looked at his watch, "Okay, thanks. Now I have someone *important* to talk to before my lecture."

So close and yet so far. No royal treatment for me: just dirt and mud. And then the next thing I heard was Murray holding forth. He was telling a group of his cronies everything I had told him. "Susskind says this, and Susskind says that. We have to learn Susskind's String Theory." And then Murray gave his big talk, the last of the conference if I recall correctly. Although String Theory was only a small part of the lecture, it had received Murray's blessing. The whole thing was quite a roller-coaster ride.

Although Murray did not work on String Theory, his mind was open to new ideas, and he played a significant role in encouraging others. With no question, he was one of the first to recognize String Theory's potential importance both as a theory of hadrons and later as a theory of Planck-scale phenomena.

String Theory comes in many versions. The versions that we knew about in the early seventies were mathematically very precise — too precise. Although it is absolutely clear from the modern vantage point that hadrons are strings, the theory would have to be modified several ways before it could describe real baryons and mesons.

Three huge problems plagued the original String Theories. One was so bizarre that conservative physicists, particularly the S-matrix enthusiasts, found it a source of humor. It was the problem of too many dimensions. String Theory, like all physical theories, takes place in space and time. Before Einstein, space and time were two separate things; but under the influence of Minkowski, the two merged into space-time, the four-dimensional world in which every event has a location in space and a moment in time. Einstein and Minkowski turned time into the "fourth dimension." But time and space are not entirely similar. Even though the Theory of Relativity sometimes mixes space and time in various mathematical transformations, time and space are different. They "feel" different. For this reason, instead of describing space-time as being four-dimensional, we say it is three-plus-one-dimensional to indicate that there are three dimensions of space and one of time. Is it possible to have more dimensions of space? Yes, that's commonplace in modern physics. It's not too hard to imagine moving around in more, or for that matter fewer, than three dimensions. Edwin Abbott's famous nineteenth-century book *Flatland* describes life in a world of only two dimensions of space. But a world with more (or fewer) than one time dimension is incomprehensible. It doesn't seem to make any sense. So, for the most part, when physicists want to play with the number of dimensions of space-time, they work with 3+1, 4+1, 5+1, or any number of space dimensions, but with only one time dimension.

Physicists have always hoped that someday they would be able to explain why space has three dimensions and not two or seven or eighty-four. So, in theory, string theorists should have been delighted to discover that their mathematics works out consistently only in a very particular number of dimensions. The trouble was that the number was 9+1 dimensions, not 3+1. Something very subtle goes wrong with the mathematics unless the number of space dimensions is nine — three times as many space dimensions as the world we actually live in! It seemed the joke was on string theorists.

As a physics teacher, I hate to tell students something important and then say I can't explain it. It's too advanced. Or it's too technical. I spend a lot of time figuring out how to explain difficult things in elementary terms. One of my biggest frustrations is that I have never succeeded in finding an elementary explanation of why String Theory is happy only if the number of dimensions is 9+1. Nor has anyone else. What I will tell you is that it has to do with the violent jittery quantum motion of a string. These quantum fluctuations can pile up and get completely out of control unless some very delicate conditions are met. And those conditions are met only in 9+1 dimensions.

Being off by a factor of three in cosmology was not so bad in those days, but it was very bad in particle physics. Particle physicists were used to high accuracy in their numbers. There is no number that they were more confident of than the number of dimensions of space. No amount of experimental uncertainty could explain the loss of six dimensions. It was a debacle. Space-time, then and now, is 3+1-dimensional with no uncertainty.

Being wrong about the dimensionality of space was bad enough, but to compound the problem, the nuclear force law between hadrons came out all wrong. Instead of producing the kind of short-range forces that exist between particles in the nucleus, String Theory gave rise to long-range forces, which looked, for all the world, almost exactly like electric and gravitational forces. If the nuclear short-range force was adjusted to be the right strength, the electric force would be about one hundred times too strong, and the gravitational force would be too strong by a stupendous factor — about 10^{40}. Identifying these long-range forces with the real gravitational and electric forces was out of the question, but only if one wanted to use the same strings to describe hadrons.

All forces in nature — be they gravitational, electric, or nuclear — have the same origin. Think of an electron circling a central nucleus. From time to time it emits a photon, and where does the photon go? If the atom is excited, the photon can escape while the electron jumps to a lower-energy orbit. But if the atom is already in its lowest-energy state, the photon cannot carry off any energy. The only alternative for the photon is to be absorbed, either by another electron or by the charged nucleus. Thus, in a real atom, the electrons and nucleus are constantly tossing photons back and forth in a kind of atomic juggling act. This "exchange" of particles, in this case the photon, is the source of all forces in nature. Force — whether it is electric, magnetic, gravitational, or any other — ultimately traces back to Feynman "exchange diagrams," in which quanta hop from one particle to another. For the electric and magnetic force, photons are the exchanged quanta; for the gravitational forces the graviton does the job. You and I are kept anchored to the earth by gravitons jumping between earth and our bodies. But for the forces binding protons and neutrons into nuclei, the exchanged objects are pions. If one goes deeper into the protons and neutrons, quarks are found tossing gluons between them. This connection between force and exchanged "messenger" particles was one of the great themes of twentieth-century physics.

If the origins of the nuclear, electromagnetic, and gravitational forces are so similar, how is it that the results are so different? The electromagnetic and gravitational forces are long range, long enough for gravity to keep the planets in orbit, while the nuclear force becomes negligible when the nuclear particles are separated by the diameter of a single proton. If you are thinking that the difference has to do with some property of the messengers — graviton, photon, pion, and gluon — you're absolutely right. The determining factor for the range of a particular force is the mass of the messenger: the lighter the messenger, the longer the range. The reason that gravity and electric forces are long range is that the graviton and photon are massless. But the pion is not massless; it is almost three hundred times heavier than the electron. The result of this much mass carried by the messenger is that, like an overweight athlete, it can't jump more than a small distance to bridge the gap between distant particles.

String Theory is also a theory of forces. Let's go back to the dance of the strings. As before, two lines of dancers approach each other. This time, instead of temporarily joining to form a single string, they do a different dance. Before they meet, one of the strings splits off a few of its members to form a short third string. The third string then runs over to the other group and joins up with it. All in all the two initial groups of dancers exchange a short string and, in so doing, a force is produced between the two groups.

From a distance the world sheet describing this exchange dance would look like the letter H, but under the microscope the lines that form it would be revealed as plumbing. The crossbar of the H is the world sheet of an exchanged string that jumps across the space between the vertical legs and creates a force between them. In those first days of String Theory, those of us who hoped to explain everything about hadrons were delighted with the possibility of explaining the nuclear force that binds protons and neutrons to form nuclei.

Unfortunately our hopes were soon shattered. When the calculations were done, the force law between strings looked nothing like the real forces holding the nucleus together. Instead of the short-range force of nuclear physics, we found long-range forces that more closely resembled electric and gravitational forces, as I mentioned earlier. The reason was not hard to find. Among the particle-like vibrating strings, there were two particular objects with a very special property — one, an open string of the kind that describes mesons, and the other, a closed glueball. Both of them had the exceptional trait of having absolutely no mass, just like the photon and the graviton! When juggled between other particles, they create forces almost exactly like the electric force between charged particles and the gravitational forces between masses. The open string mimicked the photon, but to me the biggest surprise was the similarity between the closed glueball and the mysterious elusive graviton. This would have been a source of unbounded joy if we were trying to create a new theory of gravity and electricity, but that was far from our goal. We were in the business of describing nuclear forces and had unquestionably failed. An impasse had been reached.

String Theory had one more difficulty. It was either a "theory of everything" or a "theory of nothing." The original goal of String Theory was to describe hadrons — nothing more. The electron, photon, and graviton were to remain point particles. Experiments over the years had taught us that electrons and photons, if they had any size at all, were vastly smaller than hadrons. They might as well be mere points as far as anyone could tell. On the other hand, hadrons are obviously not points. A point cannot be spun about an axis. When I think of spinning an object, I think of a pizza chef spinning a wad of dough or a basketball player spinning the ball on his finger. But you can't spin an infinitely small point. Hadrons can be easily spun: the excited spinning hadrons were encountered regularly in many accelerator laboratories. Hadrons must be more like a wad of dough than a mathematical point. But no one has ever spun a photon or an electron.[4]

Real hadrons can and do interact with point particles. A proton can absorb and emit a photon in the same way that an electron does. But when we tried to develop the theory so that the stringlike hadrons could interact with photons, all hell broke loose. One mathematical contradiction after another obstructed our attempts.

The obvious idea occurred to many people. The vibrating strings were certainly not points, but we had always assumed that the string ends were pointlike quarks. Why not allow all the electric charge to reside on the quarks? After all, making a point charge interact with a pointlike photon was child's work. But as we know, the best-laid plans sometimes go awry. The mathematics just wouldn't hold together.

The problem was that the strings of String Theory have an exceptionally violent case of the quantum jitters. The very high-frequency quantum fluctuations are so wild and out of control that the quark ends are most likely found far away *at the very edges of the universe.* It sounds crazy but the bits of string jitter and jiggle so violently that if you looked very quickly, you would discover that they are infinitely far away!

Let me try to explain this extraordinary unintuitive behavior of strings. The easiest way is to imagine a guitar string. The guitar string is somewhat different from the strings that we have to deal with in String Theory. For one thing their ends are held in place by the rest of the guitar. But that is not an important consideration right now. The important point is that both kinds of strings can vibrate in a wide variety of patterns. The guitar string can vibrate as a whole, forming a long, bowlike shape that oscillates as one long string. In that configuration, the string sounds the fundamental note.

But as every guitar player knows, a string can be made to vibrate in higher-pitched *harmonics* or *modes of oscillation*. These are vibrations

4. This can be confusing because electrons and photons have a property called spin. But the spin of an elementary particle is not due to the kind of rotational motion that a basketball, a wad of dough, or a hadron can have. In particular, the spin of an electron can never be changed: it is always one-half of Planck's constant. A basketball or hadron can be spun up to increase its angular momentum.

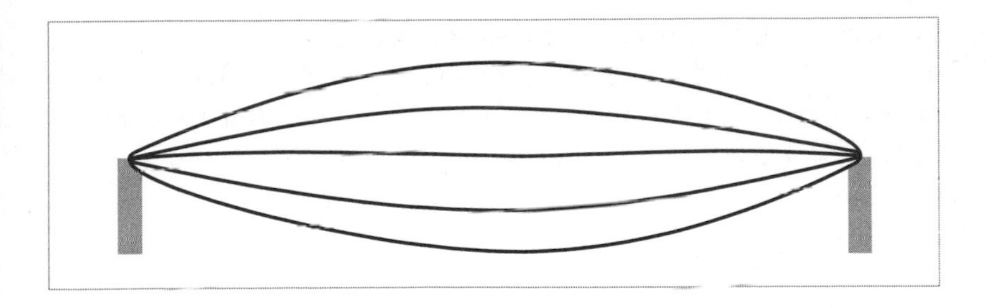

in which the string vibrates in sections as if it were multiple shorter strings. For example, in the first harmonic the two halves of the string move separately.

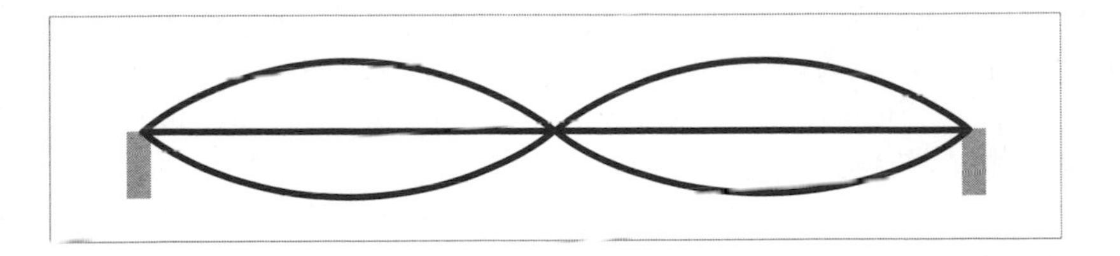

In principle an ideal infinitely thin string could oscillate in an infinite number of harmonics at higher and higher frequency, although in practice, friction and other contaminating influences damp these vibrations almost before they get started.

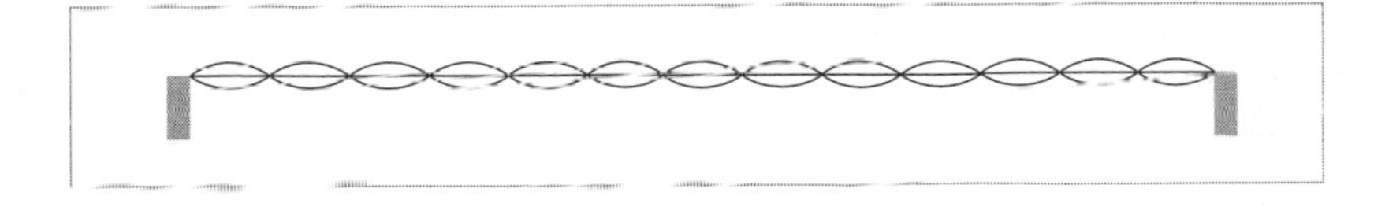

Now let's recall the quantum mechanics lesson of chapter 1. Every oscillation has a bit of jittery zero-point motion that is unavoidable. This has a very dramatic consequence for a perfectly ideal string: all of the possible vibrations, all infinitely many modes of oscillation simultaneously vibrate in a mad symphony of pure noise. All the various oscillations mount up in their influence on the location of bits of the string and would cause it to vibrate to infinite distance.

Why doesn't this insane oscillation happen to a real guitar string?

The reason is that an ordinary string is made of atoms spaced out along the string. There is no meaning to vibrations in which the string vibrates in more pieces than the number of atoms it contains. But a mathematically ideal string, not made of atoms but continuous along its length, would vibrate in this uncontrolled way.

Perhaps the most amazing mathematical miracle of String Theory is that if everything is set up just right, including the fact that the number of space-time dimensions is ten, the wild vibrations of different strings precisely match each other and cause no harm. Your strings and my strings might wildly oscillate to the ends of the universe, but if the world is ten-dimensional, these oscillations are miraculously undetectable.

But this miracle of the strings works only if everything in the world is made of strings. If the photon were a point particle and the proton a string, a horrible clash would occur. For this reason the only things that strings can interact with are other strings! This is what I meant when I said that String Theory is either a theory of everything or a theory of nothing.

Violently jittering strings, fluctuating out to the boundaries of the universe, seemed so dismal a prospect that I gave up thinking about the inflexible mathematics of String Theory for more than a decade. But in the end this berserk behavior became the basis for one of the most exciting and strange developments of modern theoretical physics. In chapter 10 we will meet the Holographic Principle, which states that the world is a kind of quantum hologram at the boundaries of space. In part it was inspired by the extreme jitters of strings. But the Holographic Principle is a feature of the quantum mechanics of gravity, not nuclear physics.

Some theories are so mathematically precise that they are inflexible. That's a good thing if the theory is successful. But if it doesn't quite work, the inflexibility may become a liability. The String Theory that existed in the seventies, eighties, and most of the nineties could not interact with any objects that were not strings. If the purpose was to describe hadrons, the theory was not promising: it involved too many dimensions, massless gravitons and photons, and the impossibility of interacting with smaller objects. String Theory was in trouble, at least in its guise as a theory of hadrons. Nevertheless, there was no denying that hadrons behaved like stretchable strings with quarks at the end. In the thirty-five years since the discovery of String Theory, the stringy nature of hadrons has become a well-established experimental fact. But in the meantime String Theory found another life for itself. How String Theory was reborn as a fundamental theory unifying quantum mechanics and gravity is the subject of the next chapter.

CHAPTER EIGHT

Reincarnation

Although the String Theory of hadrons had failed in its most precise mathematical form, some brave souls saw opportunity in the train wreck. "If the mountain won't come to Mohammad, Mohammad will go to the mountain." If you can't make a String Theory of hadrons because the theory persists in behaving like gravity, then let gravity be described by String Theory. Why not use it to describe everything: gravity, electromagnetic forces, quarks, and all the rest? Problems two and three from the last chapter are gone; the predicted range of forces now fits the reality, and everything is made of strings. The inflexibility of the theory becomes an asset. A radically new vision of the world made up of one-dimensional threads of energy, fluctuating wildly out to the edges of the universe, would replace an older vision of matter made of point particles.

To give you a picture of what this transformation of String Theory meant, let's talk about size scales. Hadrons have a size somewhere between 10^{-13} and 10^{-14} centimeters. There is some variation, but mesons, baryons, and glueballs are all in this range. While 10^{-13} centimeters may seem terribly small, 100,000 times smaller than an atom, by the standards of modern particle physics, it's very large. Accelerators have long probed objects 1,000 times smaller and are beginning to get down to 10,000 times smaller.

The natural size of a graviton is vastly smaller. Gravitons, after all, are the result of mixing gravity together with the quantum nature of matter. And whenever you work on the quantum level, you will always discover exactly what Planck discovered in 1900: the natural unit of length is the minuscule Planck length — 10^{-33} cm. Physicists expect that the size of a graviton is about that small.

Just how much smaller is a graviton than a proton? If the graviton were expanded until it was as big as the earth, the proton would become about 100 times bigger than the entire known universe! Using exactly the same String Theory that had failed as a theory of hadrons, string theorists such as John Schwarz and Joël Sherk were proposing to leapfrog completely over that vast range of scales. Like MacArthur's frog leap across the Pacific, it was either a very bold, heroic move or a very foolish one.

If the range of forces was no longer a problem, the dimensionality of space was; mathematical consistency still required nine dimensions of space and one more for time. But in the new context, this turned out to be a blessing. The list of elementary particles of the Standard Model — the particles that are supposed to be points — is a long one. It includes thirty-six distinct kinds of quarks, eight gluons, electrons, muons, tau leptons[1] and their antiparticles, two kinds of W-bosons, a Z-boson, the Higgs particle, photons, and three neutrinos. Each type of particle is distinctly different from all the others. Each has its own particular properties. You might say they have their own personalities. But if particles are mere points, how can they have personalities? How can we account for all their properties, their *quantum numbers*, such as spin, isospin, strangeness, charm, baryon number, lepton number, and color?[2] Particles evidently have a lot of internal machinery that can't be seen from a distance. Their pointlike, structureless appearance is surely a temporary consequence of the limited resolving power of our best microscopes, i.e., particle accelerators. Indeed the resolving power of an accelerator can be improved only by increasing the energy of the accelerated particles, and the only way to do that is to increase the size of the accelerator. If, as most physicists believe, the internal machinery of elementary particles were revealed only at the Planck scale, it would be necessary to build an accelerator at least as big as our entire galaxy! So we go on

1. Electrons, muons, tau particles, and neutrinos are all examples of particles that physicists call leptons. The term simply refers to fermions that do not have strong interactions like quarks.

2. These are, of course, the whimsical names that particle physicists invented for the various properties of particles that were discovered during a forty-year period ending in the early 1970s.

thinking of particles as points, despite the fact that they obviously have so many properties.

But String Theory is not a theory of point particles. From a theorist's point of view, String Theory provides plenty of opportunity for particles to have properties. Among other things strings can vibrate in many different quantized patterns of vibration. Anyone who ever played the guitar knows that a guitar string can vibrate in many harmonics. The string can vibrate as a whole or it can vibrate in two pieces with a node in the middle. It can also vibrate in three or any number of separate sections, thus producing a series of harmonics. The same is true of the strings of String Theory. The different patterns of vibration do produce particles of different types, but this in itself is not enough to explain the difference between electrons and neutrinos, photons and gluons, or up-quarks and charmed-quarks.

Here's where string theorists made brilliant use of what had previously been their greatest embarrassment. The sow's ear — too many dimensions — was turned into a silk purse. The key to the unexplained diversity of elementary particles — their electric charge, color, strangeness, isospin, and more — is very likely the extra six dimensions that previously dogged our efforts to explain hadrons!

At first sight there doesn't seem to be an obvious connection. How does moving around in six extra dimensions explain electric charge or the difference between quark types? The answer lies in the profound changes in the nature of space that Einstein explained with his General Theory of Relativity — the possibility that space, or some part of space, can be *compact.*

Compactification

The easiest examples of *compactification* are two-dimensional. Once again let's imagine that space is a flat sheet of paper. The paper could be unbounded, an infinite sheet that stretches endlessly in every direction. But there are other possibilities. When discussing Einstein's and Friedmann's universes, it was necessary to conceive of a two-dimensional space with the shape of a 2-sphere — a closed-and-bounded space. No matter what direction you travel in, you eventually come back to the starting point.

Einstein and Friedmann were imagining space to be a gigantic sphere, big enough to move around in for billions of years without encountering the same galaxy or star twice. But now imagine shrinking the sphere smaller and smaller until it is far too small to hold a human being or even a molecule, an atom or even a proton. If the 2-sphere is shrunken to microscopic proportions, it becomes hard to distinguish it from a point — *a space with no dimensions* to move in. This is the simplest example of hiding dimensions by *compactifying*, or shrinking, them.

Can we somehow choose the shape of a two-dimensional space so that it looks for most purposes like a one-dimensional space? Can we effectively hide one of the two dimensions of the sheet of paper? Indeed, we can. Here's how you do it: Start with an imaginary infinite sheet of paper. Cut out an infinite strip a few inches wide. Let's say the strip is along the x-axis. The tip of your pencil can move forever along the x-axis, but if you move it in the y-direction, you'll soon come to one of the edges. Now take the strip and bend it into a cylinder so that the upper and lower edges are joined in a smooth seam. The result is an infinite cylinder that can be described as compact (finite) in the y-direction, but infinite in the x-direction.

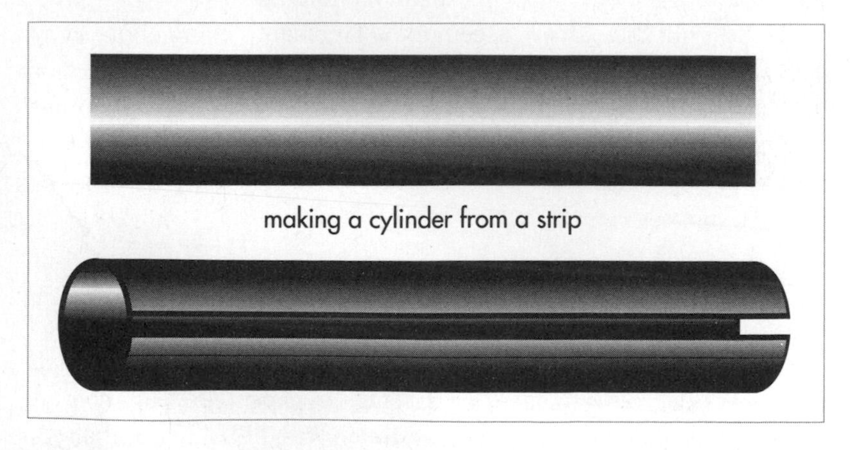

making a cylinder from a strip

Let's imagine such a space, but instead of making the y-direction a few inches in circumference, let's take it to be a micron (one ten-thousandth of a centimeter). If we looked at the cylinder without a microscope, it would look like a one-dimensional space, an infinitely thin "hair." Only if we look through a microscope will it reveal itself to be two-dimensional. In this way a two-dimensional space is disguised as one-dimensional.

Suppose we further reduce the size of the compact direction all the way down to the Planck length. Then no existing microscope would be able to resolve the second dimension. For all practical purposes the space would be one-dimensional. This process of making some directions finite and leaving the rest infinite is called compactification.

Now let's make things a little harder. Take three-dimensional space with three axes: x, y, and now z. Let's leave the x- and y-directions infinite but roll up the z-axis. It's harder to visualize, but the principle is the same. If you move in the x- or y-direction, you go on forever, but moving along z brings you back to the start after a certain distance. If that distance were microscopic, it would be hard to tell that the space was not two-dimensional.

We can go a little further and compactify both the z- and the y-direction. For the moment completely ignore the x-direction and concentrate on the other directions. One thing you can do with two directions is to roll them into a 2-sphere. In this case you can move forever along the x-direction, but moving in the y- and z-direction is like moving on the surface of a globe. Again, if the 2-sphere were microscopic, it would be hard to tell that the space was not one-dimensional. So you see we can choose to hide any number of dimensions by rolling them up into a small compact space.

The 2-sphere is only one way to compactify two dimensions. Another very simple way is to use a torus. Just as the 2-sphere is the surface of a ball, the torus is the surface of a bagel. There are lots of other shapes you could use, but the torus is the most common.

Let's return to the cylinder and imagine a particle moving on it. The particle can move up and down the infinite x-axis exactly as if the space were only one-dimensional. It has a velocity along the x-direction. But moving in the x-direction is not the only thing the particle can do: it can also move along the compact y-direction, circling the y-axis endlessly. With this new motion the particle has velocity along the hidden microscopic direction. It can move in the x-direction, the y-direction, or even with both motions simultaneously, in a helical (corkscrew) motion, winding around y while traveling along x. To the observer who cannot resolve the y-direction, that additional motion represents some new peculiar property of the particle. A particle moving with velocity along the y-axis is different from a particle with no such motion, and yet the origin of this difference would be hidden by the smallness of y. What should we make of this new property of the particle?

The idea that there might be an extra, unobserved direction to space is not new. It goes all the way back to the early years of the twentieth century, shortly after Einstein completed the General Theory of Relativity. A contemporary of Einstein's named Theodor Franz Eduard Kaluza began to think about exactly this question — how would physics be influenced if there were an extra direction of space? At that time the two important forces of nature were the electromagnetic force and the gravitational force. In some ways they were similar, but Einstein's theory of gravity seemed to have a much deeper origin than Maxwell's theory of electromagnetism. Geometry itself — the elastic, bendable properties of space — were what gravity was all about. Maxwell's theory just seemed like an arbitrary "add-on" that had no fundamental reason in the scheme of things. The geometry of space was just right to describe the properties of the gravitational field and no more. If the electric and magnetic forces were somehow to be united with gravity, the basic geometric properties of space would have to be more complex than envisioned by Einstein.

What Kaluza discovered was amazing. If one additional direction of space were added to the usual 3+1 dimensions, the geometry of space would encompass not only Einstein's gravitational field but also Maxwell's electromagnetic field. Gravity and electricity and magnetism would be unified under a single all-encompassing theory. Kaluza's idea was brilliant and caught the attention of Einstein, who liked it very much. According to Kaluza, particles could move not only in the usual three spatial dimensions but also in a fourth, hidden dimension. However, the theory had one obvious, enormous problem. If space has an extra dimension, why don't we notice it? How is the extra fourth dimension of space hidden from our senses? Neither Kaluza nor Einstein had an answer. But in 1926 the Swedish physicist Oscar Klein did have an answer. He added the new element that made sense out of Kaluza's idea: the extra dimension must be rolled up into a tiny compact space. Today theories with extra compact dimensions are known as Kaluza Klein theories.

Kaluza and Klein discovered that the gravitational force between two particles was modified if both particles moved in the additional direction. The astonishing thing was that the extra force was identical to the electric force between charged particles. Moreover, the electric charge of each particle was nothing but the component of momentum in the extra dimension. If the two particles cycled in the same direction around the compact space, they repelled each other. If they moved in opposite directions, they attracted. But if either of them did not cycle in the compact direction, then only the ordinary gravitational attraction affected them. This smells like the beginnings of an explanation of why some particles (the electron, for example) are electrically charged, while other, similar particles (neutrinos) are electrically neutral. Charged particles move in the compact direction of space, while those without charge have no motion in this direction. It even begins to explain the difference between the electron and its antiparticle, the positron. The electron cycles around the compact direction one way, say, clockwise, while the positron moves counterclockwise.

Another insight was added by quantum mechanics. Like all other cyclic or oscillating motions, the motion around the compact y-axis is quantized. The particle cannot cycle the y-axis with an arbitrary value of the y-momentum: it is quantized in discrete units just like the motion of a harmonic oscillator or the electron in Bohr's theory of the atom. This means that the y-motion, and therefore, the electric charge, cannot be any old number. Electric charge in Kaluza's theory is quantized: it comes in integer multiples of the electron charge. A particle with charge twice or three times the electron would be possible but not one with charge ½ or .067 times the electron charge. This is a highly desirable state of affairs. In the real world no object has ever been discovered carrying a fractional electric charge: all electric charges are measured in integer multiples of the electron's charge.

This was a spectacular discovery, which largely lay dormant for the rest of Kaluza's life. But it's the heart of our story. The Kaluza theory is a model of how properties of particles can arise from extra dimensions of space. Indeed, when string theorists discovered that their theory required six extra dimensions of space, they seized on Kaluza's idea. Just roll up the extra six directions in some manner and use the motion in the new directions to explain the internal machinery of elementary particles.

String Theory is richer in possibilities than theories of point particles. Returning to the cylinder, let's suppose a small closed string is moving on a cylinder. Start with a cylinder of circumference large enough to be visible to the naked eye. A tiny closed string moves on the cylinder pretty much the same way a point particle would. It can move along the length of the cylinder or around it. In this respect it is no different from the point particle. But the string can do something else that the point cannot. The string can wind around the cylinder just like a real rubber band can be wrapped around a cardboard cylinder. The wound string is different from the unwound string. In fact the rubber band can be wound any number of times around a cardboard cylinder, at least if it doesn't break. This gives us a new property of particles: a property that depends not only on the compactness of a dimension but also on the fact that particles are strings or rubber bands. The new property is called the winding number, and it represents the number of times the string is wound around the compact direction.

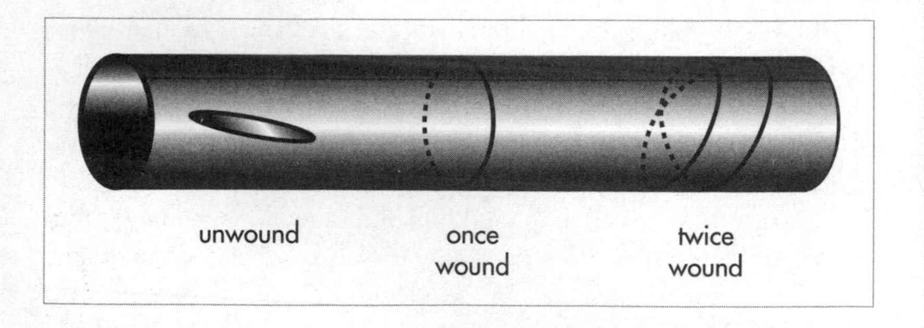

The winding number is a property of a particle that we could not understand if our microscope was not strong enough to resolve the tiny distance around the compact direction. So you see, the extra dimensions needed by String Theory are a blessing, not a curse: they are essential for understanding the complex properties of elementary particles.

A two-dimensional cylinder is easy to visualize, but a nine-dimensional space with six dimensions wrapped up into some kind of six-dimensional tiny space is beyond anyone's powers of visualization. But making pictures in your head or on paper is not the only way to understand the mind-boggling six-dimensional geometry of String Theory. Geometry can often be reduced to algebra just the way you learned in high school when you used an equation to represent a straight line or a circle. Still, even the most powerful methods of mathematics are barely enough to scratch the surface of six-dimensional geometry.

For example, the number of possible ways String Theory allows of rolling up six dimensions runs into the millions. I won't try to describe them other than to give you the special mathematical name for these spaces; they are called *Calabi Yau spaces,* after the two mathematicians who first studied them. I don't know why mathematicians were interested in these spaces, but they came in handy for string theorists. Fortunately for us the only thing we have to know is that they are extremely complicated, with hundreds of "donut holes" and other features.

Back to the two-dimensional cylinder. The distance around the cylinder is called the *compactification scale.* For a cardboard cylinder it might be a few inches, but for String Theory it should most likely be a few Planck lengths. You might think that this scale is so small that it doesn't matter for anything we care about, but that's not so. Although we can't actually see such small scales, they nonetheless have their effect on ordinary physics. The compactification scale in Kaluza's theory fixes the magnitude of the electric charge of a particle like the electron. It also fixes the masses of many of the particles. In other words the scale of compactification determines various constants that appear in the ordinary Laws of Physics. Vary the size of the cylinder, and the Laws of Physics change. Vary the values of scalar fields as in chapter 1, and the Laws of Physics change. Is there a connection? Absolutely! And we will return to it.

To specify the cylinder you need to specify only one parameter, the scale of compactification, but other shapes require more. For example, a torus is determined by three parameters. See if you can visualize them. First there is the overall size of the torus. Keeping the shape fixed, the torus can be magnified or shrunk. In addition, the torus can be "thin" like a narrow ring or "fat" like an overstuffed bagel. The parameter that determines the fatness is a ratio: the ratio of the size of the hole to the overall size. For the thin ring, the overall size and the size of the hole are about the same, so the ratio is near one. For the fat torus the hole is much smaller than the overall size, and the ratio is correspondingly small. There is one more quantity, which is harder to picture. Imagine taking a knife and cutting the ring, not in half but just so that it can be opened up to a section of a cylinder. Now twist one end of the cylinder, keeping

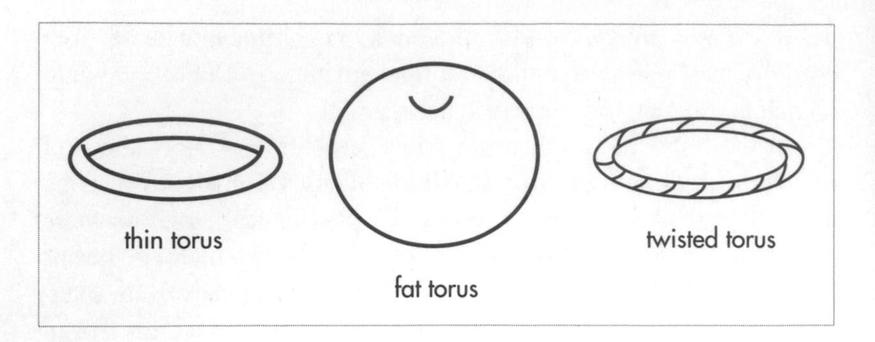

the other end fixed. Finally, reconnect the ends of the cylinder so that it becomes a ring but with a twist. The angle of the twist is another variable. If you can't picture it, that's okay. You won't need to.

Mathematicians call these parameters that determine the size and shape of the torus moduli (plural of *modulus*). A torus has three moduli. The cylinder, or more accurately, the circular cross section of the cylinder, has only one modulus. But a typical Calabi Yau space has hundreds. Perhaps you can see where this is going, but if not, I will spell it out. It is leading us to a Landscape — and an incredibly complicated one at that.

One very important issue is whether the size and shape of the compact component of the space can vary from one point to another. Visualize a clumsily constructed cylinder. Suppose that as you move along the length of the cylinder, the circumference of the cross section varies: here the cylinder is narrow, there it's wider.

Keep in mind that even if the cylinder is extremely thin, far too thin to detect its compact dimension, the size of that dimension determines various coupling constants and masses. Evidently we have made a world where the Laws of Physics can vary from point to point. What does an ordinary physicist who cannot see the small dimension make out of all this? She says: "Conditions are varying from point to point. It seems that some kind of scalar field controls the value of the electric charge and mass of particles, and it can vary from point to point." In other words, the moduli form some kind of Landscape — a Landscape of hundreds of dimensions.

A Calabi Yau space is vastly more complicated than the circular cross section of the cylinder, but the principle is the same: the size and shape of the compact space can vary with position just as if there were hundreds of scalar fields controlling the Laws of Physics! Now we begin to see why the Landscape of String Theory is so complicated.

The Elegant Supersymmetric Universe?

The real underlying principles of String Theory are largely shrouded in mystery. Almost everything we know about the theory involves a very special portion of the Landscape where the mathematics was amazingly simplified by a property that was mentioned in chapter 2 called supersymmetry. The supersymmetric regions of the Landscape form a perfectly flat plain at exactly zero altitude, with properties so symmetric that many things can be deduced without having complete mastery of the entire Landscape. If one were looking for simplicity and elegance, the flat plain of supersymmetric String Theory, a.k.a. superString Theory, is the place to look. In fact until a couple of years ago it was the only place string theorists looked. But some theoretical physicists are finally waking up and trying to wean themselves off the elegant simplifications of the superworld. The reason is simple: the real world is not supersymmetric.

The world of experience that includes the Standard Model and a small cosmological constant is not located on this plain of zero altitude. It is somewhere in the rough, textured regions of the Landscape with hills, valleys, high plateaus, and steep descents. But there is at least some reason to think that our valley is close to the supersymmetric part of the Landscape and that there may be remnants of the mathematical supermiracles that would help us understand features of the empirical world. One example that we will encounter in this section involves the mass of the Higgs boson. In a real sense the discoveries that made this book possible are all about the initial timid explorations away from the safety of the supersymmetric plain.

Supersymmetry is all about the distinctions and similarities of bosons and fermions. As with so much else in modern physics, the principles trace back to Einstein. The year 2005 marks the one hundredth anniversary of the *anno mirabilis* — the "miracle year" — of modern physics. Einstein set two revolutions in motion that year and completed a third.[3]

3. The revolution that Einstein completed in 1905 was the molecular theory of matter. In his paper on Brownian Motion, he established beyond a doubt that molecules existed and determined just how big and numerous they are.

It was of course the year of the Special Theory of Relativity. But what many people don't know is that 1905 was much more than the "relativity year." It also marked the birth of the photon, the start of modern quantum mechanics.

Einstein received only one Nobel Prize in physics, although I think it's fair to say that almost every prize given after 1905 was in one way or another a tribute to his discoveries. The prize was ultimately awarded not for relativity but for the photoelectric effect. This, the most radical of Einstein's contributions, is where the idea that light is composed of discrete quanta of energy was first argued. Physics was ready for the Special Theory of Relativity. In fact it was overdue. But the photon theory of light was a bolt from the blue. As noted previously, Einstein argued that a beam of light, usually thought of as a pure wave phenomenon, had a discrete, or grainy, structure. If the light had a definite color (wavelength), then the photons would all be marching in lock step, each photon identical to every other photon. Particles that can all be in the same quantum state like photons are called bosons, after the Indian physicist Satyendra Nath Bose.

Almost twenty years later, building on Einstein's work, Louis de Broglie closed the cycle by showing that electrons, the quintessential particles, have a wavelike side to them. Like other waves, electrons can reflect, refract, diffract, and interfere. But there is a fundamental difference between electrons and photons: unlike photons, no two electrons can ever occupy the same quantum state. The Pauli exclusion principle ensures that each electron in an atom must exist in its own quantum state and that no other electron can ever shoulder its way into an already occupied state. Even outside an atom, two otherwise identical electrons cannot have the same position or the same momentum. Particles of this kind are called fermions, after the Italian physicist Enrico Fermi, although they should be called paulions. Of all the particles of the Standard Model, about half are fermions (electrons, neutrinos, and quarks) and half are bosons (photons, Z- and W-bosons, gluons, and Higgs bosons).

Fermions and bosons play very different roles in the grand design. Ordinarily we think of matter as made up of atoms, and that means electrons and nuclei. To review, the nuclei at one level are bunches of protons and neutrons stuck together by means of the nuclear force, but at a deeper level, the protons and neutrons are made of even smaller building blocks — quarks. All of these particles — electrons, protons, neutrons, and quarks — are fermions. Matter is made of fermions. But without bosons the atoms and nuclei as well as the protons and neutrons would just fall apart. It's the bosons, especially photons and gluons, hopping back and forth between fermions that create the attraction holding everything together. Although both fermions and bosons are critically important to making the world what it is, they were always thought of as very different kinds of creatures.

But sometime in the early 1970s, as a consequence of discoveries by string theorists, physicists began to play around with a new mathematical idea, the idea that fermions and bosons are not really so different. The idea is that all particles come in precisely matched pairs, identical twins that are the same in every way except that one is a fermion and one is a boson. This was a wild hypothesis. If true in the real world, it would mean that physicists had completely missed half the particles of nature, had failed to discover them in their labs. For example, this new principle would require that there is a particle exactly like the electron — same mass and charge — except that it would be a boson. How could we have failed to detect it in particle physics laboratories like SLAC and CERN? Supersymmetry implies that a fermion twin for the photon, massless and with no electric charge, would have to exist. Similarly, boson partners for electrons and quarks would be required. A complete world of "opposites" was mysteriously missing if this new idea were right. In fact the whole exercise was only a mathematical game, a pure theoretical exploration of a new kind of symmetry that a world — some world not our own — might possess.

The identical twins don't exist. Physicists didn't blunder and miss a whole parallel world. Why, then, is there interest in such idle mathematical speculation — interest that has only intensified over the last thirty years? Physicists are always interested in possible mathematical symmetries, even if the only question is why they aren't part of nature. But also both the real world and the physicist's description of it are full of symmetry. Symmetry is one of the most far-reaching and powerful weapons in the arsenal of theoretical physics. It permeates every branch of modern physics, particularly those that have to do with quantum mechanics. In many cases all we really know about a physical system is its symmetries, but so powerful are these symmetries that they sometimes tell us almost everything we want to know. Symmetries are often at the core of what physicists find esthetically pleasing about their theories. But what are they?

Let's begin with a snowflake. Every child knows that no two snowflakes are exactly alike, but they nevertheless all share a certain feature: a symmetry. Just looking at a snowflake, you can easily see its symmetry. Take the snowflake and rotate it by an angle. It will look different, tilted. But if you rotate it by exactly 60 degrees, it will look unchanged. A physicist would say that rotating a snowflake through 60 degrees is a symmetry.

Symmetries are concerned with the operations you can do to a system that don't change the results of experiments. In the case of the snowflake, the operation is to rotate it by 60 degrees. Here is another example: suppose we build an experiment to determine the acceleration due to gravity at the earth's surface. The simplest version would be to drop a rock from a known height and time its fall. The answer is approximately ten meters per second per second. Notice that I didn't bother to tell you where I dropped the rock, whether it was in California or Calcutta. To a pretty good approximation, the answer is the same everywhere on the earth's surface: the result of the experiment would not change if you moved the entire experiment from one point of the earth's surface to another. In physics jargon moving or displacing something from one point to another is called a translation. So we speak about the symmetry of the earth's gravitational field "under translation." Of course some annoying irrelevant effect might contaminate the symmetry. For example, if we did the experiment just above a very large and heavy mineral deposit, we would get a slightly bigger result. In that case we would say that the symmetry is only approximate. Approximate symmetries are also called *broken symmetries.* The presence of the localized mineral deposit "broke the translation symmetry."

Can the symmetry of the snowflake be broken? No doubt some snowflakes are not perfect. If the flake formed in less than ideal conditions, one side might be a little different from the others. You might still be able to detect the overall hexagon shape of the crystal, but it would have an imperfect, or broken, symmetry.

In outer space, far from any contaminating influences, we might measure the gravitational force between two masses and deduce Newton's law of gravity. No matter where we did the experiment, we would get the same answer. So Newton's law of gravity has translation invariance.

To measure the force law between two objects, you have to separate them along some direction in space. For example, you might measure the force by separating the masses along the x-axis or along the y-axis or any other direction for that matter. Could the force between two objects depend on the direction along which they are separated? In principle, yes, but only if the laws of nature were different from what they are. It's a fact of nature that Newton's law of gravity — force proportional to the product of masses and inversely proportional to the square of the distance — is the same in every direction. If the whole experiment were rotated in space to realign the separation, the results would be unchanged. This insensitivity to direction is called *rotation symmetry.* Translation and rotation symmetries are two of the most basic properties of the world we live in.

Look at yourself in the mirror as you get dressed in the morning. Your mirror image looks exactly the same as you do. The mirror image of your pants is exactly the same as your real pants. The mirror image of your left glove is the same as your left glove.

Wait — that's wrong. Look again. The mirror image of your left glove is not at all like the real left glove. It's identical to the real right glove! And the image of the right glove is actually a left glove.

Look a little more closely at your own mirror image. It's not you. The freckle on your left cheek is on the right cheek of the mirror image. If you could open up your chest cavity, you would find something really

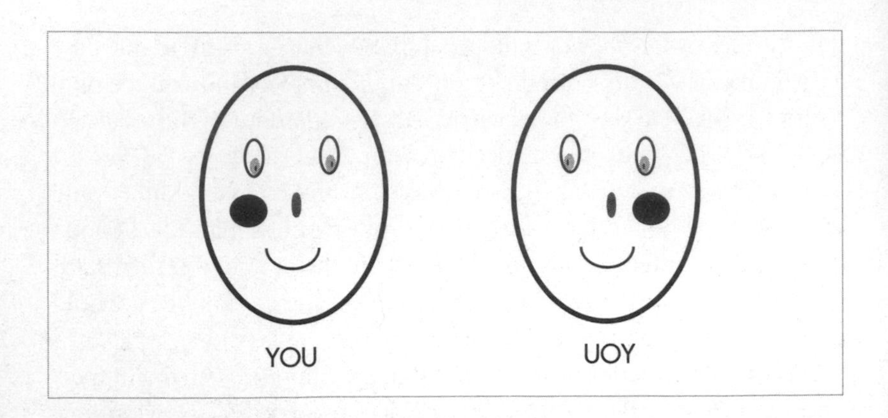

strange about the image. Its heart would be on the right side. It's like no real human being. Let's call it a namuh.

Let's imagine that we had a futuristic technology that allowed us to build up any object we pleased by assembling atoms, one by one. Here is an interesting project. Let's build a human whose mirror image is exactly like you — heart on the left, freckle on the left, and so on. Now the original would be a namuh.

Does the namuh work? Will it breathe? Will its heart work? If we feed it will it metabolize the sugar that we give it? Mostly the answers are yes. It will work as well as you do. But we will find a problem with its metabolism. It just won't process ordinary sugar.

The reason is that sugar comes in two types just like gloves — left-handed and right-handed. Humans can metabolize only sugar. A namuh can metabolize only ragus. The ordinary sugar molecule comes in two varieties — the kind you eat and its mirror image, which you can't. The technical names for sugar and ragus are of course levulose (left-handed sugar molecules that we eat) and dextrose (right-handed sugar that we don't metabolize). Namuhs will work just as well as humans but only if you also replace everything in their environment, including their food, by reflected counterparts.

Replacing everything by its mirror image is called *reflection symmetry* or sometimes *parity symmetry*. The implication of this symmetry is probably clear, but let's spell it out. If everything in the world is replaced by its mirror image, the world will behave in a completely unchanged way.

In actual fact reflection symmetry is not exact. It is an example of a broken symmetry. Something causes the mirror image of a neutrino to be many times heavier than the original. This infects other particles but only in the minutest way. It is as if nature's mirror were a slightly distorted mirror — a funhouse mirror that gives a distorted image, but only very slightly. The distortion is so insignificant for ordinary matter that snamuh would not notice at all. But high-energy elementary particles do notice the distortion and behave in ways that their mirror images don't. However, let's ignore the distortion for the time being and pretend that reflection symmetry is an exact symmetry of nature.

What do we mean when we say that "a symmetry relates particles"? Simply stated, it means that for each type of particle there is a partner or a twin with closely related properties. For reflection symmetry it means that if a left glove is possible, so is a right glove — if levulose can exist, so can dextrose. And if reflection symmetry were not broken, this would apply to every elementary particle. Every particle would have an identical but reflected twin. When reflecting humans to make snamuh, every elementary particle must be replaced by its twin.

Antimatter is another manifestation of symmetry called charge *conjugation symmetry*. Once again, the symmetry involves replacing everything by something else: in this case every particle is replaced by its antiparticle. That replaces a positive electric charge, such as the proton, by a negative charge — the antiproton. Likewise, a positron replaces every ordinary negative electron. Hydrogen atoms become antihydrogen atoms consisting of a positron and an antiproton. Such antiatoms have actually been made in the laboratory — just a handful, not enough to make even an antimolecule. But no one doubts that antimolecules are possible. Antihumans are possible but don't forget to feed them antifood. In fact you'd best keep away from antipeople. When matter meets antimatter it explosively annihilates into a burst of photons. The explosion that would occur if you inadvertently shook hands with an antihuman would be similar to a nuclear bomb explosion.

As it turns out charge conjugation symmetry is also a slightly broken symmetry. But as with reflections the effect would be completely insignificant except for very high-energy particles.

. . .

Now back to fermions and bosons. The original String Theory, the one that Nambu and I discovered, is called *bosonic String Theory* because all the particles that it describes are bosons. This was not a good thing for describing hadrons: after all, the proton is a fermion. Neither would it be good if the goal of String Theory were a theory of everything. Electrons, neutrinos, quarks — they are all fermions. But it didn't take long before new versions of String Theory were discovered that contained fermions as well as bosons. And one of the remarkable mathematical properties of these so-called superString Theories is supersymmetry — the symmetry between bosons and fermions that requires particles to come in exactly matched pairs, a boson for every fermion and vice versa.

Supersymmetry is, again, an indispensable and very powerful mathematical tool for string theorists. Without it the mathematics is so hard that it is very difficult to be certain of the consistency of the theory. Almost all of the reliable work on the subject assumed the world was supersymmetric. But as I emphasized, supersymmetry is most definitely not an exact symmetry of nature. At best it is a fairly badly broken symmetry — the kind of distorted symmetry that a badly warped funhouse mirror would imply. In fact no superpartner has ever been discovered for any of the known elementary particles. If a boson existed with the same mass and charge as the electron, it would have been discovered long ago. Nevertheless, if you go to your Web browser and look up research papers in elementary-particle physics, you will find that since the mid-1970s the overwhelming majority have something to do with supersymmetry. Why is this? Why didn't theoretical physicists toss out supersymmetry in disgust and, with it, superString Theory? The reasons vary.

The subject that was once called theoretical high-energy physics, or theoretical elementary-particle physics, long ago split into two separate disciplines called *phenomenology* and *theory*. If you type in the URL *http://arXiv.org* on your Web browser, you'll find the site where physicists post their research papers. Separate subdisciplines are listed by subject: nuclear theory, condensed-matter physics, and so on. If you look for high-energy physics (hep), you will find four separate archives, only two of which are devoted to the theoretical aspects of the subject that concern us (the other two are for experimental physics and computer simulations). One archive (hep-ph) is for phenomenology (the meaning of which will become clear shortly). The other (hep-th) is for the more theoretical and mathematical papers. Open them both. You'll find that the hep-ph papers are all about the problems of conventional-particle physics. They sometimes refer to experiments, either past or future, and the results of the papers are often numbers and graphs. By contrast, the hep-th papers are largely about String Theory and gravity. They are more mathematical, and for the most part, they have little to do with experiments. However, in the last few years, the lines between these sub-subdisciplines have been blurred a good deal, which I think is a good sign.

But in both archives you will find that many of the papers have something to do with supersymmetry. Each camp has its own reasons for this. For the hep-th people, the reasons are mathematical. Supersymmetry leads to amazing simplifications for problems that are much too hard otherwise. Remember that in chapter 2 I explained that the cosmological constant is guaranteed to be zero if particles come in superpartner pairs. This is one of many mathematical miracles that occur in supersymmetric theories. I won't describe them, but the main point is that supersymmetry so simplifies the mathematics of quantum field theory and String Theory that it allows theorists to know things that would otherwise be far beyond calculation. The real world may not be supersymmetric, but there may be interesting phenomena that can be studied with the aid of supersymmetry: phenomena that ordinarily would be far too hard for us to comprehend. An example is black holes. Every theory that includes the force of gravity will have black holes among the objects it describes. Black holes have very mysterious and paradoxical properties that we will come upon later in this book. Speculations about these paradoxes have been much too complicated to check in an ordinary theory. But, as if by magic, the existence of superpartners makes black holes easy to study. For the string theorist this simplification is essential. The mathematics of the theory as it is now practiced relies almost wholly on supersymmetry. Even many old questions about the quantum mechanics of quarks and gluons become easy if we add superpartners. The supersymmetric world is not the real world (at least in our pocket universe), but it is close enough to hold many lessons about elementary particles and gravity.

Although the ultimate aims of hep-phers and hep-thers may be the same, phenomenologists have a different immediate agenda from string theorists. Their goal is to use the older methods and sometimes the new ideas of String Theory to describe Laws of Physics as they would have been understood for most of the twentieth century. They are usually not trying to build a theory from some first principles, a theory that would be mathematically complete, nor are they expecting to discover the ultimate theory. Their interest in supersymmetry is as a possible approximate or broken symmetry of nature, something that can be discovered in future laboratory experiments. For them the most important discoveries would be of a whole family of new particles — the missing superpartners.

Broken symmetries, remember, are not perfect. In a perfect mirror, an object and its mirror image are identical except that left and right are interchanged, but in a funhouse mirror the symmetry is imperfect. It may be good enough to recognize the image of an object, but image and object are distorted versions of each other. The image of a thin man might be a fat man who, if he were real, would have twice the weight of his thin twin.

In the funhouse that we call our universe, the mathematical supersymmetry "mirror" that reflects each particle into its superpartner is badly distorted, so badly distorted that the superpartners are like a very fat image. If they exist at all, they are many times heavier — more massive — than the known particles. No superpartner has ever been discovered: not the electron's mate or the photon's or the quarks'. Does that mean that they don't exist and that supersymmetry is an irrelevant mathematical game? Perhaps so, but it could also mean the distortion is enough to make the superpartners so heavy that they are beyond the reach of current particle accelerators. If for some reason the superpartners were all a bit too massive — a couple of hundred times the proton mass — they could not be discovered until the next round of accelerators is built.

The superpartners all have names similar to their ordinary twins. It's not too hard to remember them if you know the rule. If the ordinary particle is a boson like the photon or the Higgs boson, then the name of the fermion twin is obtained by adding *ino*. Thus the photino, the Higgsino, the Zino, and the gluino. If, on the other hand, the original particle is a fermion, then you just add the letter s at the beginning: selectron, smuon, sneutrino, squark, and so on. This last rule has generated some of the ugliest words in the physicist's vocabulary.

There is always a strong tendency to hope that new discoveries are "just around the corner." If the superpartners fail to show up at a hundred times the mass of the proton, then estimates will be revised, and an accelerator will have to be built in order to discover them at a thousand proton masses. Or at ten thousand proton masses. But is it all wishful thinking and no more? I don't think so. There is a deep puzzle about the Higgs particle to which supersymmetry may hold the key. The problem is closely connected with the "mother of all physics problems" as well as with the surprising weakness of gravity.

The same quantum jitters that can create an enormous vacuum energy can also have an effect on the masses of elementary particles. Here's the reason. Suppose a particle is placed into the jittery vacuum. That particle will interact with the quantum fluctuations and disturb the way the vacuum jitters in the immediate vicinity of the particle. Some particles will damp the jittery behavior; others will increase it. The overall effect is to modify the energy due to the jitters. This change in the jitter energy, due to the presence of the particle, must be counted as part of its mass (remember $E = mc^2$). A particularly violent example is the effect on the mass of the Higgs boson. Physicists know how to estimate the additional extra mass of the Higgs, and the result is almost as absurd as the estimate for the vacuum energy itself. The vacuum jitters in the neighborhood of the Higgs boson ought to add enough mass to the Higgs boson to make it as heavy as the Planck mass!

Why is this so problematic? Although theorists usually focus on the Higgs particle, the problem really infects all the elementary particles, with the exception of the photon and graviton. Any particle placed in a jittery vacuum will suffer an enormous increase in its mass. If all the particles had their masses increased, all matter would become heavier, and that would imply that the gravitational forces between objects would become stronger. It takes only a modest increase in the strength of gravity to render the world lifeless. This dilemma is conventionally called the Higgs mass problem and is another fine-tuning problem that belabors theorists' attempts to understand the Laws of Physics. The two problems — the cosmological constant and the Higgs mass — are in many ways very similar. But what do they have to do with supersymmetry?

Remember that in chapter 2 I explained that an exact twinning of fermions and bosons would cancel the fluctuating energy of the vacuum. Exactly the same is true of the extra, unwanted mass of particles. In a supersymmetric world the violent effects due to quantum fluctuations would be tamed, leaving the particle masses undisturbed. Moreover, even a distorted supersymmetry would alleviate the problem a great deal if the distortion were not too severe. This is the primary reason that elementary-particle physicists hope that supersymmetry is "just around the corner." It should be noted, however, that distorted supersymmetry cannot account for the absurdly small value of the cosmological constant. It's just too small.

The problem of the Higgs mass is similar to the problem of vacuum energy in another way. Just as Weinberg showed that life could not exist in a world with too much vacuum energy, heavier elementary particles would also be disastrous. Perhaps the explanation of the Higgs mass problem lies not in supersymmetry but rather in the enormous diversity of the Landscape and the anthropic need for the mass to be small. In a few years we may know whether supersymmetry really is just around the corner or if it is a mirage that keeps receding as we approach it.

One question that seems not to have been asked by theoretical physicists is: "If supersymmetry is such a wonderful, elegant, mathematical symmetry, why isn't the world supersymmetric? Why don't we live in the kind of elegant universe that string theorists know and love best?" Could the reason be anthropic?

The biggest threat to life in an exactly supersymmetric universe doesn't have to do with cosmology but, rather, with chemistry. In a supersymmetric universe every fermion has a boson partner with exactly the same mass, and therein lies the trouble. The culprits are the superpartners of the electron and the photon. These two particles, called the selectron (ugh!) and the photino, conspire to destroy all ordinary atoms.

Take a carbon atom. The chemical properties of carbon depend primarily on the valence electrons — the most loosely bound electrons in the outermost orbits. But in a supersymmetric world, an outer electron can emit a photino and turn into a selectron. The massless photino flies off with the speed of light, leaving the selectron to replace the electron in the atom. That's a big problem: the selectron, being a boson, is not blocked (by the Pauli exclusion principle) from dropping down to lower energy orbits near the nucleus. In a very short time, all the electrons will become selectrons trapped in the innermost orbit. Good-bye to the chemical properties of carbon — and every other molecule needed for life. A supersymmetric world may be very elegant, but it can't support life — not of our kind, anyway.

If you go back to the physics archive Web site, you will find two other archives, one called General Relativity and Quantum Cosmology, the other Astrophysics. In these archives supersymmetry plays a much less prominent role. Why should a cosmologist pay any attention to supersymmetry if the world is not supersymmetric? To paraphrase Bill Clinton, "It's the Landscape, stupid." Although a particular symmetry may be broken to a greater or lesser degree in our little home valley, that doesn't mean that the symmetry is broken in all corners of the Landscape. Indeed, the portion of the String Theory Landscape that we know most about is the region where supersymmetry is exact and unbroken. Called the *supersymmetric moduli space* (or supermoduli space), it is the portion of the Landscape where every fermion has its boson and every boson has its fermion. As a consequence, the vacuum energy is exactly zero everywhere on the supermoduli space. Topographically, that makes it a plain at exactly zero altitude. Most of what we know about String Theory comes from our thirty-five-year exploration of this plain. Of course this also implies that some pockets of the megaverse will be supersymmetric. But there are no superstring theorists to enjoy it.

The Magical Mystery aMazing M-Theory

By 1985 String Theory — now called superString Theory — had five distinct versions.[4] They differed in a number of ways. Two had open strings (strings with two ends) as well as closed strings — three did not. The names of the five are not particularly enlightening, but here they are.

4. The *super* refers to the fact that these theories all had fermion-boson twins and were, therefore, supersymmetric.

The two with open strings are called Type Ia and Ib String Theories. The remaining three with only closed strings are known as Type IIa, Type IIb, and Heterotic String Theories. The distinctions are too technical to describe without boring the reader. But one thing that they have in common is far more interesting than any of the differences: although some have open strings and some don't, all five versions have closed strings.

To appreciate why this is so interesting, we need to understand a very disappointing failure of all previous theories. In ordinary theories — theories such as Quantum Electrodynamics or the Standard Model — gravity is an optional "add-on." You can either ignore gravity or add it into the brew. The recipe is simple: take the Standard Model and add one more particle, the graviton. Let the graviton be massless. Also add some new vertices: any particle can emit a graviton. That's it. But it doesn't work very well. The mathematics is intricate and subtle, but at the end of the day, the new Feynman diagrams involving gravitons make hash out of the earlier calculations. Everything comes out infinite. There is no way to make sense of the theory.

In a way I think it is a good thing that the simple procedure failed. It contains no hint of an explanation of the properties of particles, it has no explanation of why the Standard Model is special, and it explains nothing about the fine-tuning of the cosmological constant or the Higgs mass. Frankly, if it worked, it would be very disappointing.

But the five String Theories are very clear on this point: they simply cannot be formulated without gravity. Gravity is not an arbitrary input — it is an inevitable outcome. String Theory, in order to be consistent, *must* include the graviton and the forces that it mediates by exchange. The reason is simple. The graviton is a closed string, the lightest one. Open strings are optional, but closed strings are always there. Suppose we try to create a theory with only open strings. If we succeed we will have a String Theory without gravity. But we will always fail. The two ends of an open string can always find each other and join to form a closed string. Ordinary theories are consistent *only* if gravity is left out. String Theory is consistent *only* if it includes gravity. That fact, more than any other, gives string theorists confidence that they are on the right track.

The four theories labeled Types I and II were first discovered in the

1970s. Each had fatal defects, not in their internal mathematical consistency, but in the detailed comparison with experimental facts about particles. Each described a possible world. They just did not describe our world. Thus, enormous excitement ensued when the fifth version was discovered in Princeton, in 1985. The Heterotic String Theory appeared to be the string theorist's dream. It looked enough like the real world to perhaps be the real thing. Success was declared imminent.

Even then there were reasons to be suspicious of the strong claims. For one thing, there was still the problem of too many dimensions: nine of space and one of time. But theorists already knew what to do with the extra six dimensions: "Compactify!" they said. But there are millions of possible choices among Calabi Yau spaces. Moreover, every one of them gives a consistent theory. Even worse, once a Calabi Yau manifold was chosen, there were hundreds of moduli associated with its shape and size. These, too, had to be fixed by hand. Furthermore, the known theories were all supersymmetric: in each case the particles came in exactly matched pairs, which we know does not fit our reality.

Nevertheless, string theorists were so blinded by the myth of uniqueness that throughout the 1980s and early 1990s they continued to claim that there were only five String Theories. In their imagination the Landscape was very sparse: it had only five points! This of course was nonsense, since each compactification came with many moduli that could be varied; but still, physicists clung to the fiction that there were only five theories to sort through. Even if there were only five possibilities, what principle picked, from among them, the one that describes the real world? No ideas surfaced. But in 1995 came a breakthrough, not in finding the right version to describe the world, but in understanding the connections among the various versions.

University of Southern California, 1995

Every year in late spring or early summer, the world's string theorists convene for their annual jamboree. Americans, Europeans, Japanese, Koreans, Indians, Pakistanis, Israelis, Latin Americans, Chinese, Muslims, Jews, Christians, Hindus; believers and atheists: we all meet for a week to hear one another's latest thoughts. Almost all of the four or five hundred participants know one another. The senior people are generally old friends. When we meet we do what physicists always do: give and listen to lectures on the latest hot topics. And have a banquet.

The year 1995 was memorable, at least to me, for two reasons. First, I was the after-dinner speaker at the banquet. The second reason was an event of momentous importance to the people assembled there: Ed Witten gave a lecture reporting spectacular progress that turned the field in totally new directions. Unfortunately Witten's lecture went right past me, not because I couldn't get there, but because I was happily daydreaming about what I would say in the after-dinner speech.

What I wanted to talk about that evening was an outrageous hypothesis: a guess about how today's physics might have been discovered by very smart theorists even if physics had been deprived of any experiment after the end of the nineteenth century. The purpose was partly to amuse but also to bring some perspective to what we (string theorists) were attempting. I will come back to it in chapter 9.

What my daydreaming caused me to miss was a new idea that would become central to my conception of the Landscape. Ed Witten, not just a great mathematical physicist but also a leading figure among pure mathematicians, had long been the driving force behind the mathematical development of String Theory. He is a professor (some would say "*The* Professor") and leading light at the intellectually supercharged Institute for Advanced Study. More than anyone, Witten has single-mindedly driven the field forward.

By 1995 it was becoming clear that the vacuum described by String Theory was far from unique. There were many versions of the theory, each one leading to different Laws of Physics. This was not seen as a good thing but rather as an embarrassment. After all, ten years earlier, the Princeton string theorists had promised not only that the theory was almost unique but also that they were about to find the one true version that describes nature. Witten's primary objective had been to prove that all but one version was mathematically inconsistent. But instead he found a Landscape, or more accurately, the portion of the Landscape at zero altitude, i.e., the supersymmetric part of the Landscape. Here's what happened:

Imagine that physicists had discovered two theories of electrons and photons: the usual Quantum Electrodynamics but also a second theory.

In the second theory electrons and positrons, instead of moving freely through three-dimensional space, could move only in one direction, let's say the x-direction. They were simply unable to move in any other direction. Photons, on the other hand, move in the usual way. The second theory was an embarrassment. As far as physicists could tell, it was mathematically every bit as consistent as the Quantum Electrodynamics that ruled the real world of atoms and photons, but it had no place in their view of the real world. How could there be two theories, equally consistent, with no way to explain why one should be discarded while the other describes nature? They hoped and prayed that someone would discover a flaw, a mathematical contradiction, that would eliminate the unwanted theory and give them reason to believe the world is the way it is because no other world is possible.

Then, while attempting to prove that the second theory was inconsistent, they hit upon some interesting facts. Not only did they find no inconsistency, but they also began to understand that the two theories were both part of the same theory. The second version, they realized, was merely a limiting version of the usual theory in a region of space with an enormously large magnetic field — a super MRI machine. As any physicist will tell you, a very strong magnetic field will constrain charged particles to move in only one direction: along the magnetic lines of force. But the motion of an uncharged particle, like the photon, is unaffected by the field.[5] In other words, there is only one theory, one set of equations, but two solutions. Even better, by varying the magnetic field continuously, a whole family of theories interpolates between the two extremes. The fictitious physicists had discovered a continuous Landscape and set about to explore it. Of course they still had no idea what mechanism might choose among the continuum of solutions — why the world of reality has no background magnetic field. They hoped to explain that later.

This is exactly the situation that Witten left us with after his 1995 talk in Los Angeles. He had discovered that all five versions of String Theory were really solutions to a single theory: not many theories, but

5. Placing a magnet under a sheet of paper and then sprinkling iron filings on the sheet allows the lines of force to be easily seen. The filings line up along the field lines and form filaments.

many solutions. In fact they all belonged to a family that includes one more member, a theory that Witten called M-theory. Moreover, the six theories each correspond to some extreme value of the moduli: some distant limiting corner of the Landscape. As in the example of the magnetic field, the moduli could be continuously varied so that one theory morphed into any of the others! "One theory — many solutions": that became the guiding slogan.

There are many conjectures about what M stood for. Here are some of the possibilities: mother, miracle, membrane, magic, mysterious, and master. Later, matrix was added. No one seems to know for sure what Witten had in mind when he coined the term *M-theory*. Unlike the previously known five theories, the new cousin is not a theory with nine space dimensions and one time. Instead, it is a theory with ten dimensions of space and one of time. Even more alarming, M-theory is not a theory of strings: the basic objects of M-theory are membranes, two-dimensional sheets of energy that resemble elastic sheets of rubber instead of one-dimensional rubber bands. The good news was that M-theory seemed to provide a unifying framework in which the various String Theories appear when one or more of the ten dimensions of space are rolled up by compactification. This was real progress that held the promise of a unifying foundation for String Theory. But there was also a down side. Almost nothing was known about the mix of eleven-dimensional general relativity with quantum mechanics. The mathematics of membranes is horribly complicated, far beyond that of strings. M-theory was as dark and mysterious as any quantum theory of gravity had ever been before String Theory made its appearance. It seemed as if we had taken one step forward and two steps backward.

It didn't stay that way for long. By the next string meeting, in the summer of 1996, I had the pleasure of reporting that three of my friends and I had uncovered the secret of M-theory. We had found the underlying objects of the theory, and the equations governing them were incredibly simple. Tom Banks, Willy Fischler, Steve Shenker, and I discovered that the fundamental entities of M-theory were not membranes but simpler objects, basic "partons" of a new kind. In some ways similar to Feynman's old partons, these new constituents had an astonishing ability to assemble themselves into all kinds of objects. The graviton itself, once thought to be the most fundamental particle, was a composite of many partons. Assemble the same partons in a different pattern, and membranes emerged. In another pattern they formed a black hole. The detailed equations of the theory were far simpler than the equations of String Theory, simpler even than the equations of general relativity. The new theory is called Matrix Theory or sometimes M(atrix) Theory to emphasize its connection to M-theory.

Witten was not the first to speculate about a connection between an eleven-dimensional theory and String Theory. For years a number of physicists had tried to call attention to an eleven-dimensional theory with membranes in it. Mike Duff at Texas A&M (now of Imperial College, London) had had most of the ideas years earlier, but string theorists weren't buying it. Membranes were just too complicated, the mathematics too poorly understood, for Duff's seminal idea to be taken seriously. But Witten being Witten, string theorists latched on to M-theory and never let go.

What is it about M-theory that so captured the imagination of theoretical physicists? It is not a String Theory. No one-dimensional filaments of energy inhabit this world of eleven space-time dimensions. So why, all of a sudden, did string theorists become interested in two-dimensional sheets of energy — membranes, as they are called? The answers to these riddles lie in the subtle mysteries of compactification.

Let's return to the infinite cylinder and recall how we got to it. Beginning with an infinite sheet of paper, we first cut out an infinite strip a few inches wide. Think of the two edges as the ceiling and floor of a two-dimensional room. The room is extremely big. It goes on forever in the x-direction, but in the y-direction it is bounded above and below by the floor and ceiling. In the next step the ceiling is joined to the floor to make a cylinder.

Imagine a particle moving through the infinite room. At some point it may arrive at the ceiling. What happens next? If the strip were rolled up into a cylinder, there would be no problem: the particle would just keep going, passing through the ceiling and reappearing at the floor. In fact we don't really need to bend the paper into a cylinder; it is enough to know that every point on the ceiling is matched to a unique point on the floor so that when the particle passes through an edge it suddenly jumps to the other edge. We can roll it up or leave it flat: we need only to keep track of the rule that each ceiling point is identified with the floor point vertically below it.

Now let's get a little more advanced: our room is now three-dimensional like a real room, except for being infinite, this time in the x- *and* z-directions. But the vertical direction, y, is bounded above and below by the ceiling and floor. As before, when a particle passes through the ceiling, it reappears instantly at the corresponding point on the floor. Three-dimensional space has been compactified to two

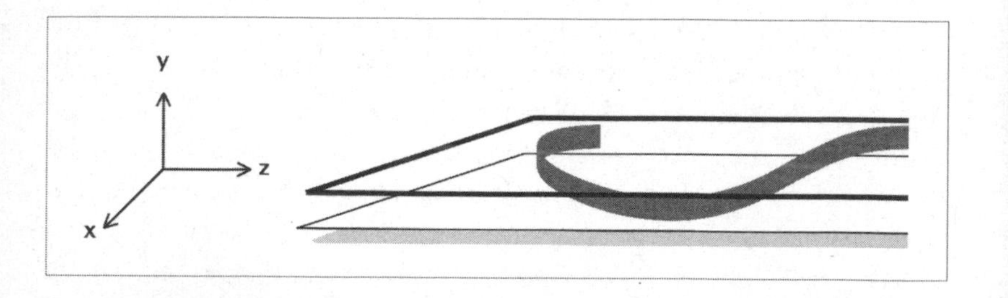

dimensions. If the height of the room — in other words, the distance around the y-direction — were shrunk to a microscopic size, the space would be practically two-dimensional.

As I said, M-theory has no strings, only membranes. So what is its connection to String Theory? Imagine a ribbon, whose width is exactly the height of the room, with its width stretched from floor to ceiling. The length of the ribbon wanders about the room following some curve inscribed on the floor. The only rule is that the upper edge of the ribbon must lie exactly above the lower edge. In fact the ribbon no more has edges than the cylinder of paper did. But it is easiest to visualize a long ribbon snaking through the infinite room with its edges following the ceiling and floor.

By now you must have a pretty good idea of how the ribbon, itself a two-dimensional membrane, mimics a one-dimensional string. If the compact direction were so small that it couldn't be seen without a microscope, the ribbon would, for all practical purposes, be a string. If the ribbon closed back on itself, it would be indistinguishable from a closed string: a Type IIa string, to be precise. That is the connection between M-theory and String Theory. Strings are really very thin ribbons or

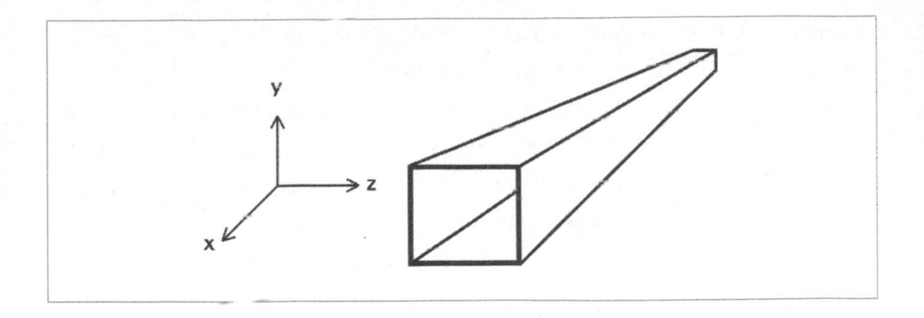

membranes that look more and more like thin strings as the distance around the y-direction shrinks. That's not so difficult.

But things can get stranger. Now let's go another step and compactify two dimensions: call them z and y. In order to visualize this, imagine the infinite room replaced by an infinite hallway. You have walls to the left and right of you and a ceiling and floor above and below. But if you look straight down the hallway, you can see forever in either direction. Again, if an object gets to the ceiling, it reappears at the floor. But what if it approaches one of the walls that bound the z-direction? You probably already know the answer: it appears at the opposite wall, directly across from the place where it touched the first wall.

Exactly the same trick can be done in the ten-dimensional space of M-theory, only this time the "hallway" extends indefinitely in eight of the ten spatial directions. As you might expect, when the width and height of the hallway get very small, the clumsy, large-scale observer thinks that he is living in a world of eight dimensions (plus one of time).

Now comes a shocking and bizarre consequence of String Theory. If the width and height of the hallway get smaller than a certain size, a new dimension grows out of nowhere. This new direction of space is none of the ones we started with. We know about it only through the indirect mathematics of String Theory. The *smaller* we make the original compact directions, the *bigger* the newly created compact direction becomes. Eventually, as the hallway is shrunken to zero height and width, the new direction grows infinitely big. Astoundingly, by shrinking away two of the space dimensions, we find nine, not eight large directions left over. This peculiar fact — that "ten minus two equals nine" — is one of the strangest consequences of String Theory. The geometry of space is not always what Euclid or even Einstein thought. Evidently at the smallest distances, space is different from anything physicists or mathematicians imagined in their wildest dreams.

By now you may be slightly confused by the precise distinction between String Theory and M-theory. String theorists are also confused (and confusing) about the terminology. For example, is the eleven-dimensional theory, which contains membranes but no strings, part of String Theory? Is the compactified version of M-theory, when it morphs into String Theory, still M-theory? I'm afraid the field has been rather imprecise on these matters. My own terminology is to call everything that grew out of the original theory of strings String Theory. That includes everything that is now called M-theory. The term *M-theory* I use when I want to emphasize the eleven-dimensional features of the theory.

The story of String Theory goes on in chapter 10, but now I want to take a time-out from the difficult technical aspects of String Theory and turn to an issue that deeply concerns every serious physicist. In fact it concerns anyone who has an interest in understanding nature at the very deepest level.

CHAPTER NINE

On Our Own?

The search for the fundamental principles of physics is a very risky business. It's like any other exploration into the deep unknown. There is no guarantee of success and plenty of possibility of becoming hopelessly lost. The physicist's guiding star has always been experimental data, but in this respect things are harder than ever. All of us (physicists) are very aware of the fact that experiments designed to probe ever deeper into the structure of matter are becoming far bigger, more difficult, and costlier. The entire world's economy for one hundred years would not be nearly enough to build an accelerator that could penetrate to the Planck scale, i.e., 10^{-33} centimeters. Based on today's accelerator technology, we would need an accelerator that's at least the size of the entire galaxy! Even if future technology could shrink it to a more manageable size, it would still need a trillion barrels of oil per second to power it.

How then can we hope to succeed? Without experimental tests and new discoveries to keep us on the right track, it may be a futile enterprise. On the other hand, some grand leap, perhaps involving String Theory, might allow us to ignore the experimental difficulties and create a theory that so accurately describes the Laws of Physics that there will be no doubt of its correctness. The truth is that we just don't know if this is possible. What we are attempting is so bold that there is no historical precedent for it. Some think it's quixotic: a fool's errand. Even those doing it are doubtful of eventual success. To divine the fundamental laws of nature that govern a world sixteen orders of magnitude smaller than any microscope will ever see is a very tall order. It will take not only cleverness and perseverance but also tremendous quantities of chutzpah.

Is the human race anywhere near being smart enough? I mean collectively, not individually. Are the combined talents of humanity sufficient to solve the great riddle of existence? Is the human mind even wired in the right way to be able to understand the universe? What are the chances that the combined and diverse intellects of the world's greatest physicists and mathematicians will be able to divine the final theory with only the absurdly limited experiments that will be possible?

It was these questions that I wanted to explore with my colleagues that evening in 1995, at the physicists' banquet. I feel they are also important to discuss in this book, if for no other reason than to give the reader an idea of the difficulties physicists will be up against in the twenty-first century. To get some perspective at that time, I indulged in a little conceit, a thought experiment. I tried to imagine a best-case scenario for how physics might have evolved if physicists of the twentieth century had been deprived of all experimental results after December 31, 1899. Most people will tell you that physics, or science in general, would quickly have bogged down. They may very well be right. But then again, they may be lacking in imagination.

The precise question I wanted to explore at the banquet was how much of twentieth-century physics could prodigiously smart theoretical physicists have discovered without any new experimental guidance. Might they have discovered all or most of what we know today? I did not claim that they would have succeeded but only that there were lines of argument that could have led them toward much of today's physics. In the rest of this chapter, I'll take you through my thinking.

The twin pillars of twentieth-century physics were the Theory of Relativity and quantum mechanics. Both were born during the first few years of the century. Planck discovered his constant in 1900, and Einstein interpreted Planck's work in terms of photons in 1905. Planck's discovery involved nothing more than the properties of heat radiation: the glow of electromagnetic radiation emitted by a hot object. Physicists call this radiation *blackbody* radiation because it would be emitted even by a perfectly black object if it is heated. For example, even the blackest of pots will glow red if heated to thousands of degrees. Physicists by the year 1900 were not only familiar with the problem of blackbody radiation, they also were deeply troubled by an apparent contradiction. The mathematical theory showed that the total amount of energy in black-body radiation was infinite. The amount of energy stored in each individual wavelength was finite, but according to nineteenth-century physics, when it was all added up, an infinite amount of energy would reside in the very short wavelengths; hence, the term *ultraviolet catastrophe.* In a sense it was a problem of the same sort as the mother of all physics problems: too much energy stored in very short wavelengths. Einstein solved it (the problem of heat radiation) with the radical but very well-motivated hypothesis that light consists of indivisible quanta. No role was played by any twentieth-century experiment.

The year of the photon was also the year of the Special Theory of Relativity. The famous Michelson Morley experiment that failed to detect the earth's motion through the ether was already thirteen years in the past when the century turned.[1] In fact it is not clear that Einstein even knew of Michelson and Morley's work. According to his own reminiscences, the main clue was Maxwell's theory of light, which dated from the 1860s. Einstein, master of thought experiments, asked himself at the age of sixteen — the year was 1895 — what a light beam would be like to someone moving alongside it with the speed of light. Even at this early age, he realized that a contradiction would result. This, not new experiments, was the soil from which his great discovery sprung.

By the end of the nineteenth century, physicists had begun the exploration of the microscopic world of electrons and the structure of matter. The great Dutch theoretical physicist Hendrik Antoon Lorentz had postulated the existence of electrons, and by 1897, the British physicist J. J. Thomson had discovered and studied their properties. Wilhelm Conrad Roentgen had made his dramatic discovery of X-rays in 1895. Following up on Roentgen's discovery, Antoine-Henri Becquerel discovered radioactivity a year later.

But some things were not known until years later. It took Robert Millikan until 1911 to determine the precise value of the electron's

1. According to nineteenth-century physics, the ether was the hypothetical elastic material that fills all space. I always picture it as colorless Jell-O. Light was supposed to be vibrating wiggles in the ether. It was expected in the nineteenth century that anyone moving relative to the ether would find that the velocity of light was altered by the observer's motion.

electric charge. And until Ernest Rutherford devised a clever experiment to probe the atom, the picture of electrons orbiting a tiny nucleus was not known, although some speculation along those lines did go back to the nineteenth century.[2] And of course the modern idea of the atom goes way back to John Dalton, in the early years of the nineteenth century.

Rutherford's discovery of the "planetary" structure of the atom — light electrons orbiting a heavy, tiny nucleus — was key. It led, in just two years, to Bohr's theory of quantized orbits. But was it absolutely necessary? I doubt it. I was surprised to learn recently that Heisenberg's first successful attempt to create a new quantum mechanics did not involve the atom at all.[3] His first application of the radical "matrix mechanics" was to the theory of simple vibrating systems, so-called harmonic oscillators. In fact the Planck-Einstein theory was understood as a theory of the harmonic oscillation (vibration) of the radiation field. That the energy of an oscillator comes in discrete jumps is the analog of Bohr's discrete orbits. It seems unlikely that Rutherford's atom was essential to the discovery of quantum mechanics.

But still there was the problem of the atom. Could its solar system–like structure have been guessed? Here I think the key would have been spectroscopy, the study of the same spectral lines that Hubble used to determine the velocity of galaxies. There was a huge amount of nineteenth-century spectroscopic data. The details of the hydrogen spectrum were well known. On the other hand, the idea that the atom consists of electrons and some object with positive electric charge had been in the air for a few years by the time 1900 rolled around. I recently learned from a Japanese friend that the first speculations of a planetary atom (electrons orbiting a nucleus) were due to a Japanese physicist, Hantaro Nagaoka. There is even a Japanese postage stamp with a picture of Nagaoka and his atom.

2. Rutherford bombarded gold atoms with alpha particles (helium nuclei), and from the deflection of the alpha particles, he deduced that the atom contained light electrons orbiting a tiny, heavy nucleus. This was the first modern particle physics experiment.

3. Abraham Pais, *Niels Bohr's Times: In Physics, Philosophy, and Polity* (Oxford: Oxford University Press, 1991).

Nagaoka's paper, available on the Internet, dates from 1903, eight years before Rutherford's experiment. Had the idea come a few years later, when more was known about quantum theory, the story might have been different. Given the wealth of spectroscopic data, the quantum behavior of oscillators, and Nagaoka's idea, would a brilliant young Heisenberg or Dirac have had the necessary eureka moment? "Ah ha, I've got it. The positive charge is at the center, and the electrons orbit around it in quantized orbits." Perhaps Bohr himself would have done it. Physicists have made much larger leaps than that: witness the General Theory of Relativity or, for that matter, the discovery of String Theory from the spectroscopy of hadrons.

And what about the General Theory of Relativity? Could it have been guessed without a twentieth-century experiment? Certainly! All that was required was Einstein's thought experiment that led to the equivalence principle. Reconciling the equivalence principle with special relativity was the path taken by Einstein.

No serious theoretical physicist today is content with two apparently incompatible theories. I am referring, of course, to quantum mechanics and the General Theory of Relativity. In the late 1920s a very similar problem existed — how to reconcile quantum mechanics with special relativity. Physicists of the caliber of Dirac, Pauli, and Heisenberg would not, and did not, rest until they saw the Special Theory of

Relativity reconciled with quantum mechanics. This would entail a relativistic quantum theory of the electron interacting with the electromagnetic field. Here I really don't have to speculate. The early development of Quantum Electrodynamics was motivated by nothing other than Dirac's desire for such a synthesis of quantum mechanics and special relativity. But would he have known that his Dirac equation was correct?

Here Pauli makes his dramatic entrance with the exclusion principle. What Pauli was motivated by was chemistry: the periodic table and how it was built up by successively placing electrons in atomic orbits. In order to understand how the electrons fill the atomic orbits and block other electrons from already filled orbits, Pauli had to invoke a new property of electrons, their so-called spin. And where did the idea of spin come from? Not from new, twentieth-century experiments, but rather from nineteenth-century spectroscopy and chemistry. The addition of the new spin degree of freedom meant that Pauli could place two electrons in every orbit, one with its spin pointing down and the other with its spin up. Thus, in helium two electrons fill the lowest Bohr orbit. This was the key to Mendeleyev's periodic table. Pauli's idea was a guess based on nineteenth-century chemistry, but Dirac's relativistic theory of the electron precisely explained this new, mysterious property of spin.

But Dirac's theory did have one serious problem. In the real world the energy associated with every particle is a positive quantity. At first Dirac's theory seemed inconsistent — it had electrons, which carried negative energy! Particles with negative energy are a very bad sign. Remember that in an atom, electrons of higher energy eventually "drop down" to orbits of lower energy by emitting photons. The electrons seek out the lowest-energy orbit that isn't blocked by the Pauli exclusion principle. But what if an infinite number of negative energy orbits were available to the electrons? Wouldn't all the electrons in the world start cascading to increasingly negative energy, giving off enormous amounts of energy in the form of photons? Indeed, they would. This potentially damning feature of Dirac's idea threatened to undermine his whole theory — unless something could prevent the electrons from occupying the negative energy states. Again Pauli saves the day. Pauli's exclusion principle would rescue Dirac from disaster. Just suppose that what we normally call vacuum is really a state full of negative-energy electrons, one in every negative-energy orbit. What would the world be like? Well, you could still put electrons into the usual positive-energy orbits, but now when an electron gets to the lowest positive-energy orbit, it is blocked from going any farther. For all intents and purposes, the negative-energy orbits might as well not exist, since an electron is effectively blocked from falling into these orbits by the presence of the so-called Dirac Sea of negative energy electrons. Dirac declared the problem solved, and so it was.

This idea soon led to something new and totally unexpected. In an ordinary atom an electron can absorb the energy of nearby photons and be "kicked up" into a more energetic configuration.[4] Dirac now showed his real brilliance. He reasoned that the same thing could happen to the negative-energy electrons that fill the vacuum; photons could kick negative-energy electrons up to positive-energy states. What would be left over would be one electron with positive energy and a missing negative-energy electron — a hole in the Dirac Sea. Being a missing electron, the hole would seem to have the opposite electric charge from the electron and would look just like a particle of positive charge. This, then, was Dirac's prediction: particles should exist identical to electrons, except with the opposite electric charge. These positrons, which Feynman would later interpret as electrons going backward in time, Dirac pictured as holes in the vacuum. Moreover, they should be created together with ordinary electrons, when photons collide with enough energy.

Dirac's prediction of antimatter was one of the great moments in the history of physics. It not only led to the subsequent experimental discovery of positrons, but it heralded the new subject of quantum field theory. It was the forerunner of Feynman's discovery of Feynman diagrams and later led to the discovery of the Standard Model. But let's not get ahead of the story.

Dirac wasn't thinking about any experiment when he discovered his remarkable equation for the relativistic quantum mechanics of electrons. He was thinking about how the nonrelativistic Schrödinger equation could be made mathematically consistent with Einstein's Special Theory of Relativity. Once he had the Dirac equation, the way lay open

4. This was known from the absorption spectrum of atoms: once again nineteenth-century physics.

to the whole of Quantum Electrodynamics. Theorists studying QED would certainly have found the inconsistencies that were papered over by renormalization theory.[5] There was no obstruction to the discovery of modern quantum field theory. And physicists would have puzzled endlessly over the enormous vacuum energy and why it didn't gravitate. We might question whether theorists would have been willing to carry on without experimental confirmation of their ideas. We might question whether young people would want to pursue such a purely theoretical enterprise. But I don't think we can question the possibility of physics progressing up to this point. Moreover, the thirty-five-year history of String Theory suggests that as long as someone will pay them, theoretical physicists will continue to push the mathematical frontiers until the end of time.

What about the nucleus, the positively charged "sun" at the center of the tiny atomic solar system? How might the proton and neutron have been deduced? The proton would not have been too difficult. Dalton in 1808 had made the first step. The mass of any atom is an integer multiple of a certain numerical value. That certainly suggests a discrete collection of basic constituents in the nucleus. Moreover, because the electric charge of a nucleus is generally smaller than the atomic number, the constituents cannot all have the same charge. The simplest possibility by far would have to be a single type of positively charged particle and a single neutral particle with practically identical masses. Smart theorists would have figured this out in no time.

Or would they? One thing might have led them astray, for how long I don't know. There was a possibility even simpler than the neutron — a possibility that required no new particle. The nucleus might be understood as a number of protons stuck together with a smaller number of electrons. For example, a carbon nucleus with six protons and six neutrons might have been mistaken for six electrons stuck to twelve protons. In fact the mass of a neutron is close to the combined mass of a

5. Quantum Electrodynamics was plagued by serious mathematical inconsistencies. The rules of calculation led to infinite answers that made no sense. A temporary cure, called *renormalization*, was cooked up in the 1950s. But it wasn't until Kenneth Wilson developed a deeper theory in the early 1970s that the problems were resolved.

proton and an electron. Of course a new type of force would have to be introduced: the ordinary electrostatic force between electron and proton would not have been nearly strong enough to tightly bind the extra electrons to the protons — and with a new force, a new messenger particle. Perhaps in the end they would have decided the neutron was not such a bad idea.

Meanwhile, Einstein had developed his theory of gravity, and curious physicists were exploring its equations. Here again, we don't need to guess. Karl Schwarzschild, even before Einstein had completed his theory, worked out the solution of Einstein's equations that we now call the Schwarzschild black hole. Einstein himself derived the existence of gravitational waves that eventually led to the graviton idea. That most certainly required no experiment or observation. The consequences of the General Theory of Relativity were worked out without appeal to any empirical proof that the theory was correct. Even the modern theory of black holes, which we will encounter in the tenth chapter of this book, only involved the Schwarzschild solution combined with primitive ideas of quantum field theory.

Could theorists have guessed the full structure of the Standard Model? Protons and neutrons, perhaps, but quarks, neutrinos, muons, and all the rest? I don't see any way that these things could have been guessed. But the basic underlying theoretical foundation — Yang Mills theory? Here I think I am on very firm ground. The experiment has been done, and the data are in. In 1953, with no other motivation than generalizing Kaluza's theory of an extra dimension, one of history's greatest theoretical physicists did invent the mathematical theory that today is called non-abelian gauge theory. Remember that Kaluza had added an extra dimension to the three dimensions of space and, in so doing, gave a unified description of gravity and electrodynamics. What Pauli did was to add one more dimension for a total of 5+1. The two extra dimensions he rolled up into a tiny 2-sphere. And what did he find? He found that the extra two dimensions gave rise to a new kind of theory, similar to electrodynamics but with a new twist. Instead of a single photon, the list of particles now had three photonlike particles. And, curiously, each photon carried charge; it could emit either of the other two. This was the first construction of a non-abelian, or Yang Mills, gauge theory.[6] Today we recognize non-abelian gauge theory as the basis for the entire Standard Model. Gluons, photons, Z-particles, and W-particles are simple generalizations of Pauli's three photonlike particles.

As I said, there was little or no chance that theorists would have been able to deduce the Standard Model with its quarks, neutrinos, muons, and Higgs bosons. And even if they had, it most likely would have been one of dozens of ideas. But I do think there is a possibility they could have found the basic theoretical ingredients.

Could they possibly have discovered String Theory? The discovery of String Theory is a good example of how the searching, probing minds of theorists often work. Again with absolutely no experimental basis, string theorists constructed a monumental mathematical edifice. The historical development of String Theory was somewhat accidental. But it easily could have arisen through other kinds of accidents. Stringlike objects play an important role in non-abelian gauge theories. Another plausible possibility is that it might have been developed through hydrodynamics, the theory of fluid flow. Think of the swirling vortex that forms when you let water drain from the sink. The actual center of the vortex forms a long, one-dimensional core that in many ways behaves like a string. Such vortices can form in air: tornadoes are an example. Smoke rings provide a more interesting example, vortex loops that resemble closed strings. Might fluid dynamics experts attempting to construct an idealized theory of vortices have invented String Theory? We will never know, but it doesn't seem out of the question. Would physicists trying to explore the quantum theory of gravity have seized on it when the fluid people found closed strings that behaved like gravitons? I think they would have.

On the other hand, a skeptic could reasonably argue that for every good idea there would have been a hundred irrelevant, wrongheaded directions pursued. With no experiments to guide and discipline theorists, they would have gone off in every imaginable direction, with intel-

6. Chen Ning Yang and Robert Mills independently worked out non-abelian gauge theory one year after Pauli's work. The only reason for not including it in my history is that Yang and Mills were partly motivated by certain empirical facts about nuclei that were not known until long after my cutoff date of January 1, 1900.

lectual chaos ensuing. How would the good ideas ever be distinguished from the bad? Having every possible idea is just as bad as having no ideas.

The skeptics have a good point; they may be right. But it is also possible that good ideas have a kind of Darwinian survival value that bad ideas don't. Good ideas tend to produce more good ideas — bad ones tend to lead nowhere. And mathematical consistency is a very unforgiving criterion. Perhaps it would have provided some of the discipline that would otherwise have come from experiment.

In a century without experiment, would physics have progressed the way I have suggested? Who knows? I don't say it would have — only that it could have. In trying to gauge the limits of human ingenuity, I am certain that we are much more likely to underestimate where the limits lie than to overestimate.

In looking back I realize that in 1995 I was guilty of a very serious lack of imagination in speaking only of the ingenuity of theorists. In trying to console myself and the other physicists at the banquet about the poor prospects for future experimental data, I badly underestimated the ingenuity, imagination, and creativity of experimental physicists. Since that time they have gone on to create the revolutionary explosion of cosmological data that I described in chapter 5. In the last chapter of this book, I will discuss some other exciting experiments that will take place in the near future, but for now let's return to String Theory and how it produces a huge Landscape of possibilities.

CHAPTER TEN

The Branes behind Rube Goldberg's Greatest Machine

We come now to the heart of the matter. The unreasonable apparent design of the universe and the appeal to some form of Anthropic Principle is old stuff. What is really new, the earthquake that has caused enormous consternation and controversy among theoretical physicists and the reason that I wrote this book, is the recognition that the Landscape of String Theory has a stupendous number of diverse valleys. Earlier theories like QED (the theory of photons and electrons) and QCD (the theory of quarks and gluons) that had prevailed throughout the twentieth century had very boring Landscapes. The Standard Model, as complicated as it is, has only a single vacuum. No choices ever have to be made, or ever can be made, about which vacuum we actually live in.

The reason for the paucity of vacuums in older theories is not hard to understand. It was not that quantum field theories with rich Landscapes are mathematically impossible. By adding to the Standard Model a few hundred unobserved fields similar to the Higgs field, a huge Landscape can be generated. The reason that the vacuum of the Standard Model is unique is not any remarkable mathematical elegance of the kind that I explained in chapter 4. It has much more to do with the fact that it was constructed for the particular purpose of describing some limited facts about our own world. They were built piecemeal, from experimental data, with the particular goal of describing (not explaining) our own vacuum. These theories admirably do the job that they were designed to do but no more. With this limited goal in mind, theorists had no reason to add loads of additional structure just to make a Landscape. In fact most physicists (with the exception of farsighted visionaries like Andrei Linde and Alex Vilenkin) throughout the twentieth century would have considered a diverse Landscape to be a blemish rather than an advantage.

Until recently string theorists were blinded by this old paradigm of a theory with a single vacuum. Despite the fact that at least a million different Calabi Yau manifolds could be utilized for compactifying (rolling up and hiding) the extra dimensions implied by String Theory, the leaders of the field continued to hope that some mathematical principle would be discovered that would eliminate all but a single possibility. But with all the effort that was spent on searching for such a vacuum selection principle, nothing ever turned up. They say that "hope springs eternal." But by now most string theorists have realized that, although the theory may be correct, their aspirations were incorrect. The theory itself is demanding to be seen as a theory of diversity, not uniqueness.

What is it about String Theory that makes its Landscape so rich and diverse? The answer involves the enormous complexity of the tiny, rolled-up geometries that hide the extra six or seven dimensions of space. But before we get to this complexity, I want to explain a simpler and more familiar example of similar complexity. In fact this example was the original inspiration for the term *Landscape*.

The term *Landscape* did not originate with string theorists or cosmologists. When I first used it in 2003 to describe the large number of String Theory vacuums, I was borrowing it from a much older field of science: the physics and chemistry of large molecules. The possible configurations of a large molecule, made of hundreds or thousands of atoms, had long been described as landscapes or, sometimes, energy landscapes. The Landscape of String Theory has much less in common with the impoverished landscapes of quantum field theory than with the "configuration space" of large molecules. Let's pursue this point before returning to the exploration of String Theory.

Begin with a single atom. Three numbers are required in order to specify the location of the atom: the coordinates of the atom along the x-, y-, and z-axes. If you don't like x, y, and z, you may use longitude, latitude, and altitude instead. Thus, the possible configurations of a single atom are the points of ordinary three-dimensional space.

The next-simplest system made of atoms is a diatomic molecule — a molecule composed of two atoms. Specifying the position of two atoms requires six coordinates: three for each atom. It would be natural to call the six coordinates x_1, y_1, z_1 and x_2, y_2, z_2, the subscripts 1 and 2 referring to the two atoms. These six numbers describe two points of three-dimensional space, but we can also combine the six coordinates to form an abstract, six-dimensional space. That six-dimensional space is the landscape describing a diatomic molecule.

Now let's jump to a molecule composed of one thousand atoms. For inorganic chemistry this would be a very large molecule, but for an organic biomolecule, it is fairly ordinary. How do we describe all the ways that the one thousand atoms can arrange themselves? This question is not entirely academic: biochemists and biophysicists who want to understand how protein molecules fold and unfold themselves think in terms of a molecular landscape.

Evidently, to specify the configuration of all one thousand atoms, we need to give three thousand numbers, which we can think of as the coordinates of a three-thousand-dimensional landscape: a landscape of possible molecular "designs."

The collection of atoms has potential energy that varies as the atoms are moved around. For example, in the case of the diatomic molecule, if the two atoms are squeezed together, the potential energy becomes large. If the atoms move apart, they will eventually reach a point of minimum energy. Of course it is much more difficult to visualize the energy of one thousand atoms, but the principle is the same: the potential energy of the molecule varies as we move across the landscape. As in chapter 3, if we think of potential energy as altitude, the landscape will have a rich topography with mountains, valleys, ridges, and plains. It shouldn't come as a surprise that the stable configurations of the molecule correspond to the bottoms of valleys.

The striking thing is that the number of these valleys is enormous: it grows exponentially with the number of atoms. For a large molecule the number of isolated valleys is way beyond millions or billions. The landscape of a molecule with one thousand atoms can easily have 10^{100}

valleys. What does all of this have to do with the Landscape of vacuums and String Theory? The answer is that, like a molecule, a compactification of String Theory has a great many "moving parts." Some of those parts we have already met. The compactification moduli were the quantities that determine the sizes and shapes of the various geometric features of Calabi Yau manifolds. In this chapter we are going to explore some additional moving parts and see why the Landscape is so complex and extraordinarily rich.

D-Branes

In chapter 8 I described how Ed Witten's 1995 idea combined the multitude of String Theories into one big M(aster)-theory. But that theory had one serious problem: it needed new objects, objects that String Theory had not previously predicted. The theory would have to work something like this: each one of the String Theories must contain previously unsuspected objects deeply hidden in its mathematics. The fundamental strings of one version were not the same objects as the fundamental strings of another version. But as the moduli varied — as one moved through the Landscape — the new objects of version A would morph into the old objects of version B. One example that we have already seen is how the membranes of M-theory morph into the strings of Type IIa theory. Witten's ideas were attractive — even compelling — but the nature of the new objects and their mathematical place in the theory was a complete mystery. That is, until Joe Polchinski discovered his branes.

Joe Polchinski has the good looks and sunny disposition of "the boy next door." Commenting about food, Joe once said, "There are only two kinds of food — the kind you put chocolate sauce on and the kind you put ketchup on." But the boyish exterior hides one of the deepest and most powerful minds to attack the problems of physics in the last half century. Even before Witten introduced his M-theory, Joe had been experimenting with a new idea in String Theory. More or less as a mathematical game, he postulated that there could be special places in space where strings could terminate. Picture a child holding the end of a jump rope and shaking it to make waves. The waves travel down to the far end of the rope, but what happens next depends on whether the far end is free to flop around or is attached to some anchor. Before Polchinski, open strings always had free ends — the floppy option — but Joe's new idea was that there could be anchors in space that held string ends from flopping. The anchor could be a simple point in space: that would be more or less like a hand rigidly constraining the end from moving. But there are other possibilities. Suppose the end of the rope were attached to a ring that could slide up and down a pole. The end would be partly fixed but partly free to move. Although attached to the pole, the end would be free to slide along a line — the pole itself. What ropes with poles can do, so can strings, or so Polchinski reasoned. Why not have special lines in space to which string ends can attach? Like the rope and pole, the string end would be free to slide along the length of the line. The line might even be bent into a curve. But points and lines don't exhaust the possibilities. The string end could be attached to a surface, a kind of membrane. Free to slide in any direction along the surface, the string end could not escape from the membrane.

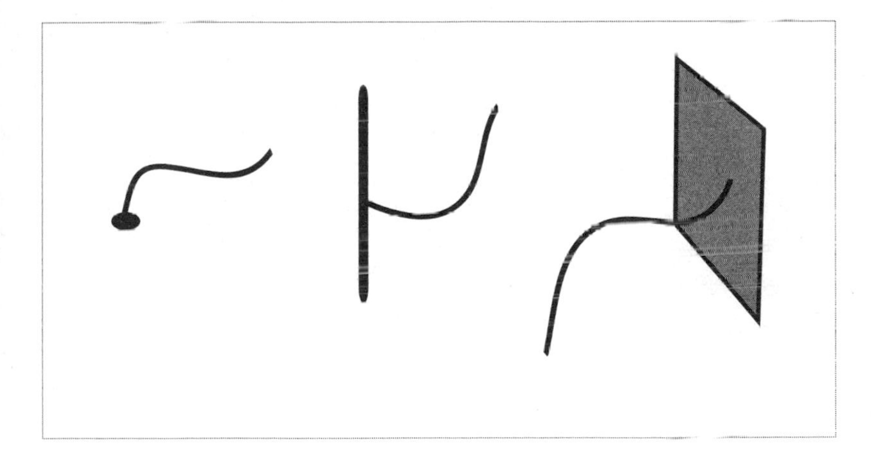

These points, lines, and surfaces where strings could end needed a name. Joe called them Dirichlet-branes or just D-branes. Peter Dirichlet was a nineteenth-century German mathematician who had nothing whatever to do with String Theory. But 150 years earlier he had studied the mathematics of waves and how they reflected off fixed objects. By all rights the new objects should be called Polchinski-branes, but the term *P-branes* was already in use by string theorists for another kind of object.

Joe is a good friend of mine. Over a period of twenty-five years we had worked closely together on a number of physics projects. The first I heard of D-branes was over coffee in Quackenbush's Intergalactic Café and Espresso bar in Austin, Texas. I think the year was 1994. The idea seemed amusing but not the stuff of revolutions. I wasn't alone in underestimating their importance. D-branes were not high on the to-do list of anyone at that time — maybe not even Joe's list. It wasn't until shortly after Witten's 1995 lecture that D-branes exploded into the consciousness of theoretical physicists.

What is the connection with Witten's lecture? A few months later, in November, Joe wrote a paper that has had tremendous repercussions throughout all areas of theoretical physics. The new objects that Witten needed were exactly Joe's D-branes. Armed with D-branes, physicists could now complete Witten's project of replacing several apparently different theories by one single theory with many solutions.

Branes of Every Dimension

What's so special about strings? What is it about one-dimensional filaments of energy that makes string theorists so certain that they are the building blocks of all matter? The more we learn about the theory, the more certain we are becoming that nothing is very special about them. In the previous chapter we encountered the Magical Mystery aMazing eleven-dimensional M-theory. That theory doesn't have strings at all. It has membranes and gravitons but no strings. As we saw, the strings appear only when we compactify M-theory, and even then the strings are just limits of ribbonlike membranes that become truly stringlike only when the compact dimension shrinks to a vanishing size. In other words, String Theory is a theory of strings only in certain limiting regions of the Landscape.

In a world with three space dimensions there are three types of objects that string theorists call branes. The simplest is a point particle. Since a point has no extension in any direction, it is common to think of the point as a *zero-dimensional space*. Life on a point would be very dull; there are no directions to explore. String theorists refer to point particles as 0-branes, the 0 representing dimensionality of the particle. According to the String Theory lingo, a 0-brane, on which strings can terminate, is called a D0-brane.

After the 0-branes come the 1-branes, or strings. A string has extension in only one direction. Living on a string is still very limiting, but at least you would have one dimension in which to move. There are two kinds of 1-branes in String Theory — the original strings and D1-strings: the one-dimensional objects where the ordinary strings can end.

Finally there are 2-branes, or membranes — flexible sheets of matter. Life is infinitely more varied on a 2-brane but still not as interesting as in three-dimensional space. In fact we could call our three-dimensional world a 3-brane, but unlike the 0-, 1-, and 2-branes, we cannot move the 3-brane around in space. It *is* space. But suppose we lived in a world with four space dimensions. The extra direction of space would allow a 3-brane freedom to move. In a world with four space dimensions, it is possible to have 0-, 1-, 2-, and 3-branes.

How about in the 9+1 dimensional world of String Theory? It is possible that branes might exist all the way from 0-branes to 8-branes. This in itself does not mean that a given theory actually has such objects. That depends on the basic constituents of matter and how they can be assembled. But it does mean that there are enough dimensions to contain such branes. The ten space directions of M-theory are enough to contain one more kind of brane: the 9-brane.

Just because ten different kinds of branes can fit into the ten dimensions of space, it doesn't mean that M-theory actually has all of them as possible objects. In fact M-theory does not. It is a theory of gravitons, membranes, and 5-branes. No other branes exist. To explain why would take us far afield into the abstract mathematics of supersymmetric general relativity, but we don't need to go there: it's enough to know that eleven-dimensional supergravity (that's 10+1-dimensional) is a theory of membranes and 5-branes interacting gravitationally by tossing gravitons back and forth.

The ten-dimensional String Theories each have a variety of D-branes. One version — Type IIa String Theory — has even-dimensional branes: D0, D2, D4, D6, and D8. Type IIb theory has the odd-dimensional branes: D1, D3, D5, D7, and D9.

Just as you could attach more than one rope to the same pole, any number of strings can terminate on a D-brane. In fact a single string can have both its ends attached to the same D-brane just like both ends of the jump rope could be attached to the same pole. These segments of string would be free to move along the brane, but they couldn't get off it. They are creatures confined to live out their lives on the D-brane.

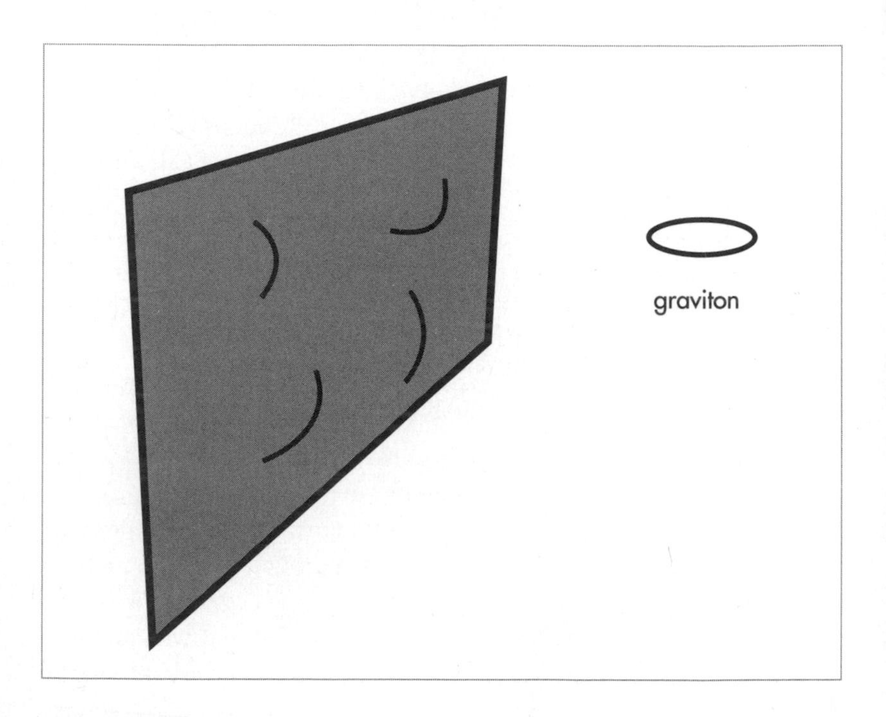

The thing that makes these small segments of string so interesting is that they behave just like elementary particles. Take for example D3-branes. The short strings, with both ends attached to the brane, are free to move throughout the three-dimensional volume of the D-3 brane. They can come together, attach to form a single segment, vibrate, and disconnect. They move and interact just like the particles that String Theory was originally cooked up to explain. But now they live on a brane.

The D-brane is a model of a world with elementary particles behaving much like the real elementary particles. The only thing missing on the D-brane is gravity. That's because the graviton is a closed string — a string with no ends. A string with no ends would not be stuck to the brane at all.

Could the real world (with the exception of the graviton) of electrons, photons, and all the other elementary particles — as well as atoms, molecules, people, stars, and galaxies — all take place on a brane? To the majority of theorists working on these problems, it seems the most likely possibility.

Branes and Compactification

All kinds of things can be done with branes. Take a D2-brane — a membrane — and curve it into a 2-sphere. You've made a balloon. The trouble is that the tension of the membrane makes it quickly collapse like a punctured balloon. You could shape the D2-brane to form the surface of a torus, but this, too, would collapse.

But now imagine a brane that is stretched from one end of the universe to the other. The simplest example to visualize is an infinite D1-brane stretched right across the universe like an infinite cable. An infinite D-brane has no way to shrink and collapse. You can imagine that two cosmic giants hold its ends in place, but since the D-brane is infinite, the giants are infinitely far away

There is no need to stop at D1-branes: an infinite sheet stretched across the universe is also stable. This time we would need many giants to hold the edges in place, but again, they would be infinitely far off. The infinite membrane would be a world with elementary particles that might resemble a "flatland" version of our own universe. You might think the creatures on the membrane would have no way of telling that more dimensions exist, but that would not be quite right. The giveaway would be the properties of the gravitational force. Remember that gravity is caused by gravitons jumping between objects. But gravitons are closed strings without ends. They have no reason to stick to the brane. Instead, they freely travel through all of space. They can still be exchanged between objects on the brane but only by traveling out into the extra dimensions, then back to the brane. Gravity would be like a science-fiction "message" telling the flatland creatures that there are more dimensions out there and that they are imprisoned on a two-dimensional surface.

The "unobserved" dimensions of gravity would in fact be easy to detect. When objects collide they can radiate gravitons, just as when electrons collide, they radiate photons. But typically the radiated gravitons will fly off into space and never return to the brane. Energy would be lost from the brane in this way. The flatland creatures would discover that energy doesn't get converted to heat, potential energy, or chemical energy: it just disappears.

Now imagine that space has more dimensions than the usual three. Infinite D-3 branes could be stretched through space in the same way, and on a 3-brane all the usual things of our world could exist — except that gravity would be all wrong. The gravitational force law would reflect the fact that the graviton moves through more dimensions. Gravity would be "diluted" by spreading out in the extra dimensions. The result would be calamitous. Gravity would be much weaker, and galaxies, stars, and planets would be poorly held together. In fact gravity would be too weak to hold us to the earth even if the earth were somehow kept together.

Let's take the extra dimensions — the ones that we can't explore but the graviton can — and roll them up into a microscopically small compact space. The three dimensions of ordinary experience form an infinite room, but the other directions have walls, ceilings, and floors. The points on opposite walls or on ceiling and floor are matched just as I described in chapter 8.

To help visualize, let's return to the example in which we compactified three-dimensional space by rolling up one direction. Beginning with an infinite room, each point of the ceiling was identified with the point on the floor directly beneath it. But now the floor has a carpet that stretches to infinity in infinite directions. The carpet is a D-brane. Imagine the carpet-brane slowly moving through the vertical dimension. It slowly rises from the floor like a magic carpet in the Arabian Nights. It continues to levitate and rise until it just touches the ceiling. And abracadabra — zap! The carpet instantly reappears at the floor.

The graviton is still not attached to the carpet-brane, but now it can't get very far away. There is very little room for it to move in the extra dimension. And if the extra dimension is microscopically small, then it is hard to tell if the graviton is off the brane. The result: gravity is almost exactly as it would be if, like everything else, the graviton moved on the brane. And of course there is nothing new if we replace the membrane with a D3-brane in a higher dimensional space. A D3-brane in the

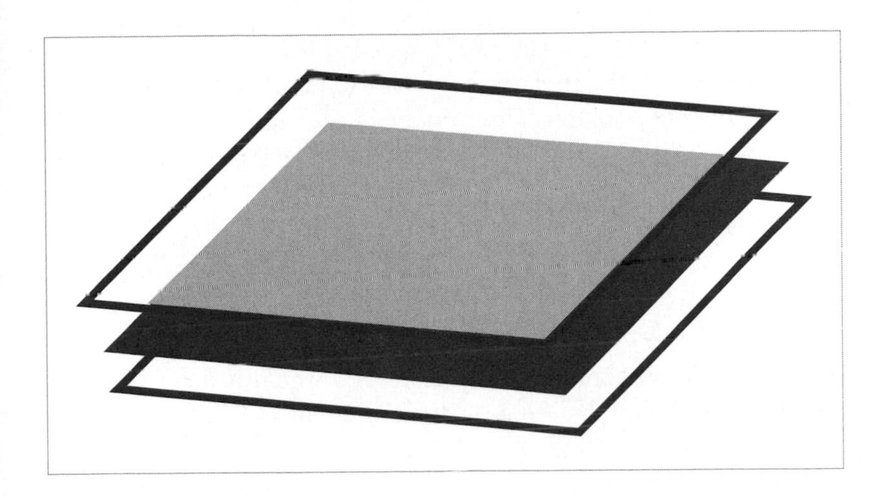

nine-dimensional space of String Theory would be very similar to our world if the extra six dimensions were tightly rolled up.

Most string theorists think we really do live on a brane-world, floating in a space with six extra dimensions. And perhaps there are other branes floating nearby, microscopically separated from us but invisible (to us) because our photons stick to our own brane, and theirs stick to their brane. Though invisible, these other branes would not be impossible to detect: gravity, formed of closed strings, would bridge the gap. But isn't that exactly what dark matter is: invisible matter whose gravitational pull is felt by our own stars and galaxies? Polchinski's D-branes open up all sorts of new directions. From our point of view, a universe with many brane-worlds living peacefully side by side is just one more possibility that can be found in the Landscape. Calabi Yau spaces of incredible complexity, hundreds of moduli, brane-worlds, fluxes (yet to come): the universe is starting to look like a world that only Rube Goldberg's mother could love. To paraphrase the famous experimental physicist I. I. Rabi, "Who ordered all that stuff?"[1]

But by no means have we exhausted all the gimmicks and gadgets with which Rube can play. Here's another: in addition to floating in the

1. Rabi commented about the newly discovered muon (a particle similar to the electron but two hundred times heavier), "Who ordered that?" No doubt he was referring to the seeming arbitrariness of the elementary particles.

compact space, branes can also be wrapped around the compact directions. The simplest example is to go back to the infinite cylinder and wind a D1-brane around it. This would look the same as winding an ordinary string around the cylinder, except the string is replaced by a D1-brane. This object, from a distance, would look like a point particle on a one-dimensional line. On the other hand, suppose the compact space were an ordinary 2-sphere. You could try to wrap a string or D1-brane around the equator of the sphere like a belt around the middle of a fat man. But the belt could slip off the spherical fat man. A string or D1-brane wrapped on a sphere is not stable — it would not stay there for long. In the words of the physicist Sidney Coleman, "You can't lasso a basketball."

What about the torus — the surface of a bagel? Can a D1-brane be wrapped on the torus in a stable way? Yes, and in more ways than one. There are two ways to "belt the bagel." One way is to run the belt through the hole. Try it. Take a bagel or donut and run a string through the hole. Wrap it around and tie it. The string can't come off. Can you see the other way to belt the torus?

The deciding factor is the "topology" of the torus. Topology is the mathematical subject that distinguishes spheres from tori (plural of *torus*) and more complicated spaces. An interesting extension of the torus is a surface with two holes in it. Take a lump of clay and mold it into a ball. The surface is a sphere. Now poke a hole through it so that it resembles a donut: the surface is a torus. Next, poke a second hole. The surface is a two-holed generalization of a torus. There are more ways that you could wind a D1-brane on the two-hole torus than on the one-hole torus. A mathematician would call the sphere a *zero genus surface*, the torus a *genus one surface*, and the two-hole torus a *genus two surface*. Obviously you can poke any number of holes to make surfaces of any genus. The higher the genus, the more ways there are to wrap branes.

Having nine space dimensions, String Theory has six extra dimensions to hide by compactification. Six-dimensional spaces are vastly more complicated than two-dimensional spaces. Not only can you wrap D-1 branes but also there are higher dimensional versions of donut holes that allow you to wrap D2-, D3-, D4-, D5-, and D6-branes in hundreds of ways.

So far we have mainly thought about branes one at a time. But in fact you can have stacks of them. Think of the carpet in an infinite room. But why not have two carpets, one lying on top of the other? In fact it is possible to stack them up like stacks of carpets in a Persian bazaar. Just as the carpets could float freely of one another, a stack of D-branes can separate into several freely floating branes. But the D-branes are a bit like sticky carpets. If you bring them together, they will stick, forming a compound brane. This gives Rube Goldberg more options in designing his machine. He can place several carpet stacks at different heights in the room. He has new flexibility to make worlds with all sorts of properties. In fact with five carpets, stuck together in a stack of two and a stack of three, he can make a world with Laws of Physics that have many similarities to the Standard Model!

The locations of branes in the compact space are new variables to add to the moduli when counting the possibilities for creating a universe. From a distance, when the compact directions are microscopic — too small to see — the brane positions just appear to be additional scalar fields that define the Landscape.

Fluxes

Fluxes have emerged as one of the most important ingredients in the Landscape. They, more than anything else, make the Landscape prodigiously large. Fluxes are a bit more abstract, and harder to visualize, than branes. They are interesting new ingredients, but the bottom line is simple. From a distance they just look like even more scalar fields. The most familiar examples of fluxes are the electric and magnetic fields of Faraday and Maxwell. Faraday was not a mathematician, but he had a powerful ability to visualize. He must almost have been able to see the electromagnetic fields in his experimental apparatuses. His picture of the field of a magnet was *lines of force* emanating out of the North Pole and flowing back into the South Pole. At every point in space, the lines of force specify the direction of the magnetic field, while the density of the lines (how close together they are) specifies the field's intensity.

Faraday pictured the electric field in the same way — lines flowing out of positive charges and into negative charges. Picture an imaginary sphere surrounding an isolated charged object with lines of electric force flowing out and receding off to infinity. The lines of force must pass through the sphere. These imaginary lines passing through the sphere are an example of the electric flux through a surface.

There is a measure of the total amount of flux passing through a surface. Faraday pictured it as the number of lines of force passing through the surface. Had he known calculus, he might have described it as a surface integral of the electric field. The idea of the *number of lines* was an even better one than Faraday knew. The flux through a surface happens to be one of those things that modern quantum mechanics tells us is quantized. Like photons, the unit of flux cannot be subdivided into fractions. Indeed, the flux cannot vary continuously but must be thought of in terms of discrete lines, so that the flux through any surface is an integer.

Ordinary electric and magnetic fields point along directions of three-dimensional space, but it is also possible to think of fluxes that point along the six compact directions of space. In a six-dimensional space the mathematics of fluxes is more complicated, but you can still think of lines or surfaces of force, winding their way over a Calabi Yau space, and passing through its donut holes.

To go more deeply into flux on a Calabi Yau space would require a good deal of modern geometry and topology. But the important conclusions are not so hard. As in the case of magnetic fields, the flux through the various donut holes is quantized. It is always an integer multiple of some basic flux unit. This means that to specify the flux completely all you need to specify is a number of integers — how many units of flux there are through each hole in the space.

How many integers are needed to describe the flux on a Calabi Yau space? The answer depends on the number of holes the surface has. Calabi Yau surfaces are far more complicated than a simple torus and typically have several hundred holes. Thus, hundreds of flux integers are part of the description of a point on the Landscape!

Conifold Singularities

Thus far a typical setup can involve a few hundred moduli to fix the size and shape of the compact space, some branes located at various positions on the space, and now an additional few hundred flux integers. What more can we provide for Rube?

There are many more things to play with, but to keep this book of manageable size I will explain only one more — the conifold singularity. A soccer ball is a sphere. If you ignore the texture and seams on the surface, it is smooth. An American football, by contrast, is smooth everywhere except at the ends, where it comes to points. An infinitely sharp point somewhere on a smooth surface is an example of a *singularity*. In the case of the football, the singularities are called conical singularities. The pointy shape of the ends is like the tip of a cone.

Singularities in higher dimensional spaces — places where the space is not smooth — are more complicated. They have more complex topology. The conifold is one such singularity that can exist on a Calabi Yau space. Although complicated, as its name suggests, it is similar to the tip of a cone. For our purposes we can think of the conifold as a pointy conical place in the geometry.

An interesting thing happens when you combine conifolds and fluxes on the same Calabi Yau space. The flux exerts a force on the tip of the cone, stretching it out into a long, narrow neck like the snout of an anteater. In fact you can have more than one conifold singularity so that the space becomes spiky with pointy tips sticking out like some six-dimensional sea urchin.

Now Rube has the parts. What kind of screwball machine can he build? The possibilities are enormous, but I'll describe one machine called the KKLT construction, named for the first initials in the last names of its founders.[2] KKL and T began with a Calabi Yau space. There are millions to choose from. Just take your pick. Somewhere on the space there is a snouty conifold singularity. Next KKLT filled the various holes with fluxes: an integer for each hole. All of this meant specifying about five hundred parameters — moduli and fluxes. The result is a valley on the Landscape, but not like any we have talked about yet. This point is the Death Valley of the Landscape — not because it is hot — but because it is below sea level. The altitude is negative. This of

2. KKLT stands for Kachru Kallosh Linde and Trivedi. Shamit Kachru, Renata Kallosh, and Andrei Linde are professors at Stanford University. Sandip Trivedi is a professor at the Tata Institute, in India.

course means that the vacuum energy, and therefore, the cosmological constant, is negative — the wrong sign for our universe. Instead of giving rise to a universal repulsion, it would cause a universal cosmic attraction. Instead of accelerating the expansion of the universe, it would hasten the tendency to collapse.

But KKLT had one more Rube Goldberg trick. They added an antibrane — an anti-carpet-brane. D-branes are like particles. Just as every particle has its antiparticle, every brane has its antibrane. Like ordinary particles, if a brane and an antibrane come together, they can annihilate in an explosion of energy. But KKLT put only antibranes into their construction.

As it turns out the antibrane experiences a force that pulls it toward the tip of the conifold singularity. That is the only possible location for the antibrane. The mass of the extra antibrane adds just enough energy to make the altitude positive. Thus, by a mix of a little bit of everything, KKLT discovered a point on the landscape, a valley really, with a small positive cosmological constant — the first of its kind.

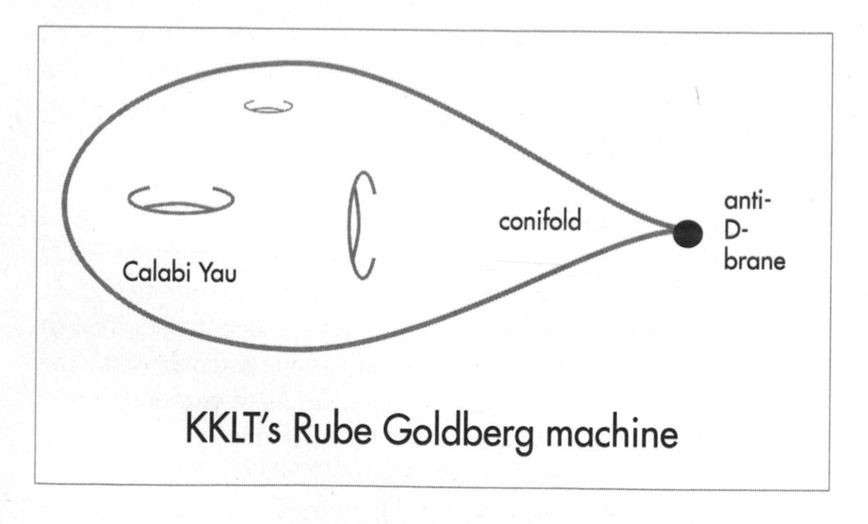

KKLT's Rube Goldberg machine

The importance of the valley that KKLT discovered is not that it closely resembles our own valley. It has no Standard Model of particle physics, and in the original form, it didn't have the ingredients to describe Inflation. Its significance is that it is the first successful attempt to depart from the supersymmetric plain and find a valley "above sea level." It was proof of a principle — that the String Theory Landscape has valleys with a small positive cosmological constant.

The KKLT machine does have a feel of Rube Goldberg complexity, but it has one feature that Rube never would have permitted. It has one part that serves two purposes. The antibrane not only raises the energy and makes the cosmological constant positive, but it also does another important job. Our world, the world that we live in, is not supersymmetric. There is no massless fermion partner of the photon, no bosonic identical twin for the electron. Before putting the antibrane down the conifold throat, the KKLT construction was still supersymmetric. But the antibrane warps the funhouse mirror so that supersymmetry is broken. That's very un-Goldberg-like — to use one part to do two jobs.

The KKLT point on the Landscape is not our world. But it may not be so hard to build in the Standard Model by including a few more branes. Somewhere out of the way of the antibrane, five additional D-branes could provide the extra ingredients.

The "Discretuum" of Bousso and Polchinski

What KKLT found was not a single valley but rather a vast collection of valleys. As I mentioned at the start of chapter 7, Polchinski and Raphael Bousso, then a postdoc at Stanford, had already explained the basic idea in a largely ignored paper. To understand how compactification can lead to tremendous numbers of vacuums, Bousso and Polchinski had concentrated on a single Calabi Yau geometry and asked how many ways there are to fill hundreds of donut holes in the space with fluxes.

Let's suppose that the Calabi Yau manifold has a topology that is rich enough to allow five hundred distinct donut holes through which the fluxes wind. The flux through each hole must be an integer, so a string of five hundred integers has to be specified.

Theoretically there are no limits on the size of the integers, but in practice we would not want to put too much flux through any hole. The effect of a very large flux would be to stretch out the size of the manifold to proportions that might be dangerous. So let's make some limits. Suppose that the flux integers are constrained to be no bigger than nine. Then each of the fluxes is an integer between zero and nine. How many possibilities does that add up to?

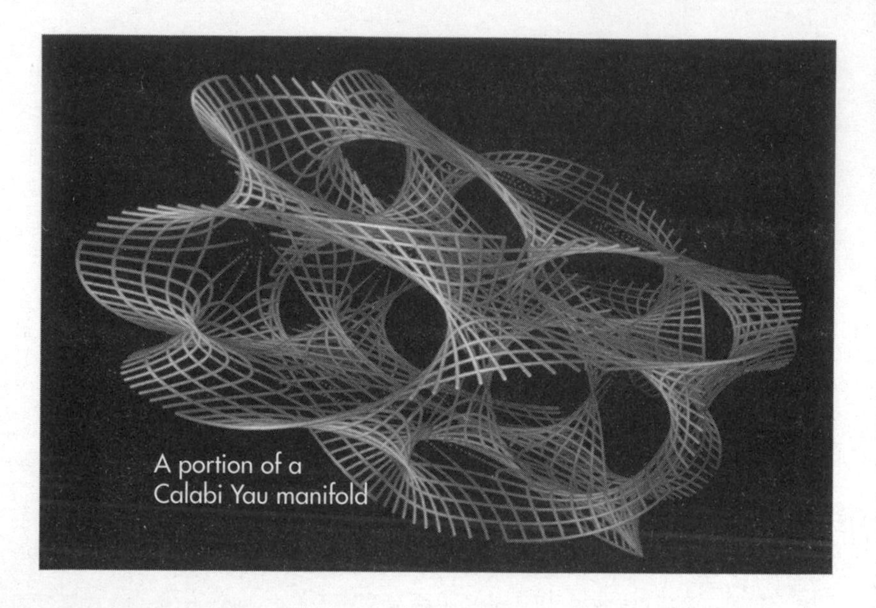
A portion of a Calabi Yau manifold

Let's begin with an easier example. Suppose there is only one hole to deal with instead of five hundred. If the flux through the hole can be any integer between zero and nine, then there are ten possibilities — zero, one, two, three, four, five, six, seven, eight, and nine. The point is that each of these possibilities defines a potential vacuum, an environment with its own laws and, most important, its own vacuum energy. Although ten vacuums are a lot from the perspective of ordinary twentieth-century quantum field theory, it is hardly a promising number for overcoming the incredible improbability of 119 vanishing decimal entries. But let's continue.

Suppose there are two holes, each of which can have flux between zero and nine. Then the number of possible configurations is 10^2, or one hundred. This is slightly better but still far too modest. But note that each time a hole is added, the number of possibilities grows ten times larger. Six holes give a million possibilities; twelve holes give one trillion. With five hundred holes we get the stupendously large number of configurations 10^{500}. Moreover, each valley in this gargantuan list has some vacuum energy, and no two of them are likely to have exactly the same value.

Let's make a chart to show all the possible values of the cosmological constant. Take a sheet of paper and draw a horizontal axis. Halfway along the line mark a point and call it zero. At the right mark a point and call it one. The value one stands for the benchmark value of the vacuum energy — one Unit. Now start marking all the points that correspond to the vacuum energies of the 10^{500} valleys. With a very sharp pencil you may be able to make one thousand random dots before they begin to run into one another and make a continuous line.

To do better get a bigger sheet. With a sheet as big as the Empire State Building, you may get a million randomly placed dots before they start to touch. With a page the size of the galaxy, perhaps 10^{24} points can be shown. None of these numbers is anywhere near 10^{500}. Even if you spaced the points a Planck length apart and made the sheet as big as the known universe, you would get only a measly 10^{60} points. The number 10^{500} is so staggeringly large that I can't think of any way of graphically representing that many points.

The word for *all* possible numbers in a given range is *continuum*. The points on our vacuum energy plot do not really form a continuum, but they are so dense that for practical purposes every number is represented. To describe such a terrifically large and dense set of values, string theorists like Bousso and Polchinski coined the word *discretuum* — discrete but almost a continuum.

But the real point is that with that many randomly chosen values for the cosmological constant, there will be a huge number in the tiny "window of life" that Weinberg calculated. No fine-tuning is needed to make sure of that. Of course it will be only a minute fraction of the valleys that are in the anthropic window of opportunity — roughly one out of 10^{120}.

The growth of the Landscape over the years since String Theory was discovered has been a source of anguish for most string theorists. During the happy early days, when the impoverished Landscape was composed of only a single point or, at most, a number that could be counted on the fingers of one hand, string theorists were overjoyed when they found that the few known theories were really just different solutions of a single theory. But as this consolidation was taking place, another more ominous trend was developing that horrified many string theorists. The number of distinct solutions was expanding into an unimaginably large Landscape. But I suspect that, in time, these same string theorists will begin to see the Landscape as the single most important and convincing feature of their theory. One might ask, "Haven't we just substituted one impossible problem for another? No longer do we have to wonder why the cosmological constant is so fine-tuned. Perhaps it's true that the Landscape is so prodigious that you can find whatever you are looking for. But what principle of physics picks out our benign valley from 10^{500} others?" The answer that we will come to in the next chapter is that nothing does. As we will see, the question is the wrong one.

CHAPTER ELEVEN

A Bubble Bath Universe

It's one thing to argue that theory gives rise to many possibilities for the Laws of Physics, but it's quite another to say that nature actually takes advantage of all the possibilities. Which of the many possible environments materialized as real worlds? The equations of physics undoubtedly have solutions describing giant spherical shells of stainless steel orbiting around massive "stars" made of pure gold. In a theoretician's sense such solutions of the equations *exist*. But are there any such objects in the universe? Probably not, and the reasons are historical. Nothing about the way the universe evolved — nothing about Big Bang cosmology — could explain how such objects could ever have formed. Mathematical existence is not the same as physical existence obviously. Discovering that String Theory has 10^{500} solutions explains nothing about our world unless we also understand how the corresponding environments came into being.

Some physicists believe that there must be a *vacuum selection principle* that singles out a unique point in the Landscape — presumably our point. Such a principle, if it exists, might be mathematical — perhaps a proof that only one of the many apparent solutions of String Theory is really consistent. But, if anything, the mathematics of String Theory has gone the other way — toward greater and greater nonuniqueness. I have heard some say that the vacuum selection principle must be cosmological; the birth of the universe could happen only in a unique way leading to an equally unique environment. But the vacuum selection principle is a lot like the Loch Ness monster: it's often claimed to exist, but no one has ever seen it. Consequently, many of us are beginning to suspect that it doesn't exist at all. Even if such a mechanism does exist, the chances that the resulting Laws of Physics would be fine-tuned to the incredible precision needed for our existence would still be negligible. My own feeling is that a true vacuum selection principle would most likely be a disaster.

What is the alternative? The answer is that nature somehow makes use of all the possibilities. Is there a natural mechanism that would have populated a megaverse with all possible environments, turning them from mathematical possibilities to physical realities? This is what an increasing number of theoretical physicists believe — myself included. I call this view the *populated Landscape.*[1]

In this chapter I will explain the main idea of the populated Landscape viewpoint: that mechanisms that rely on well-tested physical principles gave rise to a huge, or even an infinite, number of pocket universes, with each and every valley in the Landscape being represented.

The mechanisms that underlie the populated Landscape rely only on the principles of general relativity and very conventional applications of quantum mechanics. To understand how the Landscape becomes populated, we will need to examine two very basic physics concepts. The first is *metastability of the vacuum.* It refers to the fact that the properties of the vacuum can suddenly change with little or no warning. The second concept is that *space clones itself.*

Stability and Metastability

In Kurt Vonnegut's darkly hilarious science-fiction satire *Cat's Cradle,* physicist Felix Hoenikker discovers a new form of solid water called ice-nine. The crystalline structure of ice-nine is somewhat different from usual ice — a new way of stacking up the cannon balls, so to speak — and as a result the new crystal lattice is so stable that it doesn't melt until the temperature reaches 114 degrees Fahrenheit. In Vonnegut's fantasy the reason that the waters of the earth previously remained liquid is that a tiny seed of the new crystal would be needed to "teach" the water molecules how to reassemble themselves into the more stable ice-nine

1. The populated Landscape is entirely familiar to cosmologists such as Linde, Vilenkin, and Guth, who have embraced it or something like it for many years.

lattice. Once such a tiny "teacher crystal" was introduced, the surrounding water would coagulate around it, forming a rapidly expanding bubble of ice-nine. Until Hoenikker's playful little experiment, no one had ever made a crystal of ice-nine, so the earth's H_2O was uncorrupted by the more deadly cousin of ordinary ice.

Uncorrupted, that is, until a chip of Hoenikker's new stuff gets into the hands of "Papa" Monzano, president and dictator of San Lorenzo. Papa ends his own life by swallowing a bit of it, thereby destabilizing his own fluids. They turn into the lethal ice-nine, and his whole body freezes in a fraction of a second. When Papa's fortress collapses and falls into the sea, his ice-nine-laden corpse follows and starts a chain reaction. A crystal expands at breakneck speed and immediately freezes every bit of water on earth, thus ending all life.

Ice-nine is of course a fiction. There is no form of water that is solid above 32 degrees Fahrenheit. *Cat's Cradle* is really a cautionary tale about the madness and instability of a world full of nuclear weapons. But, although fictional, the ice-nine story is based on serious principles of physics and chemistry: in particular, the concept of metastability.

Stability implies a degree of resistance to sudden, unpredictable change. A pendulum hanging in the vertical position is very stable. Instability is the opposite: a pencil standing on its tip will fall in an unpredictable direction. Metastability is something in between. Some systems have the remarkable property that they appear stable for long periods of time but eventually undergo very sudden and unforeseen catastrophic changes. These systems are called metastable.

In the real world a closed tank of liquid water at room temperature is stable. But in the fictitious world of Felix Hoenikker and Papa Monzano, it is only metastable. Real water can also be metastable, but not at room temperature. Surprisingly, if it is carefully cooled below freezing or heated above boiling, water can remain liquid for a long time until it suddenly changes to ice or steam. Even odder, the vacuums of String Theory are often metastable. But before plunging into metastable water or even empty space, I want to explain a simpler example of metastability.

Some things are flat out impossible. No matter how long you wait, they will never happen. Others are just very unlikely, but if you wait long enough, they will eventually occur. Here is something that according to *classical* physics is impossible. Once again let's imagine a small ball rolling on a simple, one-dimensional landscape. In fact it's not rolling; it is trapped at the bottom of a valley between two high mountains. There's another lower valley on the other side of one of the mountains, but the ball is stuck where it is. In order to roll over the mountain and get to the lower valley, it would have to have enough kinetic energy to compensate for the extra potential energy at the summit. Standing still, it lacks the energy to climb even a short way up the hill. Getting to the other side without a push is not only unlikely, it's just plain impossible. This is an example of perfect stability.

But now let's add a bit of heat. The ball might be exposed to some random collisions with the molecules of a warm gas. It has the thermal jitters. If we wait long enough, at some point an unusually energetic molecule, or random succession of collisions, will give it enough of a kick to go over to the other side and down to the lower valley. The probability that such a random accident will occur in the next hour may be exceedingly small. But no matter how small, as long as the probability is not zero, given enough time the ball will eventually go over the barrier and land in the lower valley.

But wait! We've neglected the quantum jitters. Even without any heat — even at a temperature of absolute zero — the ball fluctuates because of the quantum jitters. You may suspect that even in the absence of thermal energy, quantum fluctuations would eventually kick the ball over the hill. If so, you're right. A quantum-mechanical ball trapped in an energy valley is not completely stable; it has a small probability of appearing on the other side of the mountain. Physicists call this weird, unpredictable quantum jump *quantum tunneling*. Typically, quantum tunneling is a very unlikely event that can take as long as the proverbial room full of monkeys who are randomly banging away at typewriters would take to write a Shakespeare play.

Systems of this type, which are not truly stable but can last for an extremely long time, are metastable. There are many examples of metastability in physics and chemistry: systems that seem stable but eventually tunnel to new configurations without warning. In Vonnegut's satire ordinary water at room temperature is metastable. Sooner or later a tiny crystal of ice-nine will form, even if only by the random motion of molecules, and then a chain reaction will rearrange the metastable liquid water into the more stable ice-nine. As we will soon see, real examples involving nothing more than ordinary ice and water also exist. But most important for this book, vacuums can be metastable. Bubbles of space with strangely different properties can spontaneously appear and grow, much like ice-nine did in *Cat's Cradle*. This is how the Landscape becomes populated and the universe becomes diverse.

A Real Ice Catastrophe

Water freezes solid at a temperature of 32 degrees Fahrenheit or, what is the same thing, zero degrees on the centigrade scale. However, you can cool very pure water to a lower temperature without its becoming solid if you do it slowly and very carefully. Liquid water below freezing temperature is called *supercooled*.

Supercooled water, just below the usual freezing temperature, can last for a very long time. But introducing a small bit of ordinary ice will cause the water suddenly to crystallize around it and form a rapidly growing chunk of ice. Just as ice-nine destroyed the world, the real ice chunk will quickly take over the entire body of water.

Putting the ice crystal into the supercooled water is very similar to giving the rolling ball a shove over the nearby hill. It's the event that pushes the system "over the edge." In the case of the rolling ball, the shove has to be strong enough to push it over the barrier. A tiny push won't do it. The ball will just roll back to the starting position. The same is true for the supercooled water. If the ice crystal is smaller than a certain critical size, it will just melt back into the surrounding liquid. For example, an ice crystal just a few molecules in diameter will not grow and take over.

But even without someone adding a chip of ice, the supercooled water will not last forever. The reason is that the molecules in the liquid are constantly fluctuating, bouncing off one another and rearranging themselves. This motion is due to both the thermal jitters and the quantum jitters. Every once in a while, by accident, a group of molecules will arrange themselves into a small crystal. Most of the time the crystal will be so small that it will quickly melt into its surroundings.

Very rarely, however, a larger crystal will spontaneously form by random accident. Then the crystal will explosively grow, and everything will freeze. The phenomenon is called bubble nucleation, the growing ice crystal being thought of as an expanding bubble. A very similar thing will happen to water that is superheated above the boiling point. The only difference is that a bubble of steam will spontaneously nucleate and grow.

The boundary between the solid ice and the liquid water (or the steam and water) is called a *domain wall.* It is like a membrane between the two different phases.[2] In fact the domain wall has properties of its own, for example, surface tension that tries to shrink the bubble. Another example of a domain wall is the boundary between ordinary water and air. As a child I was fascinated by the trick of floating a steel sewing needle on the surface of a cup of water. The domain boundary separating air and water is like a skin stretched over the liquid. It has surface tension, which actually has to be pierced in order for an object to penetrate it.

A vacuum with a positive cosmological constant is a lot like a supercooled or superheated liquid. It is metastable and can decay by nucleating bubbles. Every vacuum corresponds to a valley in the Landscape with a particular altitude or energy density. However, although the vacuum may seem quiet and featureless to our coarse senses, quantum fluctuations continually create tiny bubbles of space whose properties correspond to neighboring valleys. Usually the bubbles quickly shrink and disappear. But if the neighboring valley has lower altitude, then every so often a bubble will appear that is large enough to start growing. Will it take over everything? We will see shortly.

The domain walls that separate a bubble from its environment are two-dimensional surfaces that resemble membranes. These are not the first membranes we have encountered. In chapter 10 we learned about Polchinski's D-branes. In many cases the domain walls are nothing but the membranelike D2-branes.

Cloning Space

One thing is missing in the analogy between the cosmic bubbling of pocket universes and the bubbling of ice crystals in supercooled fluids:

2. Physicists and chemists refer to ice, liquid water, and steam as three different phases of water: the solid, liquid, and gaseous phases.

namely, the tendency for space to expand. Each point on the Landscape has a cosmological constant. Recall that a positive cosmological constant means a universal repulsion, which causes matter to separate. A modern general relativist would say that space itself is expanding, or inflating, and that matter is simply being carried along for the ride.

Long ago, when Einstein was still experimenting with the cosmological constant, the Dutch astronomer Willem de Sitter began the study of inflating space. The space or, more accurately, space-time that de Sitter discovered, and which carries his name, is the solution to Einstein's equations when there is no energy or gravitating matter other than the ubiquitous vacuum energy of empty space, i.e., a cosmological constant. Like Einstein, de Sitter assumed the cosmological constant was positive. What he found was an inflating space that grows exponentially with time. Exponential expansion means that in a certain time interval space doubles; then, in the next time interval, it doubles again; and then again. It grows to two times — four times — eight times — sixteen times its original size, in the same way that compound interest expands money. At a rate of 5 percent interest, your capital would double in fourteen years. The cosmological constant is like the interest rate: the bigger the cosmological constant, the faster the space inflates. Like any expanding space, de Sitter space satisfies Hubble's Law — velocity proportional to distance.

We have used the analogy of an expanding rubber balloon to visualize a growing universe. But de Sitter space is different from the rubber of an exponentially expanding balloon in an important way. In the case of a balloon, the rubber — the fabric of the balloon — becomes increasingly stretched, stressed, and thinned out by the expansion. Eventually it reaches its limits, and the balloon pops. But the fabric of de Sitter space never changes. It is as though the rubber molecules were continually giving birth to new rubber molecules in order to fill the spaces created by the expansion. Think of the rubber molecules as cloning themselves to fill in the gaps.

Of course no real rubber molecules are being continuously created. Space itself is reproducing to fill the gaps. One might say that space is cloning itself — each small volume giving birth to offspring, thereby growing exponentially.

Suppose an observer in de Sitter space, moving with the general expansion, looked around at her surroundings: what would she see? You might expect that she would see the universe changing with time, getting bigger and bigger. Surprisingly, that's not the case. All around her she would see space flowing away according to Hubble's Law: the close things moving slowly, the distant things moving faster. At some distance the fluid of space would be rushing away so rapidly that the recessional velocity would become equal to the speed of light. At even farther distances the outgoing points would recede with an even greater velocity! Space in those regions would be flowing away so fast that even light signals, emitted straight toward the observer, would be swept away. Because no signal can travel faster than light, contact with these distant regions is completely cut off. The farthest points that can be observed, i.e., the point where the recessional velocity is the speed of light, is called the *horizon*, or more properly, the *event horizon*.

The concept of a cosmic event horizon — an ultimate barrier to our observations or a point of no return — is one of the most fascinating consequences of an accelerating universe. Like the horizon of the earth, it is by no means an end of space. It is merely the end to what we can see. When an object crosses the horizon, it says good-bye forever. Some objects may even have initially formed beyond the horizon. The observer can never have any knowledge of them. But if such objects are permanently beyond the limits of our knowledge, do they matter at all? Is there any reason to include the regions outside the horizon in a scientific theory? Some philosophers would argue that they are metaphysical constructions that have no more business in a scientific theory than the concepts of heaven, hell, and purgatory. Their existence is a sign that the theory has unverifiable and, therefore, unscientific elements to it — or so they say.

The trouble with that view is it does not permit us to appeal to a vast and diverse megaverse of pocket universes, an idea which does have explanatory power: most importantly here, the power to explain the anthropic fine-tuning of our region of space. We will see shortly that all the other pockets are in the mysterious ghostly portions of space out beyond our horizon. Without the idea of a megaverse of pockets, there is no natural way to formulate a sensible Anthropic Principle. My own view of this dilemma will be explained in the next chapter, but I will state it briefly here. I believe this whole discussion is based on a fallacy. In a universe governed by quantum mechanics, the apparent ultimate barriers are not so ultimate. In principle, objects behind horizons are quite within our grasp, but only in principle. More on this in the next chapter.

Curiously, in a universe accelerating under the influence of a cosmological constant, the distance to the event horizon never changes. It is fixed by the value of the cosmological constant — the larger the cosmological constant, the smaller the distance to the horizon. The observer lives in an unchanging world of finite radius bounded by her horizon, but in exactly the way that the earth's horizon eludes anyone who tries to approach it, the cosmic horizon of de Sitter space can never be reached. It's always a finite distance away, but when you approach it, there's nothing there! However, if we could get outside the de Sitter space — watch it from a distance, so to speak — we would see the whole space exponentially growing with time.

Metastable de Sitter Space

I want to return to the topic of metastable substances, but with a new twist — suppose the substance in question is inflating. To help visualize the expanding metastable substance, imagine an infinite shallow lake of supercooled water. To simulate the cloning of space, the bottom of the lake could be filled with small feeder pipes that continuously provide new supercooled water. In order to make room for the new fluid, the water spreads out horizontally — any two molecules getting farther apart because new molecules come to fill in the growing space between them. If boats were floating on the lake, they would separate and lose contact. The lake inflates just like de Sitter space.

In that inflating body of supercooled water, crystals of ice will randomly nucleate from time to time. If they are large enough, they will grow and become expanding ice islands. But because they are being carried along with the spreading fluid, the growing islands may separate so fast that they never meet one another. The regions between the islands inflate and prevent the entire lake from becoming solid ice. The space between islands eternally grows, remaining liquid, even though the islands of ice also grow indefinitely. Nevertheless, any observer floating along with the flow will end up surrounded by ice: given enough time, a tiny crystal of ice will eventually nucleate in the person's vicinity and swallow her up. This is a somewhat paradoxical but, nonetheless, correct conclusion: there is always plenty of liquid water but any given bit of it is sooner or later engulfed in ice.

What I have described is a precise analogy for the phenomenon called *Eternal Inflation*: growing islands of alternate vacuum in a sea of eternally inflating space. It is by no means a new idea. My colleague at Stanford, Andrei Linde, is one of the great thinkers who have pioneered many of the modern ideas of cosmology. For as long as I've known Andrei — certainly since he came to the United States from Russia, about fifteen years ago — he has been preaching the doctrine of an eternally inflating universe, constantly spinning off bubbles of many kinds.[3] Alexander Vilenkin is another Russian-American cosmologist who has determinedly tried to push cosmology in the direction of a superinflating megaverse of enormous diversity. But, for the most part, physicists have ignored these ideas, at least until very recently. What is shaking up the field right now is the realization that String Theory — our best guess for a theory of nature — has features that mesh very well with these older ideas.

The combination of general relativity, quantum mechanics, and an initial high-density universe, together with the Landscape of String Theory, suggest that an eternally inflating, metastable universe may be inevitable.

Eternal Inflation

If you purchased this book hoping to find the ultimate answer to how the universe began, I am afraid you will be disappointed. Neither I nor anyone else knows. Some think it began with a singularity, an infinitely violent state of infinite energy density. Others, notably Stephen Hawking and his followers, believe in a quantum tunneling from nothing. But however it began, we know one thing. At some time in the past, the universe existed in a state of very large energy density, probably trapped in an inflationary expansion. Almost all cosmologists will agree that a history of rapid exponential inflation is very likely the explanation of many

3. In Linde's original work he used the term *Self-Reproducing Universe*. I have used *Eternal Inflation* because it seems to be more common in the current literature on the subject.

puzzles of cosmology. In chapter 5 we learned about the observational basis for this belief. It seems all but certain that the *observable* history of our universe began about fourteen billion years ago at a point in the Landscape with enough energy density to inflate our patch of space by at least 10^{20} times. That is probably an enormous underestimate. The energy density during this period was very large — how large we don't know for sure but vastly larger than anything we can make in the laboratory, even during the most violent collisions of elementary particles in the largest accelerators. It appears that at that time, the universe was not quite trapped in a valley of the Landscape but was resting on a slightly tilted plateau. As it inflated, our pocket of space (the observable universe) slowly rolled down the shallow tilt, toward a sudden, steep ledge, and when it got to the edge of the ledge, it quickly descended, converting potential energy to heat and particles. This event, which created the material of the universe, is called *reheating*. Finally, the universe rolled down to our present valley, with its tiny anthropic cosmological constant. That's it: cosmology as we know it was a brief roll from one value of the vacuum energy to another. All the interesting things happened during this transitory period.

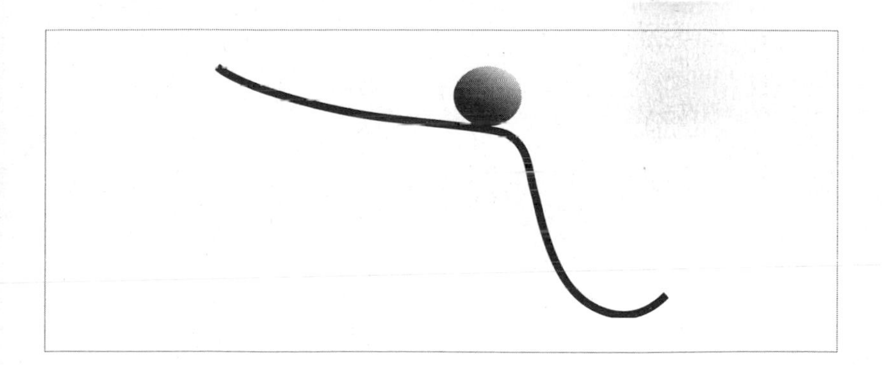

How did our pocket universe get up on the ledge? That's what we don't know. But it is mighty convenient that it started where it did. Without the Inflation caused by the energy density on the ledge, the universe could not have evolved to the large, matter-filled universe we see around us: a universe big enough, smooth enough, and with density contrasts just right for our own existence.

The problem with a theory that places us on the ledge at the very beginning is that this is just one of a stupendous number of starting points. Its only distinction is that it provides a potentially successful beginning for a universe with a chance for life to evolve. Arbitrarily placing the universe at such a lucky spot on the Landscape would defeat the goal of explaining the world without an intelligent designer. But as I will explain, a theory with an enormous Landscape has no choice. To my mind it is completely inevitable — mathematically certain — that some parts of space will evolve to find themselves at the lucky spot. But not everyone agrees.

The Princeton cosmologist Paul Steinhardt, in a critique of the Anthropic Principle, says: "The Anthropic Principle makes an enormous number of assumptions — regarding the existence of multiple universes. . . . Why do we need to postulate an infinite number of universes with all sorts of different properties just to explain our one?" The answer is that we don't need to postulate them. They are unavoidable consequences of well-tested, conventional principles of general relativity and quantum mechanics.

It is ironic that Steinhardt's own work contained the original germ of the Eternal Inflation idea, including the arguments that I find so inevitable. The bubbling up of an infinity of pocket universes is as certain as the bubbling of an opened bottle of champagne. There are only two assumptions: the existence of a Landscape and the fact that the universe started with a very high density of energy, i.e., that it started at high altitude. The first may prove to be no assumption at all. The mathematics of String Theory seems to make the Landscape unavoidable. And the second — high-energy density — is a feature of every scientific cosmology that begins with the Big Bang. Let me explain why I, together with most other cosmologists, find Eternal Inflation to be such a compelling idea.

The ideas that I am going to tell you about are not my own. They were pioneered by the cosmologists Alan Guth, Andrei Linde, Paul Steinhardt, and Alexander Vilenkin and owe a large debt to the seminal work of one of the great physicists of my generation, Sidney Coleman. Let us begin with a universe, or perhaps just a patch of space, located at an arbitrary point in the Landscape with only the simple requirement that the energy density is rather large. Like any mechanical system it will begin to evolve toward regions of lower potential energy. Think of rolling a bowling ball from the top of Mount Everest. What is the likelihood that it will roll all the way to sea level without getting stuck somewhere? Not too good. Far more likely it will come to rest in some local valley not far from the mountain. The initial conditions — exactly where it began and with what velocity it started to roll — hardly matter.

As with bowling balls, so it goes with the patch of space we are following: it will most likely plop into some valley, where it will begin to inflate. A stupendous volume of space will be cloned, all located in the same valley. There are of course lower valleys, but to get to them the universe would have to climb over mountain passes at elevations higher than the starting valley, and it cannot do so because it doesn't have the energy. So it just sits there and inflates forever.

But we've forgotten one thing. The vacuum has the quantum jitters. Just like the thermal jitters of supercooled water, the quantum jitters cause small bubbles to form and disappear. The interior of these bubbles may lie in a neighboring valley, with smaller altitude. This bubbling is constantly going on, but most of the bubbles are too small to grow. The surface tension on the domain walls separating the bubble from the rest of the vacuum squeezes them away. But, as in the supersaturated case, every now and then a bubble forms that is big enough to start growing.

The mathematics describing this bubble formation in an inflating universe has been known for many years. In 1977 Sidney Coleman and Frank De Luccia wrote a paper that was to become a classic. In their paper they calculated the rate at which such bubbles would appear in an inflating universe, and although the rate could be very small — very few bubbles per unit volume — it most certainly is not zero. The calculations used only the most trustworthy, well-tested methods of quantum field theory and are considered by modern physicists to be rock solid. Thus, unless there is something terribly wrong, the inflating vacuum will spin off growing bubbles located in neighboring valleys.

Do bubbles collide, eventually coalescing so that all of space winds up in some new valley? Or does the space between the bubbles expand too quickly to allow the islands to merge? The answer depends on the competition between two rates — the rate of bubble formation and the rate that space reproduces, or the rate of cloning. If the bubbles form very quickly, they will rapidly collide and merge, the whole space moving to some new point on the Landscape. But if the rate at which space reproduces is greater than the rate at which bubbles form, cloning wins, and the bubbles never catch up with one another. Like the islands of ice in the inflating supercooled lake, the bubbles evolve in isolation, eventually moving beyond one another's horizons. The majority of space continues to eternally inflate.

Which wins, bubble nucleation or space cloning? Generally the answer is not even close. Bubble nucleation, like all other tunneling processes, is rare and improbable. Typically, a very long time will elapse before a bubble large enough to expand accidentally nucleates. On the other hand, the cloning of space, i.e., the exponential growth due to vacuum energy, is extremely rapid if the cosmological constant is not ridiculously small. In all but the most contrived examples, space continues to clone itself exponentially, while islands or bubbles slowly nucleate in neighboring valleys of the Landscape. By a very wide margin, the cloning of space wins the competition.

Let's take a look inside one of the bubbles. What do we find? Typically we will find ourselves located in a valley with altitude somewhat lower than at the start. Space inside the bubble will also be inflating. I'm not speaking about the ordinary growth of the bubble but about the cloning of space inside the bubble. Thus, we start over again. A new patch of space is now located in a new valley. But there are still other, lower valleys. Inside the original bubble, a next-generation bubble may form in another nearby valley with a lower altitude. And if that bubble is bigger than critical size, it begins to grow — a growing bubble inside another bubble.

As a rule, I don't like biological analogies in physics. People tend to take them too literally. I'm going to use one now but *please* don't get the idea that I think that universes or black holes or electrons are alive, engage in Darwinian competition, or have sex.

Think of the megaverse as a colony of organisms that reproduce by cloning. In order to avoid confusion, let me emphasize: the organisms are not living creatures — they are reproducing regions of space. Because clones are identical to their parents, we can think of them as occupying the same valley in the Landscape. We might even think of a

Landscape of biological designs: different valleys corresponding to different species. Don't worry about the organisms getting in each other's way. In this fictitious world there is always room for more. When a bubble forms with different properties from the parent, the offspring occupies a new neighboring valley. Because the space inside the bubble also inflates, the offspring begins both the cloning process and the process of inhabiting new valleys by creating next-generation bubbles. In this way the metaphorical colony begins to spread out over the Landscape. The fastest reproducers are the patches of space at the highest altitude where the cosmological constant is largest. In these regions of the Landscape, cloning takes place especially rapidly, and the population of the high altitudes grows fastest. But the high-altitude organisms also feed the lower altitudes so that the population in the nether regions also grows with time.[4] Eventually every niche in the Landscape will become populated, not just once but with an exponentially increasing population. The only thing wrong with the analogy is that real organisms compete and kill each other when their valley becomes overcrowded. No mechanism for competition between pocket universes exists, so the population of every valley continues to increase indefinitely. One might think of these organisms as being totally invisible to one another in every way.

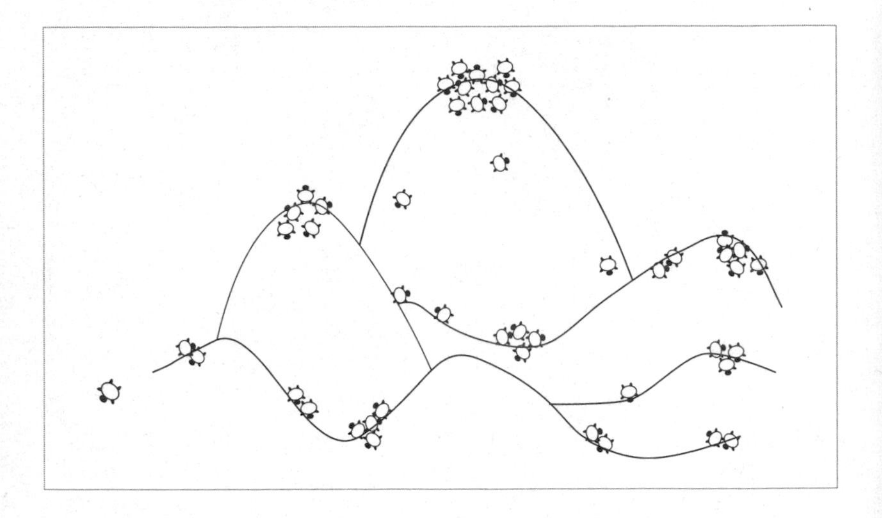

How do bubbles die? If a bubble appears with exactly zero cosmological constant, it will fail to inflate and cease reproducing. The only such vacuums are the supersymmetric parts of the Landscape. Thus, the supersymmetric region of the Landscape is the graveyard of universes in more than one sense. Ordinary life cannot exist in supersymmetric environments, but more to the point, reproduction of bubbles ceases.

While analogies often capture some truth in a way that is easy to understand, they are always misleading in other ways. The analogy between Eternal Inflation and species evolution is defective in ways other

4. An interesting and important question is whether there is any upward mobility on the Landscape. In other words can one of our fictitious organisms climb up to a higher altitude? The answer implied by the standard rules of quantum mechanics is yes — anything that can happen can also happen in the reverse direction.

than the lack of competition. Darwinian evolution depends on continuity between generations. Offspring closely resemble their parents. If we had a sequence of photographs of all the generations of apes, beginning with the "missing link" five million years ago and ending with you or me, we could line them up in a row to see how rapidly evolution did its work. If we ignore those changes that distinguish individuals at a given time, we would find that the changes from one generation to another are far too tiny to detect. Only the accumulated change over thousands of generations would be noticeable, and then just barely. The same would be true of all types of life. Big structural changes take place very infrequently, and when they do they almost always lead to evolutionary dead ends. Anyone born with two heads, three legs, or no kidneys won't survive for more than a short time (except in modern-day hospitals), and in any case, such creatures are extremely unlikely to be successful in the Darwinian dating game.

The contrast with evolution on the cosmic Landscape could not be bigger. The change that takes place when a bubble nucleates in an inflating region of space is not incremental like biological evolution. Think geologically: neighboring valleys are dissimilar. Aspen valley, in the Colorado Rockies, at altitude eight thousand feet, is more than two thousand feet lower than Twin Lakes, just on the other side of Independence Pass. They are dissimilar in other ways as well. If there is a valley so similar to Aspen that we would hardly notice the difference, chances are it's a long way off.

That's the way the cosmic Landscape is. The altitude of neighboring valleys is not particularly similar. If the neighbors differ in the composition of branes or fluxes, it will lead to differences in the list of elementary particles, the constants of nature, and even the dimensionality of space. When a parent vacuum gives birth to an offspring bubble, the results typically will be a monstrous mutation rather than a small incremental change.

Is Eternal Inflation, with its wildly prolific creation of bubble-worlds of every possible kind, a wild phantasmagoric hallucination? I don't think so. The exponential expansion of space seems rock solid; no cosmologist questions it. The possibility of more than one valley is in no way unusual, nor is the assumption that an inflating region will spin off bubbles of lower altitude. Everyone agrees.

What is new is that String Theory gives rise to an exponentially large number of valleys with a tremendous variety of environments. Many physicists are very alarmed by this idea. But even here, most serious string theorists admit that the reasoning looks solid.[5]

Let's consider the last stages of cosmic evolution just before our patch settles into the conventional era of Inflation followed by reheating and eventually life. Where did we come from before miraculously appearing on the inflationary ledge? Most likely the answer is a neighboring valley at a higher elevation. How does that valley differ from our own? String Theory gives answers: fluxes had other values, branes were in different locations, and the moduli of the compactification were different. Perhaps in the passage over the mountain toward the ledge, branes annihilated one another and rearranged, fluxes shifted, and the sizes and shapes of several hundred moduli changed to something new to give a new Rube Goldberg machine. And with the new arrangement came new Laws of Physics.

A Paradoxical Relation between Child and Parent

Einstein's General Theory of Relativity can lead to consequences that defy our usual abilities to visualize geometric relations, black holes being a prime example. Another extremely interesting curiosity involves the geometry inside bubbles that form in an inflating space. From the outside the bubble appears to be an expanding sphere bounded by a domain wall or membrane. The energy that is released by the changes inside the bubble is converted to kinetic energy of the domain wall, which quickly accelerates. After a short time the bubble will be expanding with almost the speed of light. One would expect that an observer inside the bubble would experience a finite world that at every instant is bounded by a growing wall. But that's not at all what he sees. The view from inside the bubble is very surprising.

In chapter 5 we encountered the three basic kinds of expanding universes: the closed-and-bounded universe of Alexander Friedmann, the

5. I don't mean to imply that there is universal agreement. At least one very experienced, highly regarded string theorist, Tom Banks, has argued that the reasoning concerning the Landscape is suspect.

flat universe, and the infinite open universe with negative curvature. All of the standard universes are homogeneous, and none of them has an edge or wall. One might think that an inhabitant in the interior of a bubble would observe the expanding domain wall and conclude that he didn't live in any of the standard universes. Surprisingly, this is incorrect: that inhabitant of the bubble would observe an *infinite* open universe with negatively curved space! How a finite expanding bubble can look like an infinite universe from the inside is one of those mysterious paradoxes of non-Euclidean Einsteinian geometry.

I will try to give you some idea of how the paradox is resolved. Let's start with a map of the earth. Because the earth's surface is curved, it cannot be drawn on a flat plane without distortion. For example, in a Mercator projection, Greenland looks almost as big as North America and a good deal bigger than South America and Africa. Of course it is nowhere near as big as these continents. But to flatten the earth's surface, a great deal of stretching is necessary.

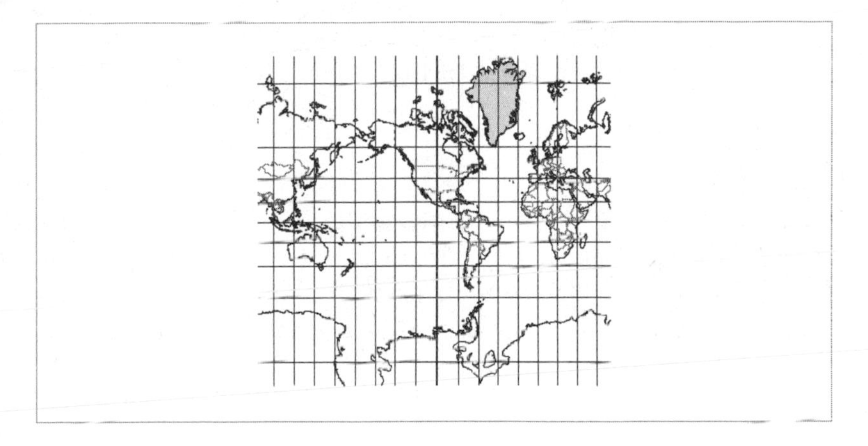

The same is true if we try to flatten a negatively curved surface so that it can be drawn on a plane. It's not easy to draw such a space, but fortunately, a famous artist has already done the work. The artist M. C. Escher's famous woodcut *Limit Circle IV* is nothing but a uniformly negatively curved space drawn on a flat piece of paper. All the angels are the same size, and so are all the devils. They can roughly be thought of as galaxies. But to flatten the space, the center has to be stretched, and the distant parts must be compressed.

Limit Circle IV, by M. C. Escher

In fact the distance from the center of the space to the boundary is infinite. An infinite number of devils (or angels) must be crossed in order to get to the edge. Because each devil is the same size as all the others, the distance is also infinite. Nevertheless, the whole infinite space appears as the interior of a circle when it is flattened onto a plane. With this in mind it is not so difficult to picture the infinite geometry fitting into a finite bubble.

What is especially strange is that if the astronomer wanted to study the expanding domain wall, he would always find it infinitely far away. Interior to the bubble, the geometry of space is unbounded, despite the fact that at any moment of time an exterior observer sees the bubble as a bounded sphere. It's not that an astronomer inside the bubble can't detect light coming from the domain wall. But that light does not seem to be coming from a boundary of space; rather, it seems to come from a boundary of time — from what appears to be a Big Bang taking place in the past. This is a most paradoxical situation, an infinite expanding universe inside a finite expanding bubble.

Knowing that we live in an open, negatively curved universe would be a strong reason to believe that our pocket universe evolved from some point in history during which it was a bubble in an exponentially expanding space. That seems like a clear prediction, but it may be impossible to confirm. The observable universe is just too big, and so far we have seen only a tiny part of it. We simply don't see enough to know if it is curved or flat.

What about our universe today? Can expanding bubbles of some other environment form, grow, and take over our pocket universe? What would happen to us if we were swallowed by such a bubble? The answer that String Theory suggests is that we will one day be engulfed in a destructive environment, fatal to all life. Remember that all evidence points to our world having a cosmological constant — a bit of vacuum energy. There is no reason why it cannot produce a bubble with smaller energy. And we know that there are such places on the Landscape, namely, the graveyard of universes: supersymmetric regions where the cosmological constant is exactly zero. Wait long enough, and we will find ourselves in just that kind of vacuum. Unfortunately, as I explained in chapter 7, even life forms as alien as superstring theorists probably cannot survive in a supersymmetric world. A supersymmetric universe might be extremely elegant, but the Laws of Physics in such a world do not allow for ordinary chemistry. It's not just the graveyard of universes: it spells the death of all chemistry-based life.

If it is certain that we will eventually be swallowed in a hostile supersymmetric environment, how long will it take? Is it something that can happen tomorrow, next year, in a billion years? Like all jittery quantum fluctuations, the answer is that it could happen anytime. Quantum mechanics tells us only how likely it is at any given time. And the answer is that it is incredibly unlikely to happen anytime soon. Indeed, it is unlikely to happen in the next billion, trillion, or quadrillion years. The best rough estimates suggest that our world will last for at least a googolplex years and probably a lot longer![6]

6. A googol is defined as 10^{100}, i.e., one with one hundred zeros after it. A googolplex is ten to the googol power.

Two Views of History

It is hard to see how the populated Landscape viewpoint can be wrong. It follows from well-tested principles. Nevertheless, there are serious matters of concern. Perhaps the most uncomfortable issue can be summarized in the following criticism — a composite of several that I have heard:

> Isn't it true that all the other pocket universes are beyond our horizon? By definition, the horizon separates the world into those places that we can get information from and those places that are absolutely impossible to observe. Doesn't this imply that the other pockets are, *in principle, unobservable?* If that is so, what difference can they make? Why should we have to appeal to the existence of worlds that have no operational meaning to us? The populated Landscape sounds more like metaphysics than physics.

Because I think this issue is so important, the entire next chapter is devoted to it. Indeed, I could easily write a whole book on the subject of horizons, and I probably will. But for now let's just contrast two ways of describing the history of the universe. The first way actually corresponds closely to the conventional way of observing the universe. We observe the universe from within, by means of various kinds of telescopes from the surface of the earth. Even if the observations take place from space — on a satellite — the results of the observations are brought back to earth for analysis.

Observations from the earth are limited to things within our horizon. Not only can't we see anything beyond the horizon, but also nothing behind the horizon can have any influence on our observations. So why not build a theory that restricts attention to a single causal patch? This is a fine pragmatic attitude of which I heartily approve.

What is the history of the universe as seen from a typical observer's vantage point? A good starting point might be a patch of space trapped in some high-altitude valley. The enormously large vacuum energy leads to repulsive forces that are so violent that even particles like protons are ripped apart almost instantly. That primordial world is extremely inhospitable. It is also very small: the horizon is only a tiny distance away —

smaller than a proton radius — and the region accessible to the observer is microscopic, perhaps not much bigger than the Planck length. Obviously no real observer can survive in this environment, but let's ignore that.

After some time a bubble nucleates and grows, taking over the entire region accessible to the observer. The observer finds himself surrounded by an environment that's only a little friendlier: the cosmological constant is smaller, and the horizon has grown, giving a bit more room in which to wiggle around. Still, the cosmological constant in the new valley is far too big for comfort. But again a bubble grows, this time resulting in an environment with a somewhat smaller cosmological constant. Such sudden changes may happen several times. The observer sees a series of environments, none suitable for life. Finally a bubble with exactly zero vacuum energy forms — a bubble of supersymmetric vacuum. The bubble evolves to a negatively curved open world and ceases evolving. The probability of passing through one of the extremely rare life-supporting environments, on the way to the graveyard, is extremely small.

But let's suppose that a bubble of our kind of universe formed before the supersymmetric Landscape was reached. This is a very unlikely event, given how scarce such valleys are, but it can happen. Will life evolve? That depends on exactly how the patch of space got there. One possibility out of many is that it first arrived at the inflationary ledge. That's good. Inflation leads to a hospitable universe. But if the patch reached our valley from some other direction in the Landscape, all bets are off. If it failed to get hung up for a time on the ledge, the universe would probably never have produced enough heat and particles to later be the stuff of life.

From the perspective of an observer who sees a series of environments ending in the graveyard, the likelihood of life is tiny. But now let's imagine that we could get outside the universe and see it as a whole. From the perspective of the entire megaverse, the history is not a sequence or series of events. The megaverse description is a more *parallel* view of things — many pocket universes, evolving in parallel. As the megaverse evolves, pocket universes spread over the entire Landscape. It is absolutely certain that some — quite likely a very small fraction — will wind up on the ledge of life. Who cares about all the others that ended badly? Life will form where it can — and only where it can.

Once again a biological analogy may be helpful. Imagine the tree of life — every branch being a species. If you follow the tree outward from the main trunk (bacteria), randomly proceeding at every fork, you will quickly come to extinction. Every species becomes extinct; but if the rate for evolving new species exceeds the extinction rate, the tree continues to spread. If you follow any particular path to extinction, the probability of encountering intelligent life is nil. But it is practically certain that the tree will eventually sprout an intelligent branch if it grows for a long enough time. The parallel view is a much more optimistic view.

Many-Worlds

What if Germany had won World War II? Or what would life be like if the asteroid that killed the dinosaurs sixty-five million years ago had not hit the earth? The idea of a parallel world that took a different course at some critical historical junction is a favorite theme of science-fiction authors. But as real science, I have always dismissed such ideas as frivolous nonsense. But to my surprise I find myself talking and thinking about just such matters. In fact this whole book is about parallel universes: the megaverse is a world of pocket universes that become disconnected — completely out of contact — as they recede beyond one another's horizons.

I am far from the first physicist to seriously entertain the possibility that reality — whatever that means — contains, in addition to our own world of experience, alternate worlds with different history than our own. The subject has been part of an ongoing debate about the interpretation of quantum mechanics. Sometime in the middle 1950s, a young graduate student, Hugh Everett III, put forth a radical reinterpretation of quantum mechanics that he called the *many-worlds interpretation.* Everett's theory is that at every junction in history the world splits into parallel universes with alternate histories. Although it sounds like fringe speculation, some of the greatest modern physicists have been driven by the weirdness of quantum mechanics to embrace Everett's ideas — among them, Richard Feynman, Murray Gell-Mann, Steven Weinberg, John Wheeler, and Stephen Hawking. The many-worlds interpretation was the inspiration for the Anthropic Principle when Brandon Carter first formulated it, in 1974.

The many-worlds of Everett seems, at first sight, to be quite a different conception than the eternally inflating megaverse. However, I think the two may really be the same thing. I have emphasized several times that quantum mechanics is not a theory that predicts the future from the past, but rather it determines the probabilities for the possible alternate outcomes of an observation. These probabilities are summarized in the basic mathematical object of quantum mechanics — the *wave function*.

If you have learned a little bit about quantum mechanics and know that Schrödinger discovered a wave equation describing electrons, then you have heard of wave functions. I want you to forget all that. Schrödinger's wave function was a very special case of a much broader concept, and it is this more general idea that I want to concentrate on. At any given time — right now, for example — there are many things that one might observe about the world. I might choose to look out the window just above my desk and see if the moon is up. Or I might plan a two-slit experiment (see chapter 1) and observe the location of a particular spot on the screen. Yet another experiment would involve a single neutron that was prepared a certain time in the past — say, ten minutes ago. You may recall from chapter 1 that a neutron, not bound in a nucleus, is unstable. On the average (but only on the average), in twelve minutes it will decay into a proton, an electron, and an antineutrino. The observation in this case would be to determine whether, after ten minutes, the neutron has decayed or if it is still present in its original form. Each of these experiments or observations has more than one possible outcome. In its most general sense, the wave function is a list of the probabilities for all possible outcomes of all possible observations on the system under consideration. More exactly, it is a list of the square roots of all these probabilities.

The decaying neutron is a good illustration to start with. With a bit of simplification, we can suppose there are only two possible outcomes when we observe the neutron: either it has decayed or it hasn't. The list of possibilities is a short one, and the wave function has only two entries. We start with the neutron in its undecayed form so that the wave function has value one for the first possibility and zero for the second. In other words initially the probability that the neutron is undecayed is one, while the probability that it has decayed (when we start) is zero.

But after a short time, there is a small probability that the neutron has disappeared. The two entries to the wave function have changed from one and zero to something a bit less than one and something a bit more than zero. After about ten minutes the two entries have become equal. Go on for another ten minutes, and the probabilities will be reversed: the probability that the neutron is still intact will be close to zero, and the probability that it has become a proton/electron/antineutrino will be up near one. Quantum mechanics contains a set of rules for updating the wave function of a system as time unfolds. In its most general form, the system of interest is everything — the entire observable universe, including the observer doing the observations. Since there may be more than one lump of matter that might be called an observer, the theory must give rise to consistent observations. The wave function contains all of this and in a way that will prove consistent when two observers get together to discuss their findings.

Let's examine the best known of all thought experiments in physics: the famous (or should I say infamous?) Schrödinger's cat experiment. Imagine that at noon a cat is placed in a sealed box along with a neutron and a gun. When the neutron decays (randomly), the ejected electron activates a circuit that causes the gun to shoot and kill the cat.

A practitioner of quantum mechanics — call him S — would analyze the experiment by constructing a wave function: a list of the probabilities for the various outcomes. S cannot reasonably take the entire universe into account, so he limits the system to include only those things in the box. At noon only one entry would exist: "The cat is alive in the box with the loaded gun and the neutron." Then S will do some mathematics analogous to solving Newton's equations in order to find out what will happen next — say, at 12:10 p.m. But the result is not a prediction of whether the cat will be dead or alive. It is an updating of the wave function, which will now have two entries: "The neutron is intact/the gun is loaded/the cat is alive" and "The neutron has decayed/the gun is empty/the cat is dead." The wave function has split into two branches — the dead and alive branches — whose numerical values give the square roots of the probabilities for the two outcomes.

S can open the box and see if the cat is dead or alive. If the cat is alive, then S can throw away the dead-cat branch of the wave function. That branch, if advanced in time, would contain all the information about the world in which the cat was shot, but since S found the cat alive, he has no further need for this information. There is a term for this process of dropping the unobserved branches of the wave function whenever an observation is made. It is called the collapse of the wave function. It is a very convenient trick that allows the physicist to concentrate on only the things that can subsequently be of interest. For example, the live branch has information that may still interest S. If he advances this branch of the wave function a bit more in time, he would be able to determine the probability that the gun would subsequently misfire and shoot S (serves him right). The collapse of the wave function whenever an observation takes place is the primary ingredient of the famous Copenhagen interpretation of quantum mechanics championed by Niels Bohr.

But the collapse of the wave function is not a part of the mathematics of quantum mechanics. It is something extraneous to the mathematical rules, something that Bohr had to tack on in order to end the experiment with an observation. This arbitrary rule has bothered generations of physicists. A big part of the problem is that S limited his system to the things in the box, but at the end of the experiment S himself gets into the act by making the observation. It is now widely understood that a consistent description must necessarily include S as part of the system. Here is the way the new description would go:

The wave function now describes everything in the box as well as the physical lump of matter that we have been calling S. The initial wave function still has only one entry but it is now described as follows: "The live cat is in the box with the loaded gun and the neutron, and the mental state of S is blank." Time goes on, and S opens the box. Now the wave function has two entries: "The neutron is intact/the gun is loaded/the cat is alive/the mental state of S is aware of the live cat" and the second branch, "The neutron has decayed/the gun is empty/the cat is dead/the mental state of S is aware of the dead cat." We have managed to describe S's perceptions without collapsing the wave function.

But now suppose there is another observer named B. B has been out of the room while S has been doing his bizarre experiment. When he opens the door to look in what he sees is one of two outcomes. There is no point in keeping track of the unobserved branch of the wave function, so B collapses the wave function. It seems we have not avoided the extraneous operation. Evidently what we need to do is to include B in the wave function. The starting point would be a system composed of everything in the box and two lumps called S and B. The initial state is: "The live cat is in the box with the loaded gun and the neutron, the mental state of S is blank, and the mental state of B (who is out of the room) is blank." When S opens the box, the wave function develops two branches: "The neutron is intact/the gun is loaded/the cat is alive/the mental state of S is aware of the living cat/B's mental state is still blank" and "The neutron has decayed/the gun is empty/the cat is dead/the mental state of S is aware of the dead cat/B's mental state is still blank." Finally, B opens the door, and the first branch becomes: "The neutron is intact/the gun is loaded/the cat is alive/the mental state of S is aware of the living cat/B's mental state is aware of the live cat and also S's mental state." I will leave it to the reader to work out the other branch. The important thing is that the experiment has been described without collapsing the wave function.

But now suppose there is another observer called E. Never mind. You should be able to see the pattern. What is evident is that the only way to avoid wave-function collapse is to include the entire observable universe as well as all the branches of the wave function in the quantum description. That is the alternative to Bohr's pragmatic rule of terminating the story by collapsing the wave function.

In Everett's way of thinking, the wave function describes an infinite branching tree of possible outcomes. Following Bohr, most physicists have tended to think of the branches as mathematical fictions, except for the actual branch that one finds oneself on after an observation. Collapsing the wave function is a useful device to cut away all the unneeded baggage, but to many physicists this rule seems to be an arbitrary external intervention by the observer — a procedure not in any way based on the mathematics of quantum mechanics. Why should the mathematics give rise to all the other branches if their only role is to be thrown away?

According to the advocates of the many-worlds interpretation, all the branches of the wave function are equally real. At each junction the world splits into two or more alternative universes, which forever live side by side. Everett's vision was of a ceaselessly branching reality, but with the one proviso that the different branches never interact with one another after they have split. On the live-cat branch, the dead-cat branch will never come back to haunt S. Bohr's rule is just a trick to cut away all the branches, which although quite real, will have no further effect on the observer.

One other point is worth noting. By the time we get to the present stage of history, the wave function has branched so many times that there are an enormous number of replicas of every possible eventuality. Consider poor B while he is out of the room. The wave function branched when S opened the box, thus splitting all of them, including B, into two branches. The number of branches containing you, sitting and reading this book, is practically infinite. In this framework the concept of probability makes perfect sense as the relative frequency of different outcomes. One outcome is more probable than another if more branches contain it.

The many-worlds interpretation cannot be experimentally distinguished from the more conventional Copenhagen interpretation. Everyone agrees that, in practice, the Copenhagen rule correctly gives the probabilities of experimental outcomes. But the two theories profoundly disagree about the philosophical meaning of these probabilities. The Copenhageners take the conservative view that probabilities refer to the statistics of a large number of repeated experiments. Think of flipping a coin. If the coin is "fair," the probability for either outcome (heads or tails) is one-half. This means that if the coin is flipped a large number of times, the fraction of heads and tails will each be about one-half. The larger the number of trials, the closer the answer will be to the ideal half-half result. Similar things apply when rolling dice. If one rolls a single die many times, one-sixth of the time (within the margin of error) the die will show each of the six possible outcomes. Ordinarily no one would apply statistics to a single coin flip or a single roll of the die. But the many-worlds interpretation does just that. It deals with single events in a way that would seem ridiculous for coin flipping. The idea that when a coin is flipped, the world splits into two parallel worlds — a heads-world and a tails-world — does not seem to be a useful idea.

Why then are physicists so bothered by the probabilities that occur in quantum mechanics that they are driven to strange ideas like the many-worlds interpretation? Why was Einstein so insistent that "God does not play dice?" To understand the puzzlement that attends quantum mechanics, it is helpful to ask why, in a Newtonian world of absolute certainty, one would ever discuss probability at all. The answer is simple: probabilities enter Newtonian physics for the simple reason that one is almost always ignorant of the exact initial conditions of an experiment. In the coin flipping experiment, if one knew the exact details of the hand that threw the coin, the air currents in the room, and all the other relevant details, there would be no need for probabilities. Each throw would lead to a definite outcome. Probability is a convenient trick to compensate for our practical inability to know the details. It has no fundamental place in the Newtonian laws.

But quantum mechanics is different. Because of the Uncertainty Principle, there is no way to predict the outcome of an experiment — no way, in principle. The fundamental equations of the theory determine a wave function and nothing more. Probability enters the theory at the outset. It is *not* a trick of convenience used to compensate our ignorance. Moreover, the equations that determine how the wave function changes with time have no provision for suddenly collapsing the unobserved branches. The collapse of the wave function is the trick of convenience.

The problem becomes especially acute in the cosmological context. Ordinary experiments, similar to the two-slit experiment that I described in chapter 1, can be repeated over and over, just like the coin toss. In fact each photon that goes through the apparatus can be thought of as a separate experiment. There is no problem accumulating enormous amounts of statistical data. But the problem with this conception of quantum mechanics is that we cannot apply it to the great cosmic experiment. We can hardly repeat the Big Bang over and over and gather statistics on the outcomes. For this reason many thoughtful cosmologists have adopted the philosophical underpinnings of the many-worlds interpretation.

Carter's early pioneering idea for synthesizing the Anthropic Principle with the many-worlds interpretation was this: suppose the wave function includes branches not only for such ordinary things as the location of an electron, the decay or nondecay of the neutron, or the life and death of a cat, but also for different Laws of Physics. If one assumes all the branches are equally real, then there are worlds with many alternative environments. In modern language we would say there are branches (as well as real worlds) for every location on the Landscape. The rest of the story is no different than what I have explained earlier in the book, except instead of talking about different regions of the megaverse, one would talk about different branches of reality. To make the point let me quote from chapter 1 and then modify the quote with some replacements. The original quote was, "Somewhere in the megaverse, the constant equals *this* number: somewhere else it is *that* number. We live in one tiny pocket where the value of the constant is consistent with our kind of life." The modified quote is: "Somewhere in the wave function, the constant equals *this* number: somewhere else it is *that* number. We live in one tiny branch where the value of the constant is consistent with our kind of life." Although the two quotes seem very similar, they are referring to two apparently completely different ideas of alternate universes. It seems that we have more than one way to achieve the kind of diversity that would allow anthropic reasoning to make sense. I might add that different proponents of the Anthropic Principle have different opinions about which version is the one true theory of parallel universes. My opinion? I believe the two versions are complementary versions of exactly the same thing.

Let's look at the situation in a bit more detail. Earlier in this chapter I described two views of eternally inflating history, the parallel and series views. The parallel view recognizes the entire megaverse with all its multiple pocket universes, which once they are separated by horizons are out of contact. That sounds quite a lot like the many-worlds of Everett. But what about the series view?

Let's consider an example. Suppose that a bubble of space has formed, with properties associated with some valley in the landscape. It will help to have some names for the valley and its neighbors, so let's call it Central Valley. To the east and west of Central Valley lie East and West valleys, each at somewhat lower altitudes. From West Valley two other nearby valleys can be reached, one called Shangri La and the other Death Valley. Death Valley is not really a valley but rather a flat plateau at exactly zero altitude. East Valley also has a few neighbors that can be easily reached, but we won't bother naming them.

Imagine yourself in Central Valley as your pocket universe inflates. Because there are nearby lower valleys, your vacuum is metastable:

bubbles can form and engulf you. After some period you might look around and observe the properties of your environment. You may find that you are still in Central Valley. Or you may find that you have made a transition to East Valley or to West Valley. The decision as to which valley you now inhabit is determined randomly according to quantum mechanics, in much the same way that quantum mechanics determined the fate of S's cat.

Let's say that you now find yourself in West Valley. You might as well throw away the branch of your wave function that corresponds to East Valley; it is irrelevant to your future. Again, wait a while, and if you are lucky, a bubble in the pleasant life sustaining Shangri La Valley may next swallow you up. But you may also wind up in Death Valley. At each junction Bohr and his Copenhagen gang would tell you how to calculate the probability for each outcome. Then they would instruct you to collapse the wave function in order to get rid of the excess baggage of those branches that didn't correspond to your experience. This is the series view of history.

My own view should be obvious by now. The series view — staying at home in your own pocket, within your horizon, observing events, and eliminating the unobserved baggage — is the Bohr interpretation of quantum mechanics. The more expansive parallel, megaverse view of history is the Everett interpretation. I find in this correspondence a pleasing consistency. Perhaps in the end we will find that quantum mechanics makes sense only in the context of a branching megaverse and that the megaverse makes sense only as the branching reality of Everett's interpretation.

Whether we use the language of the megaverse or the many-worlds interpretation, the parallel view, together with the enormous Landscape of String Theory, provides us with the two elements that can change the Anthropic Principle from a silly tautology into a powerful organizing principle. But the parallel view relies on the reality of regions of space and time that, apparently, are permanently beyond the reach of any conceivable observation. For some people that is troubling. It troubles me. If the vast sea of pocket universes is really beyond an ultimate horizon, then the parallel view seems more like metaphysics than science. The next chapter is all about horizons and whether they are really ultimate barriers.

CHAPTER TWELVE

The Black Hole War

> "Sometimes I've believed as many as six impossible things before breakfast."
>
> — LEWIS CARROLL

> We can only look on helplessly as the heat engulfs you. Soon your precious body fluids will begin to boil and then vaporize. It will become so hot the very atoms of your being will be torn apart. But it is foretold that eventually you will be returned to us in a vaporous form of pure light and radiance.
>
> But have no fear. You will pass to the other side safely and without pain. In your present form you will be lost to us forever, never to communicate again, at least not unless we make the crossing ourselves. But, my friend, from your place, you will have no trouble seeing us as we continue on without you. Good luck.

A story of martyrdom and resurrection? A man of the cloth comforting the martyr before the auto-da-fé? The crossing of the veil that separates the living from the dead? Not at all: it is the imaginary, but entirely possible, briefing of a future star traveler, curious and brave enough to enter a giant black hole and to cross its horizon. Not a briefing by a chaplain but by the starship's resident theoretical physicist.

Or, more to the point of this book, it could be the crossing of the cosmic horizon of an eternally inflating universe. But we will come to cosmic horizons a little later.

Spiritualists believe that communication with the dead is possible: all that's required is the right *medium*, an adept at the darker sciences. You can guess what I think of such claims, but ironically, I have been one of the main combatants in a war of ideas about the possibility of communication with the undead on the other side of an event horizon. The war lasted for a quarter of a century, but now it's over.

The protagonists were Stephen Hawking and his army of general relativists on one side.[1] For the first fifteen years, it was mainly Gerard 't Hooft and myself on the other. Later a band of string theorists came to our aid.

Gerard 't Hooft is a Dutchman. If measured by the number of great contributions to physics per capita, the Dutch are surely the greatest physicists in the world. Christiaan Huygens, Hendrik Antoon Lorentz, Willem de Sitter, Heike Kamerlingh Onnes, George Uhlenbeck, Johannes Diderik van der Waals, Hendrik Gerhard Casimir, Martinus Veltman, Gerard 't Hooft are just a few of the greatest names. Lorentz and 't Hooft are arguably among the greatest historical figures of physics. To me 't Hooft, more than any living physicist, represents the spirit of Einstein, Lorentz, and Bohr. Although he is six years younger than I am, I have always been in awe of 't Hooft.

I am glad to say that 't Hooft is not only one of my heroes but that he is also a good friend. Although he is mathematically far more powerful than I am, I have always found that of all my colleagues he's the one that I feel closest to in viewpoint. Throughout the years we have often found ourselves working on the same puzzles, troubled by the same paradoxes, and with similar guesses about the resolution of these problems. I think that, like me, Gerard is a very conservative physicist who will not embrace a radical solution to a problem unless he feels that all other paths have proved futile. But then he is fearless.

If Gerard is conservative I would have to say that Stephen Hawking is the Evel Knievel of physics. Brave to the point of recklessness, Stephen is a well-known traffic menace in Cambridge, where his wheelchair is often seen careening around, way beyond safe speeds. His physics is in many ways like his wheelchair driving — bold, adventurous, audacious to the maximum. Like Evel Knievel he has had his crashes.

1. *General relativist:* an expert in Einstein's General Theory of Relativity.

Three years ago Stephen turned sixty. The party was like no other physicist's sixtieth birthday. Seminars and physics lectures — sure, plenty of them — but also music, cancan dancers, a famous rock star from U2, a Marilyn Monroe look-alike, singing physicists. It was a stupendous media event.

To give you an idea of the relationship Stephen and I have had over the years, I will quote from the birthday lecture I gave at the celebration:

> Stephen, as we all know, is by far the most stubborn and infuriating person in the universe. My own scientific relation with him I think can be called adversarial. We have disagreed profoundly about deep issues concerning black holes, information, and all that kind of thing. At times he has caused me to pull my hair out in frustration — and you can plainly see the result. I can assure you that when we began to argue more than two decades ago, I had a full head of hair.

At this point I could see Stephen in the rear of the auditorium with his impishly mischievous grin. I went on:

> I can also say that of all the physicists I have known he has had the strongest influence on me and on my thinking. Just about everything I have thought about since 1980 has in one way or another been a response to his profoundly insightful question about the fate of information that falls into a black hole. While I firmly believe his answer was wrong, the question and his insistence on a convincing answer [have] forced us to rethink the foundations of physics. The result is a wholly new paradigm that is now taking shape. I am deeply honored to be here to celebrate Stephen's monumental contributions and especially his magnificent stubbornness.

That was three years ago. At the time Stephen still believed that he was right and 't Hooft and I were wrong.

In the early days of the war there were many flip-floppers trying to position themselves to be on the winning side — whichever that would be. But Stephen, to his everlasting credit, stuck to his guns until further resistance was no longer possible. Then he gracefully and unconditionally surrendered. Indeed, had Hawking fought with less conviction, we would probably know much less than we do today.

Stephen's point of view was simple and direct. The horizon of a black hole is a *point of no return*. Anything that crosses the horizon is trapped. To cross back it would be necessary to exceed the speed of light: a total impossibility according to Einstein. People, atoms, photons, every form of signal that can carry a message are bound by Einstein's speed limit. No object or signal can pass from behind the horizon to the outside world. The black hole horizon is the wall of a perfect prison. Observers waiting outside the prison for a report from inside would wait an eternity for even a single bit of information from within. At least this was Hawking's view.

To get a good idea of how black holes work without getting into the difficult mathematics of general relativity, we need an analogy. Fortunately, we have a pretty good one that is familiar and easy to understand. I'm not sure who first used it, but I learned it, or something similar, from the Canadian physicist Bill Unruh. Let's go back to the infinite shallow lake that we used in the last chapter to illustrate an inflating universe. But now we don't need the feeder tubes that keep injecting new water into the lake. Instead we introduce a drain at the center. The drain is a hole in the lake bottom that allows the water to escape, perhaps emptying on to some deadly rocks below. Let's also introduce some boats with observers onto the lake. The observers must obey two rules. The first is that they can communicate only by means of surface waves, i.e., ripples on the surface of the lake. They can wiggle their fingers in the water in order to radiate waves. The second rule is to follow a speed limit on the lake. No boat, under any circumstances, is permitted to move through the water faster than the speed of these waves.

Let's begin with observers far from the center, where the water is hardly disturbed by the drain. It's not completely unaffected: it very slowly migrates inward but almost imperceptibly. But as we move toward the drain, the flow picks up speed, and very close to the drain the inward velocity becomes larger than the velocity of surface ripples. Waves emitted from this region get swept toward the drain even if they were directed away. Obviously any boat that unknowingly finds itself this close is doomed to be swept to its destruction. In fact there is a particular boundary where the velocity of the water exactly matches the speed of surface waves. That place is the point of no return. Once crossed, there is no getting back. Not even a message can be communicated to the outside. That point of no return is just like the horizon of a black hole, except that in the case of the black hole, space is being swept inward with the velocity of light. From behind the horizon no signal can escape without exceeding Einstein's ultimate speed limit. Now it should be clear why Stephen was certain that information falling across a black hole horizon is irredeemably lost to the outside.

Stephen himself was responsible for the weapon that was turned against him. Building on the great work of Jacob Bekenstein in the early 1970s, Stephen had shown that black holes have thermal energy — heat. They are not icily cold as physicists had assumed. It is true that the bigger the black hole, the lower its temperature, but no matter how big, there is always a residual thermal heat. For the kind of black hole that would result from the final collapse of a star, the Hawking temperature would be only about one ten-millionth of a degree above absolute zero. But it's not zero.

Hawking reasoned that a black hole, like any other object with heat content, would radiate energy. A hot poker that's been left in the flame radiates light of an orange or red color. Cooler objects radiate infrared radiation invisible to the naked eye. No matter how cold, as long as an object is not at absolute zero, it will radiate energy in the form of electromagnetic radiation. In the case of a black hole, it is called Hawking radiation. That was Hawking's great discovery.

Anything that radiates will lose energy. But mass and energy are two sides of the same thing, according to Einstein. So in time black holes lose their mass, and losing mass, they shrink until they completely evaporate away, leaving only the photons of the Hawking radiation in their place. Curiously, then, the mass of any object that falls into a black hole is inevitably radiated back out as Hawking radiation. The energy of the courageous star traveler who braved the horizon crossing eventually reappears as "pure light and radiance."

But, said Hawking, because no signal can exceed the speed of light, no information from the interior can exit the horizon along with the Hawking radiation. Such information is trapped in a shrinking ball, until poof — it just disappears when the black hole is gone.

The first I heard of this was in 1980, when Stephen, Gerard 't Hooft, and I attended a small conference in San Francisco. Both Gerard and I were disturbed by Stephen's conclusion and were certain it was wrong. But neither of us could see exactly what was incorrect in the reasoning. I had a sense of deep discomfort. A paradox of very serious magnitude had been announced by Hawking: the kind of paradox that could eventually open the door to a deeper understanding of the elusive connection between gravity and quantum mechanics.

The problem was that Hawking's conclusion violated one of the central tenets of physics. Hawking of course knew that. It was why he found the idea of *information loss* in black hole evaporation so exciting. But 't Hooft and I felt that the *conservation of information* was just too deeply built into the basic logical foundations of physics to discard it, even in the presence of such a bizarre object as a black hole. If we were right then somehow the bits of information that fall to the horizon of a black hole are radiated back out with the Hawking radiation, thus opening the way for imprisoned information to be signaled to the outside.

One should not get the idea that information comes out of the black hole in an easily accessible form. It comes out in a form that is so scrambled that in practical terms it would be impossible to unscramble. But the debate was not about practicalities. It was about laws of nature and principles of physics.

What, exactly, constitutes information, especially if it is scrambled beyond recognition? To understand the ideas involved, let's pursue the analogy with a prison. A Mafia kingpin in the "big house" wants to send a message to one of his lieutenants on the outside. He first writes out his message, "Tell the Piranha brothers to bet ten thousand on the Kid." To make it hard for the censors, he adds, at the end, a much longer fake message, let's say, the text of the *Encyclopaedia Britannica*. Next the criminal mastermind places the message on a series of cards, one letter to a card. The order of the cards retains the original message, including the interesting part and the fake addition tacked on at the end. Now he scrambles the message. The kingpin has a code for doing this. He takes the entire message and shuffles the letters, not in a random way, but according to a rule. Next he shuffles the result, again by the same rule. He does this over and over ten million times. The message is then conveyed to the lieutenant.

The cards are the analog of the Hawking photons that are radiated from the black hole.

What does the lieutenant make of it? If he doesn't know the shuffling rule, he has nothing but a random, meaningless sequence of letters that conveys no information. But the information is there, nevertheless. Given the shuffling rule, the lieutenant can unscramble by reverse-shuffling ten million times. The message reappears at the top of the deck, and the lieutenant easily picks out the relevant part. The information was there even if scrambled. Even if the lieutenant didn't have the shuffling rule — perhaps he lost it — the information was, nevertheless, in the cards.

Compare that with a different situation. This time the prison censor intercepts the message on the way out and shuffles it, but according to a rule that has some intrinsic randomness built into it. Once, twice, ten million times he shuffles. Now, even if the lieutenant knows that the letters were scrambled randomly, there is no way to recover the message. The information is truly lost. The randomness of the shuffling not only scrambled the message but also destroyed the information that it contained.

The real controversy that Hawking, 't Hooft, and I were at war over had nothing to do with the practicalities of actually reconstructing the messages from inside a black hole. It had to do with the existence of rules and the kinds of rules that nature utilizes. Gerard and I claimed that nature scrambles information but never destroys it. Stephen claimed black holes create a form of randomness — a kind of noise in the system — that degrades any information before the Hawking radiation escapes the immediate environment of the black hole. Once again the issue wasn't one of technology, but rather, it had to do with the nature of the future Laws of Physics, when quantum mechanics and gravity will both be important.

The reader may find one thing confusing, even disturbing. Doesn't quantum mechanics introduce an element of randomness into the laws of nature? Don't the quantum jitters destroy information? The reason is not simple, but the answer is no. Quantum information is not as detailed as the information in a classical sequence of symbols. But the randomness of quantum mechanics is of a very special, controlled kind. Hawking was claiming a degree of randomness above and beyond the kind allowed by the standard rules of quantum mechanics: a new kind of randomness that was catalyzed by the presence of a black hole.

Let's pursue the prison analogy a little further. Imagine that the lieutenant sent a message into the prison with some irreplaceable information. In fact we can even imagine a steady stream of information flowing in. The prison has its limits. It can't keep absorbing scraps of paper indefinitely. At some point it will have to dump them back out in the trash. According to Hawking the messages go in, the trash comes out, but inside the prison the information in the message is destroyed by this new kind of randomness. But 't Hooft and I said no: the message is in the trash. It's indestructible. We argued that the quantum bits that fall into the black hole are always there to recover — but only if you know the code.[2]

The position that 't Hooft and I held was not without its problems. We insisted that information escapes from the horizon, but how could it if that requires exceeding the speed of light? What is the mechanism? The answer must be that it never goes in.

Let's send a message into a black hole with the star traveler. According to the usual rules of the General Theory of Relativity, the message together with the traveler should sail right through the horizon. On the other hand, 't Hooft and I, in order to rescue the basic principles of quantum mechanics, were claiming that the bits of information in the message would be transferred to the outgoing Hawking radiation just before passing in to the horizon and then radiated out. It was as though the message were torn out of the hands of its messenger and put in the outgoing trash just before passing the point of no return.

This conflict of principles created a very serious dilemma. General relativity said that the bits enter the horizon and continue on their way, deep into the interior of the black hole. But the rules of quantum mechanics forbid information to be lost to the outside world. One possibility might resolve the dilemma. Let's go back to the prison analogy. Suppose that at the entrance to the prison, a guard were stationed at a Xerox machine, and that every incoming message was xeroxed — one copy going into the prison and one sent back out after nonrandom shuf-

2. A *bit* is a technical term for an indivisible unit of information — a yes-no answer to a question.

fling. That should satisfy everyone. Inside the prison they would see the message entering as if it were undisturbed on its way in. Outside the observers would find that the information was never lost. Everyone is right.

Here the problem gets interesting. A very basic principle of quantum mechanics says that a quantum Xerox machine is impossible. Quantum information cannot be faithfully copied. No matter how well the machine copies some kinds of information, it will always fail badly with other kinds. I called this the No Quantum Xerox Principle. The quantum-information experts call it the No Cloning Theorem. What it states is that no physical system can ever function completely faithfully to replicate information in a quantum world.

Here is a way to understand the No Quantum Xerox Principle. Start with a single electron. The Heisenberg Uncertainty Principle tells us that it is never possible to know both the position of the electron and its velocity. But now suppose we could quantum Xerox the electron in exactly its original state. Then we could measure the position of one copy and the velocity of the other, thereby learning the forbidden knowledge.

So, here is the new dilemma: general relativity tells us that information should fall straight through the horizon toward the deep interior of the black hole. On the other hand, the principles of quantum mechanics tell us that the same information must remain outside the black hole. And the No Cloning Theorem assures us that only one copy of each bit is possible. That's the confusing situation that Hawking, 't Hooft, and I found ourselves in. By the early nineties the situation had reached a crisis: who is right? The observer on the outside who expects the rules of quantum mechanics to be respected? For him the bits of information should be located just above the horizon, where they are scrambled and then sent back out in the Hawking radiation. Or is the observer who falls through the horizon correct in expecting the bits to accompany her into the heart of the black hole?

The solution to the paradox was eventually provided by two new principles of physics that 't Hooft and I introduced in the early 1990s. Both of them are very strange, far stranger than Hawking's idea that information can be lost; so strange that, in fact, no one else besides 't Hooft and me believed them at first. But as Sherlock Holmes once told Watson, "When you have eliminated all that is impossible, whatever remains must be the truth, no matter how improbable."

Black Hole Complementarity

With the possible exception of Einstein, Niels Bohr was the most philosophical of the fathers of modern physics. To Bohr the philosophical revolution that accompanied the discovery of quantum mechanics was all about *complementarity*. The complementarity of quantum mechanics was manifest in many ways, but Bohr's favorite example was the particle-wave duality that had been forced on physics by Einstein's photon. Is light a particle? Or is it a wave? The two are so different that they seem totally irreconcilable.

Nevertheless, light is both a wave and a particle. Or more accurately, for certain kinds of experiments light behaves like particles. A very dilute beam of light falling on a photographic plate leaves tiny black dots: discrete evidence of the indivisible particle nature of the photon. On the other hand, those dots will eventually add up to a wavy interference pattern, a phenomenon that makes sense only for waves. It all depends on how you observe the light and what experiment you do. The two descriptions are complementary, not contradictory.

Another example of complementarity has to do with Heisenberg's Uncertainty Principle. In classical physics the state of motion of a particle involves both its position and its momentum. But in quantum mechanics you either describe a particle by its position or its momentum — never both. The sentence, "A particle has a position AND a momentum," must be replaced by, "A particle has a position OR a momentum." Likewise, light is particles, OR light is waves. Whether you use one description or the other depends on the experiment.

Black hole complementarity is the new kind of complementarity that results from combining quantum mechanics with the theory of gravity. There is no single answer to the question, "Who is right? The observer who remains outside the black hole and sees all information radiated from just above the horizon? Or the observer who falls through with the bits that are heading toward the center of the black hole?" Each is right in its own context: they are complementary descriptions of two different experiments. In the first the experimenter stays outside the black hole. He may throw things in, collect photons as they come out, lower probes down to just above the horizon, observe the effects on the trajectories of particles passing near the black hole, and so on.

But in the second kind of experiment, the physicist prepares an experiment in her lab. Then, lab and all, she jumps into the black hole, crossing the horizon, while performing the experiment.

The complementary descriptions of the two experiments are so radically different that it hardly seems credible that they could both be right. The external observer sees matter fall toward the horizon, slow down, and hover just above it.[3] The temperature just above the horizon is intense and reduces all matter to particles, which are finally radiated back out. In fact the external observer, monitoring the in-falling observer, sees her vaporized and reemitted as Hawking radiation.

But this is nothing like what the freely falling observer experiences. Instead she passes safely through the horizon without even noticing it. No bump or jolt, no high temperature, no warning of any kind signals the fact that she has passed the point of no return. If the black hole is big enough, let's say, with a few million light-year radius, she would sail on for another million years with no discomfort. No discomfort, that is, until she reaches the heart of the black hole, where tidal forces — the distorting forces of gravity — eventually become so strong that . . . never mind, it's too gruesome.

3. The word *see* is being used in a physicist's sense. It means reconstructing events from the output — in this case, the Hawking radiation. Such reconstruction would be incredibly complex, but in principle, it is just as possible as seeing the ordinary world through the light emitted and reflected from objects.

Two such different descriptions sound contradictory. But what we have learned from Bohr, Heisenberg, and others after them is that apparent paradoxes of this type signal genuine contradictions only when they lead to conflicting expectations for a single experiment. There is no danger of incompatible experimental results because the freely falling observer can never communicate her safe passage to the outside. Once she has safely passed the horizon, she is permanently out of contact with all observers who remain outside the black hole. Complementarity is strange but true.

The other major revolution of the early twentieth century was Einstein's Theory of Relativity. Certain things are relative to the state of motion of the observer. Two different observers moving rapidly past each other will disagree about whether two events occurred at the same time. One observer might see two flashbulbs flash at exactly the same time. The other would see one flash take place before the other.

The Principle of Black Hole Complementarity is also a new and stronger relativity principle. Once again the description of events depends on the state of motion of the observer. Remaining at rest outside the black hole, you see one thing. Falling freely toward the interior of the black hole, you see the same events entirely differently.

Complementarity and relativity — the products of the great minds of the early twentieth century — are now being united in a radically new vision of space, time, and information.

The Holographic Principle

Perhaps the error that Hawking made is to think that a bit of information has a definite location in space. A simple example of a quantum bit is the polarization of a photon. Every photon has a screw sense to it. Imagine the electric field of a photon as it moves. The tip of the electric field moves with a helical motion — a corkscrewlike motion. Think of yourself as following behind the light ray. The corkscrewing motion can be either clockwise or counterclockwise. In the first case the photons making up the beam are called right-handed photons; in the second case, they are left-handed. It's like the direction that you would have to twist a screwdriver in order to drive a screw into the wall in front of you. Ordinary screws are right-handed, but no law of nature forbids left-handed screws. Photons come in both types. The distinction is called the circular polarization of the photon.

The polarization of a single photon is composed of a single quantum bit of information. Morse code messages could be sent in the form of a sequence of photons, the messages being coded in the sequence of polarizations instead of a sequence of dots and dashes.

What about the location of that bit of information? In quantum mechanics the location of a photon may not be definite. After all, you can't specify both the location and momentum of the photon. Doesn't that mean that the bit of information is not at a definite place?

You may not know exactly where the photon is, but you can measure its location if you choose. You just can't measure both its position and its momentum. And once you measure the photon's location, you know exactly where that bit of information is. Furthermore, in conventional quantum mechanics and relativity, every other observer will agree with you. In that sense the quantum bit of information has a definite location. At least that's what was always thought to be the case.

But the Principle of Black Hole Complementarity says that the location of information is not definite, even in that sense. One observer finds the bits making up her own body are somewhere far behind the horizon. The other sees those same bits radiated back out from a region just outside the horizon. So it seems that the idea that information has a definite location in space is wrong.

There is an alternative way to think about it. In this view the bits have a location, but they're not at all where you think they are. This is the holographic view of nature that grew out of thinking about black holes. How do holograms apply?

A picture, a photograph, or a painting is not the real world that it depicts. It's flat, not full with three-dimensional depth like the real thing. Look at it from the side — almost edge on. It doesn't look anything like the real scene viewed from an angle. In short it's two-dimensional, while the world is three-dimensional. The artist, using perceptual sleight of hand, has conned you into producing a three-dimensional image in your brain, but in fact the information just isn't there to form a three-dimensional model of the scene. There is no way to tell if that figure is a distant giant or a close midget. There is no way to tell if the figure is made of plaster or if it's filled with blood and guts. The brain is providing information that is not really present in the painted strokes on the canvas or the darkened grains of silver on the photographic surface.

The screen of a computer is a two-dimensional surface filled with pixels. The actual data that are stored in a single image are in the form of some digital information about color and intensity — a collection of bits for each pixel. Like the painting or the photo, it is actually a very poor representation of the three-dimensional scene.

What would we have to do to faithfully store the full three-dimensional data, including depth information, as well as the "blood-and-guts" data about the interior of objects? The answer is obvious: instead of a collection of pixels filling two dimensions, we would need a space-filling collection of "voxels," tiny elements that fill the *volume* of space.

Filling space with voxels is far more costly than filling a surface with pixels. For example, if your computer screen is a thousand pixels on a side, the total number of pixels is one thousand squared, or one million. But if we want to fill a volume of the same size with voxels, the required number would be one thousand cubed, or one billion.

That's what makes holograms so surprising. A hologram is a two-dimensional image — an image on a piece of film — that allows you to unambiguously reconstruct full-blown three-dimensional images. You can walk around the reconstructed holographic image and see it from all sides. Your powers of depth perception allow you to determine which object in a hologram is closer or farther away. Indeed, if you move, the farther object can become the closer object. A hologram is a two-dimensional image, but one that has the full information of a three-dimensional scene. However, if you actually look closely at the two-dimensional film that contains the information, you see absolutely nothing recognizable. The image is thoroughly scrambled.

The information on a hologram, although scrambled, could be located on pixels. Of course nothing is for free. To describe the volume of space one thousand pixels on a side, the hologram would have to be composed of one billion pixels, not one million.

One of the strangest discoveries of modern physics is that the world is a kind of holographic image. But even more surprising, the number of pixels the hologram comprises is proportional only to the *area* of the region being described, not the volume. It is as though the full three-dimensional content of a region, one billion voxels in volume, can be described on a computer screen containing only a million pixels! Picture yourself in an enormous room bounded by walls, a ceiling, and a floor. Better yet, think of yourself as being in a large spherical space. According to the Holographic Principle, that fly just in front of your nose is really a kind of holographic image of data stored on the two-dimensional boundary of the room. In fact you and everything else in the room are images of data stored on a *quantum hologram* located on the boundary. The hologram is a two-dimensional array of tiny pixels — not voxels — each as big as the Planck length! Of course the nature of the quantum hologram and the way it codes three-dimensional data is very different from the way ordinary holograms work. But they do have in common that the three-dimensional world is completely scrambled.

What does this have to do with black holes? Let's place a black hole in our large spherical room. Everything — black hole, space traveler, mother ship — is stored as information on the holographic walls of the space. The two different pictures that black hole complementarity tries to reconcile are simply two different reconstructions of the same hologram by two different reconstruction algorithms!

The Holographic Principle was not widely accepted when 't Hooft and I put it forward in the early 1990s. My own view was that it was correct but that it would take many decades until we knew enough about quantum mechanics and gravity to confirm it in a precise way. But just three years later, in 1997, all that changed when a young theoretical physicist — Juan Maldacena — electrified the physics world with a paper titled "The Large N Limit of Superconformal Field Theories and Supergravity." Never mind what the words mean. Maldacena, by cleverly using String Theory and Polchinski's D-branes, had discovered a completely explicit holographic description of, if not our world, a world similar enough to make a convincing case for the Holographic Principle. Slightly later Ed Witten put his stamp of approval on the Holographic Principle with a follow-up to Maldacena's paper titled "Anti De Sitter Space and Holography." Since then the Holographic Principle has matured into one of the cornerstones of modern theoretical physics. It has been used in many ways to illuminate problems that, on the face of them, have nothing to do with black holes.

What does the Holographic Principle have to do with black hole complementarity? The answer is *everything*. Holograms are incredible scrambles of data that have to be decoded. That can be done by a mathematical algorithm or by shining laser light on the hologram. The laser light implements the mathematical algorithm.

Imagine a scene containing a large black hole and other things that might fall into the black hole as well as the radiation coming out. The entire scene can be described by a quantum hologram localized far away on some distant boundary of space. But now there are two possible ways — two algorithms — for decoding the hologram. The first reconstructs the scene as seen from outside the black hole, with the Hawking radiation carrying away all the bits that fell in. But the second reconstruction shows the scene as it would be seen by someone falling into the black hole — one hologram, but two ways to reconstruct its content.

Bubbles All around Us

It's probably too much to say that the three-dimensional world is a complete illusion. But the idea that the location of a bit of information is not necessarily where you might expect is now a widely accepted fact. What are its implications for the bubble bath universe of chapter 11? Let me remind you where we left off at the end of that chapter.

In the last chapter I explained the two views of history, one series and one parallel. According to the series view, every observer sees at most a small portion of the entire megaverse. The rest will never be seen because it is moving away so fast that light cannot bridge the gap. The boundary between what can and cannot be seen is the horizon. Unfortunately, the rest of the megaverse of pocket universes is all in this never-never land beyond the horizon. According to the classical principles of general relativity, we can wonder all we want about the existence and reality of these other worlds, but we can never know. They are irrelevant. They are meaningless in the scientific sense. They are metaphysics, not physics.

But exactly the same conclusion was incorrectly drawn about black hole horizons. Indeed, the cosmic event horizon of an eternally inflating universe is mathematically very similar to the horizon of a black hole. Let's return to the infinite lake filled with boats and observers. The black hole was very much like the dangerous drain, the horizon being the point of no return. Let's compare that situation with the eternally inflating lake, i.e., the lake fed by feeder tubes so that the floating observers all separate according to Hubble's Law. If the lake is fed at a constant rate, it provides a precise analog for Eternal Inflation.

Any particular boat will be surrounded by a boundary similar to the point of no return that surrounded the drain. Imagine a dinghy hovering around its parent vessel. If by accident or design it gets beyond the point of no return, it simply cannot get back or even communicate with the parent vessel. The only difference between the black hole horizon and the cosmic horizon of inflating space is that in one case we are on the outside looking in, and in the other case, we are inside looking out. But in every other way, the black hole and cosmic horizons are the same.

To someone outside a black hole, the events in the life of the trans-horizon explorer are behind the horizon. But those events are physics, not metaphysics. They are telegraphed to the outside in scrambled holographic code in the form of Hawking radiation. Like the prisoner's message, it doesn't matter if the code is lost — or even whether we ever had it. The message is in the cards.

Are there also "cards" coming from behind the cosmic horizon with messages from billions of pocket universes? Cosmic horizons are not nearly as well understood as black holes. But if the obvious similarity between them is any guide, cosmic horizons do yield such cards, and they are very much like the photons Hawking radiation comprises. By now you may have guessed that they are the photons of the cosmic microwave background radiation that bathe us from every direction and for all time. Messengers from the cosmic horizon, they are also coded messages from the megaverse.

George Smoot, one of the leaders in cosmic microwave detection, in an overenthusiastic moment likened a cosmic microwave map of the sky to "the face of God." I think for inquiring minds curious about the world a scrambled hologram of an infinity of pocket universes is a far more interesting and accurate image.

CHAPTER THIRTEEN

Summing Up

Slogans

One theme has threaded its way through our long and winding tour from Feynman diagrams to bubbling universes: our own universe is an extraordinary place that appears to be fantastically well designed for our own existence. This specialness is not something that we can attribute to lucky accidents, which is far too unlikely. The apparent coincidences cry out for an explanation.

An immensely popular story, not only among the general public but also many scientists, is that a "superarchitect" designed the universe with some benevolent purpose in mind.[1] The advocates of this view, intelligent design, say that it is quite scientific and perfectly fits the facts of cosmology as well as biology. The intelligent designer not only chose excellent Laws of Physics for its purpose but also guided biological evolution through its unlikely chain, from bacteria to Homo sapiens. But this is an intellectually unsatisfying, if emotionally comforting, explanation. Left unanswered are: who designed the designer, by what mechanism the designer intervenes to guide evolution, whether the designer violates the Laws of Physics to accomplish its goals, and whether the designer is subject to the laws of quantum mechanics.

One hundred and fifty years ago, Charles Darwin proposed an answer for the life sciences that has become the keystone of modern biology — a mechanism that needs no designer and no purpose. Random

1. See, for example, Paul Davies's 1983 book, *God and the New Physics* (New York: Simon and Schuster).

mutation, combined with competition to reproduce, explains the proliferation of species that eventually fill every niche, including creatures that survive by their wit. But physics, astronomy, and cosmology lagged behind. Darwinism may explain the human brain, but the specialness of the Laws of Physics has remained a puzzle. That puzzle may be yielding, at last, to physical theories that parallel Darwin's biological theory.

The physical mechanisms that I have explained in this book share two key ingredients with Darwin's theory. The first is a huge Landscape of possibilities — an enormously rich space of possible designs.[2] There are more than 10,000 species of birds, 300,000 species of beetles, and millions of species of bacteria. The total number of *possible* species is undoubtedly immeasurably larger.

Is the number of biological designs as large as the number of universe designs? That depends on exactly what we mean by a biological design. One way of listing all biological possibilities is to enumerate the ways of assigning the base pairs in a large DNA molecule. A human DNA strand has about a billion base pairs, and there are four possibilities for each. The total number of possibilities is the ridiculously large number $4^{1000000000}$ (or $10^{600000000}$). This is much bigger than the 10^{500} (similarly obtained by counting the number of ways of assigning flux integers) that string theorists guess for the number of valleys of the Landscape, but of course almost all of these do not correspond to viable life forms. On the other hand, most of the 10^{500} vacuums are also dead ends. In any case both numbers are so large that they are far beyond our powers of visualization.

The second key ingredient is a superprolific mechanism to turn the blueprint designs into huge numbers of real entities. Darwin's mechanism involved replication, competition, and lots and lots of carbon, oxy-

2. Long after I had written this chapter, while *The Cosmic Landscape* was in the final stages of editing, I happened to read an essay by Richard Dawkins titled "Darwin Triumphant" (reprinted in *A Devil's Chaplain: Reflections on Hope, Lies, Science, and Love* [New York: Houghton Mifflin, 2003]) in which Dawkins uses the term *Landscape* in exactly the way I am using it here. Some of the concepts are so similar to the ones in this book that at first I thought Dawkins must have access to my computer files. But if he did plagiarize my work, he must have solved the problem of time travel. "Darwin Triumphant" was written in 1991 and published that year in *Man and Beasts Revisited*, ed. M. H. Robinson and L. Tiger (Washington, D.C.: Smithsonian Institution Press).

gen, and hydrogen on which these mechanisms operate. Eternal Inflation also involves exponential replication — but of volumes of space.

As I discussed in chapter 11, the process of populating the Landscape does have its similarities with biological evolution, but it also has at least two very big differences. The first was discussed in chapter 11. Biological evolution along a given line of descent is through minute, undetectable changes from generation to generation. But descent through a series of bubble nucleations involves, at each stage, large changes of vacuum energy, particle masses, and the rest of the Laws of Physics. Biologically, if only such large changes were possible, Darwinian evolution would be impossible. The mutated monsters would be at such a disadvantage relative to normal offspring that their survival in a competitive world would be impossible.

How then does the megaverse become populated with diversity if biological evolution, under the same conditions, would stagnate? The answer lies in the second big difference between the two kinds of evolution: there is no competition for resources among pocket universes. It's interesting to contemplate an imaginary world in which biological evolution takes place in an environment where resources are so unbounded that there is no need for competition. Would intelligent life evolve in such a world? In most descriptions of Darwinian evolution, competition is a key ingredient. What would happen without it? Let's take a particular case, the final step in the evolution of our own species. About 100,000 years ago Cro-Magnons were in a struggle for survival with Neanderthals. Cro-Magnons won because they were smarter, bigger, stronger, or sexier. Thus, the average genetic stock of the human race was improved. But suppose resources were unbounded and that sex was unnecessary for reproduction. Would there be fewer Cro-Magnons? Not at all. Everyone who survived would survive more easily without competition. And many who did not survive would do so as well. But there would also be more Neanderthals. In fact there would be more of everyone. All populations would increase exponentially. In a world of unbounded resources, lack of competition would not have slowed the evolution of the smartest creatures, but it would have made a lot more dumb ones.

There is a third context, after physics and biology, where the same two ingredients — a Landscape and a megaverse — are essential to our existence. Planets and other astronomical bodies come in a very large number of possible designs. Hot stars, cold asteroids, giant dust clouds are just a few. Once again the Landscape of possibilities is extremely rich. Just the variation in distance from the parent star gives great diversity to planets. As for the mechanism that turns possibilities into actualities, the Big Bang, and the subsequent clumping by means of gravity, created 10^{22} planets within the observable part of our universe alone.

In each of these cases the answers to the questions of our own existence are the same. There are many creatures/planets/pocket universes and many possible designs. The numbers are so big that, statistically, some of them will be intelligent or conducive to intelligent life. Most creatures/universes/astro-bodies are dead ends from this point of view. We are just the lucky few. That is the meaning of the Anthropic Principle. There is no magic, no supernatural designer: just the laws of very large numbers.

My friend Steve Shenker, who is one of the wisest physicists I know, likes to reduce things to slogans. He feels that unless a big important idea can be encapsulated in a short phrase or two, its essence has not really been grasped. I think he is right. Here are some examples from the past.

From Newtonian mechanics:

Space and time are absolute.

Einstein and special relativity:

Space and time are relative.

and

The speed of light is an absolute constant.

Einstein and general relativity:

> The equivalence principle: gravity
> is indistinguishable from acceleration.

Quantum mechanics:

> The Heisenberg Uncertainty Principle: position and
> velocity cannot be simultaneously determined.

Cosmology:

> The Big Bang

The best scientific slogans I know don't come from physics or cosmology but from the theory of evolution:

> Survival of the fittest
> Natural selection
> The selfish gene

If this book were to be reduced to a single thought, it would be that the grand organizing principle of both biology and cosmology is:

> A Landscape of possibilities populated by
> a megaverse of actualities

There is one frustrating difference between the biological or planetary mechanism and the Eternal Inflation that populates the Landscape. In the two former cases, we can directly observe the results of the prolific mechanism of creation. We see the diversity of bio-forms all around us. Astronomical objects are a little harder to observe, but even without telescopes we can see planets, moons, and stars. But the huge sea of pocket universes created by Eternal Inflation is hidden behind our cosmic event horizon. The problem is, of course, Einstein's speed limit. If we could exceed the speed of light, there would be no problem traveling to distant pocket universes and back. We could navigate the entire megaverse. But, alas, punching a wormhole through space to a distant pocket universe is a fantasy that violates fundamental principles of physics. The existence of other pocket universes remains, and will remain, a conjecture, but a conjecture with explanatory power.

Consensus?

If the ideas that I have explained turn out to be correct, then our view of the world is about to expand far beyond the current provincial boundaries to something much grander: bigger in space, bigger in time, and bigger in possibilities. If correct, how long will it take for the paradigm to shift? Like the proverbial forest, paradigm shifts are easiest seen from a distance. While the ground is shifting, things are often too confusing, the waters too muddy, to see clearly, even a few years ahead. During those times it is almost impossible for outsiders to know whose ideas are serious and whose are fringe speculations. It's even hard for the insiders to know. My main purpose in writing this book is not primarily to convince the reader of my own point of view; scientific arguments are best fought on the pages of technical journals and the blackboards of seminar rooms. My purpose is to explain the struggle of ideas that is about to take front-and-center place in the mainstream of science so that ordinary readers can follow the ideas as they unfold and experience the drama and excitement that I feel.

The history of scientific ideas has always fascinated me. I am as interested in how the great masters came to their insights as I am in the ideas themselves. But the great masters are not all dead. The present — right now — is a marvelous time to watch the Weinbergs, Wittens,

’t Hoofts, Polchinskis, Maldacenas, Lindes, Vilenkins . . . as they struggle toward a new paradigm. As far as I can make out, here is what my most distinguished colleagues think. I will address the physicists first and then the cosmologists.

Steven Weinberg, more than any other physicist, is responsible for the discovery of the Standard Model of particle physics. Steve is not a rash man and is likely to weigh the evidence at least as carefully as anyone. His writings and lectures clearly imply that he sees the evidence, if not as definitive, then as strongly suggesting that some version of the Anthropic Principle may play a role in determining the Laws of Physics. But his own writings express regret — regret for a “paradigm lost.” In his 1992 book, *Dreams of a Final Theory*, he writes:

> Thus if such a cosmological constant is confirmed by observation, it will be reasonable to infer that our own existence plays an important role in explaining why the universe is the way it is.
>
> For what it is worth, I hope that this is not the case. As a theoretical physicist, I would like to see us able to make precise predictions, not vague statements that certain constants have to be in a range that is more or less favorable to life. I hope that string theory really will provide a basis for a final theory and that this theory will turn out to have enough predictive power to be able to prescribe values for all the constants of nature including the cosmological constant. We shall see.

Weinberg wrote these words during the afterglow of the discoveries of Heterotic String Theory and Calabi Yau compactification. But he now knows that String Theory will not be the hoped-for alternative to the Anthropic Principle.

Ed Witten is one of the greatest mathematicians in the world and a Pythagorean at heart. He has built his career around the elegant and beautiful mathematics that came out of String Theory. His ability to plumb the mathematical depths of the subject is breathtaking. Not surprisingly he is one of the most reluctant of my colleagues to give up the search for a magic, mathematical silver bullet, a bullet that will pick out a unique, consistent set of physical laws for elementary particles. If such a bullet exists, Witten has the depth and power to find it. But he has been looking for a long time with no success. Although he has done more than anyone to create the tools that are needed to explore the Landscape, I don't suppose he is at all happy about the current direction that the theory is taking.

If Witten is the driving force behind the mathematical tools of String Theory, Joe Polchinski has been the primary source of "parts" for the great machine. Joe, together with the brilliant young Stanford physicist Raphael Bousso,[3] made the first use of these parts to construct a model of the Landscape with a huge "discretuum" of vacuums. In many conversations Joe has expressed a belief that there is no alternative to the populated Landscape viewpoint.

My old comrade in arms Gerard 't Hooft has always been skeptical of String Theory's claim of approaching a Theory of Everything and recently elaborated in an e-mail message:

> Nobody could really explain to me what it means that string theory has 10^{100} vacuum states. Before you can say such a thing you must first give a rigorous definition of what string theory is, and we haven't got such a definition. Or was it 10^{500} vacua, or $10^{10000000000}$? As long as such "details" are still up in the air, I feel extremely uncomfortable with the anthropic argument.
>
> However, some form of anthropic principle I cannot rule out. After all, we live on Earth, not on Mars, Venus or Jupiter, for anthropic reasons. This, however, makes me distinguish the Discrete from the Continuous Anthr Principle. Discrete means something like: the fine-structure constant is an inverse integer, happens to be $1/137$, that gets higher order corrections. Continuous means this constant is $1/137.018945693459823497634978634913498724082734$ and so on, all of these decimals being determined by the anthr. princ. That I find unacceptable. String theory seems to be saying that the first 500 decimals are anthropic, the rest mathematic. I think it is far too early to make such speculations.

3. Today, Bousso is professor of physics at the University of California at Berkeley.

Roughly translated, what 't Hooft means by the Discrete Anthropic Principle is that the Landscape should not contain so many vacuums that every value of the constants of nature can be found. In other words, he would be less unhappy with anthropic reasoning if the number of distinct possibilities were finite as opposed to infinite.

I think it is noteworthy that, skeptical or not, Gerard neither rules out anthropic explanations nor offers an alternative explanation for the incredible fine-tuning of the cosmological constant. But about his skeptical attitude toward a final Theory of Everything, I think he is probably right.

Tom Banks is another skeptic. Tom is one of the deepest thinkers in physics and one of the most open-minded. His skepticism, like 't Hooft's, is not so much about anthropic reasoning but rather about String Theory's determination of the Landscape. Tom himself has made numerous important contributions to String Theory. But his own view is that the Landscape of metastable vacuums may just be illusory. He argues that String Theory and Eternal Inflation are simply not well enough understood to be certain that the Landscape is a mathematical reality. If certainty is the criterion, then I agree with him. But Banks feels the mathematics is not only incomplete but may actually be wrong. So far his arguments have not been persuasive, but they do raise serious concerns.

What do today's younger physicists think? By and large they are open-minded. Juan Maldacena, who is in his early thirties, has had the biggest impact on theoretical physics of anyone of his generation. It was largely his work that turned the Holographic Principle into useful science. Like Witten, he has contributed important new mathematical insight, and like Polchinski, he has had a deep impact on the physical interpretation of the mathematics. Of the Landscape he remarked, "I hope it isn't true." He, like Witten, had hoped for uniqueness, both in the Laws of Physics and in the history of the universe. Nevertheless, when I asked him if he saw any hope that the Landscape might not exist, he answered, "No, I'm afraid I don't."

At Stanford University — my home — there is pretty near unanimity on the issue, at least among the theoretical physicists: the Landscape exists. We need to become explorers and learn to navigate and to map it. Shamit Kachru and Eva Silverstein, both in their early thirties, are two of the world's young leaders. Both are busy constructing the Landscape's mountains, valleys, and ledges. Indeed, if I were to attribute to anybody the title of the Modern Rube Goldberg, it would be to Shamit. Don't get me wrong; I don't mean to say that he makes bad machines. On the contrary — Shamit has brilliantly used the complicated machine parts of String Theory better than anyone to design models of the Landscape. And the Anthropic Principle? It goes with the territory. It's part of the working assumption of all my close colleagues at Stanford, young and old.

At the other end of the country, New Jersey is the home of two of the world's great centers for theoretical physics. Princeton, with its university Physics Department and the Institute for Advanced Study, is first and foremost, but twenty miles north, in New Brunswick, is another powerhouse — Rutgers University. Michael Douglas is one of Rutgers's star attractions. Like Witten, he is both a brilliant physicist and a serious mathematician. But more important for this tale, he is a bold explorer of the Landscape. Douglas has set himself the task of studying the statistics of the Landscape rather than the detailed properties of individual valleys. He uses the laws of large numbers — statistics — to estimate which properties are most common, what percentage of valleys lies at different altitudes, and what the likelihood is that a valley that can support life exhibits approximate supersymmetry. While he prefers to use the term *statistical approach* instead of Anthropic Principle, it's probably fair to say that Douglas is on the anthropic side of the divide.

Cosmologists are equally split on the issue. Jim Peebles of Princeton University is the "grand old man" of American cosmology. Peebles has been a pioneer in every aspect of the subject. In fact in the late 1980s he was one of the very first people to suspect that cosmological data indicated the existence of something like a cosmological constant. In discussing the problems of cosmology with him, I was struck by his rather automatic acceptance that many features of the universe could be explained only by some kind of anthropic reasoning.

Sir Martin Rees, the British Astronomer Royal, is an all-out enthusiast for the Landscape, the megaverse, and the Anthropic Principle. Martin is Europe's leading cosmologist and astrophysicist. Many detailed arguments that I have used to motivate the Anthropic Principle I learned from him and from the American cosmologist Max Tegmark.

Andrei Linde and Alexander Vilenkin you have already met. Like Rees and Tegmark, they are firmly in the anthropic Landscape camp. Linde has expressed his opinion: "Those who dislike anthropic principle are simply in denial. This principle is not a universal weapon, but a useful tool, which allows us to concentrate on the fundamental problems of physics by separating them from the purely environmental problems, which may have an anthropic solution. One may hate the Anthropic Principle or love it, but I bet that eventually everyone is going to use it."

Stephen Hawking is Martin Rees's colleague at Cambridge University, but I have no doubt that his views are very much his own. Here is a quote from a lecture Stephen gave in 1999: "I will describe what I see as the framework for quantum cosmology, on the basis of M theory. I shall adopt the no boundary proposal, and shall argue that the Anthropic Principle is essential, if one is to pick out a solution to represent our universe, from the whole zoo of solutions allowed by M theory."

So it seems that Stephen and I finally agree on something.

But not all cosmologists agree. Among the best-known Americans in the field, Paul Steinhardt and David Spergel are vehement foes of anything that smells vaguely anthropic. Steinhardt, whose feelings are more or less representative, says he hates the Landscape and hopes it will go away. But like Maldacena he can find no way to get rid of it. From Steinhardt's writings (in "The *Edge* Annual Question — 2005," at www.edge.org): "Decades from now, I hope that physicists will be pursuing once again their dreams of a truly scientific 'final theory' and will look back at the current anthropic craze as millennial madness."

Alan Guth, the father of Inflation, is a fence sitter. Alan is a thorough believer in the populated Landscape. Indeed, it was he who coined the term *pocket universe*. But not being a string theorist, he takes a wait-and-see attitude toward the discretuum — in other words, he is less committed to the proposition that the number of possible vacuum environments is exponentially large. As for the Anthropic Principle, I suspect Alan is a closet believer. Whenever I see him I say, "Well Alan, have you 'come out' yet?" He invariably answers, "Not yet."[4]

I've saved for last my old friend David Gross. David and I have been good friends for forty years. During that time we have fought and argued incessantly, sometimes fiercely, but always with great respect for each other's opinions. My guess is that we will become two crusty old curmudgeons, battling to the very end. Maybe we already are.

David is, without doubt, one of the world's greatest living physicists. He is best known as one of the principal architects of Quantum Chromodynamics, i.e., the dynamics of hadrons.[5] But more important for this story, he has long been one of the most senior generals in the army of string theorists. In the mid-1980s, while a professor at Princeton, David and his collaborators Jeff Harvey, Emil Martinec, and Ryan Rohm created a sensation when they discovered Heterotic String Theory. This new version of String Theory looked much more like the real world of elementary particles than any previous version. Moreover, at about the same time Ed Witten (also at Princeton) was busy with his collaborators — Andy Strominger, Gary Horowitz, and Philip Candelas — inventing Calabi Yau compactification. When the two came together, the world of physics gasped — the results looked so realistic that it seemed only a matter of months until a definitive, final, unique theory of elementary particles would be in hand. The world held its breath — and held its breath and turned blue.

Fate was not kind. The more time that passes, the more it becomes clear that the Princeton enthusiasm was, at best, premature. But David has never given up the hope that the silver bullet will turn up and make the earlier enthusiasm justified. Myself? I suspect that, in the end, the Heterotic theory will turn out to be a very important component of

4. Bulletin: Just as I was about to finish writing this book, Guth wrote a paper (Alan Guth and David I. Kaiser, "Inflationary Cosmology: Exploring the Universe from the Smallest to the Largest Scales," *Science* 307 [2005]: 884–90) in which he said, "This idea — that the laws of physics that we observe are determined not by fundamental laws, but instead by the requirement that intelligent life can exist to observe them — is often called the anthropic principle. Although in some contexts this principle might sound patently religious, the combination of inflationary cosmology and the landscape of string theory gives the anthropic principle a viable framework." Has Alan just opened the closet door?

5. I am delighted to report that as I am writing this book, it has been announced that Gross and two others were awarded the Nobel Prize for their work on QCD.

Rube's great machine. Its resemblance to the Standard Model is impressive. But I would also guess that it is not the only component. Fluxes, branes, singularities, and other features may expand the Heterotic Landscape far beyond what the authors of the theory originally imagined.

Gross, as I said, is an extremely formidable intellectual opponent, and he is very opposed to the Anthropic Principle. Although his reasons are more ideological than scientific, they are important to discuss. What bothers him is an analogy with religion. Who knows? Maybe God *did* make the world. But scientists — real scientists — resist the temptation to explain natural phenomena, including creation itself, by divine intervention. Why? Because as scientists we understand that there is a compelling human need to believe — the need to be comforted — that easily clouds people's judgment. It's all too easy to fall into the seductive trap of a comforting fairy tale. So we resist, to the death, all explanations of the world based on anything but the Laws of Physics, mathematics, and probability.

David, along with many others, expresses the fear that the Anthropic Principle is like religion: too comforting, too easy. He fears that if we begin to open the door, even a crack, the Anthropic Principle will seduce us into a false belief and stop future young physicists from searching for the silver bullet. David eloquently quotes Winston Churchill's 1941 address to students at his own school: "Never, ever, ever, ever, ever, ever, ever, give up. Never give up. Never give up. Never give up." But the field of physics is littered with the corpses of stubborn old men who didn't know when to give up.

David's concern is very real, and I don't mean to minimize it, but I also think it's not as bad as he says. I don't for a moment worry about the younger generation lacking the moral fiber to avoid the trap. If the populated Landscape is the wrong idea, we (or perhaps I should say, they) will find it out. If the arguments that indicate the existence of 10^{500} vacuums are wrong, string theorists and mathematicians will discover it. If String Theory itself is wrong, perhaps because it is mathematically inconsistent, it will fall by the wayside and, with it, the String Theory Landscape. But if that does happen, then as things stand now, we would be left with no other rational explanation for the illusion of a designed universe.

On the other hand, if String Theory and the Landscape are right, with new and improved tools we may locate our valley. We may learn about the features in neighboring locales — including the inflationary ledge and steep downhill approach. And, finally, we may confirm that the rigorous use of mathematics leads to many other valleys, in no special way distinguished from ours except by their inhospitable environment. David has honest concerns, but to shun a possible answer because it runs counter to our earlier hopes is itself a kind of religion.

Gross has another argument. He asks, "Isn't it incredibly arrogant of us to presume that all life must be just like us — carbon based, needful of water" and so on. "How do we know that life can't exist in radically different environments?" Suppose, for example, some strange forms of life could evolve in the interior of stars, in cold dust clouds of interstellar space, and in the noxious gases that surround giant gas planets like Jupiter. In that case, the Ickthropic Principle of the codmologists would lose its explanatory power. The argument that life's need for liquid water explains the fine-tuning of the temperature would lose its force. In a similar vein, if life can form without galaxies, then Weinberg's explanation of the smallness of the cosmological constant also loses its force.

I think that the correct response to this criticism is that there is a hidden assumption that is an integral part of the Anthropic Principle, namely: *the existence of life is extremely delicate and requires very exceptional conditions*. This is not something that I can prove. It is simply part of the hypothesis that gives the Anthropic Principle its explanatory power. Perhaps we should turn the argument upside down and say that the success of Weinberg's prediction supports the hypothesis that robust intelligent life requires galaxies, or at least stars and planets.

What are the alternatives to the populated Landscape paradigm? My own opinion is that once we eliminate supernatural agents, there is none that can explain the surprising and amazing fine-tunings of nature. The populated Landscape plays the same role for physics and cosmology as Darwinian evolution does for the life sciences. Random copying errors, together with natural selection, are the only known natural explanation of how such a finely tuned organ as an eye could form from ordinary matter. The populated Landscape, together with the rich diversity predicted by String Theory, is the only known explanation of the extraordinary special properties of our universe which allow our own existence.

This is a good place for me to pause and address a potential criticism that might be leveled against this book, namely that it lacks balance. Where are the alternative explanations of the value of the cosmological constant? Aren't there any technical arguments against the existence of a large Landscape? What about other theories besides String Theory?

I assure you that I am not hiding the other side of the story. Throughout the years many people, including some of the most illustrious names in physics, have tried to explain why the cosmological constant is small or zero. The overwhelming consensus is that these attempts have not been successful. Nothing remains to report on that score.

As for serious mathematical attempts to debunk the Landscape, I know of only one. The author of that attempt is a good mathematical physicist, and as far as I know, he still believes his criticism of the KKLT construction (see chapter 10). The objection involves an extremely technical mathematical point about special Calabi Yau spaces. Several authors have criticized the criticism, but by now it may be irrelevant. Michael Douglas and his collaborators have found many examples that avoid the problem. Nevertheless, an honest assessment of the situation would have to include the possibility that the Landscape is a mathematical mirage.

Finally, as for alternatives to String Theory, a well-known one is called Loop Gravity. Loop Gravity is an interesting proposal, but it is not nearly as well developed as String Theory. In any case even its most famous advocate, Lee Smolin, believes that Loop Gravity is not really an alternative to String Theory but may be an alternative formulation of String Theory.

As much as I would very much like to balance things by explaining the opposing side, I simply can't find that other side. Opposing arguments boil down to a visceral dislike of the Anthropic Principle (I hate it) or an ideological complaint against it (it's giving up).

Two specific arguments have been the subjects of recent popular books by well-known physicists, but both have failed in my view. I'll take a moment to explain why.

The Laws of Nature Are Emergent

This is a favorite idea of some condensed-matter theorists who work on the properties of materials made of ordinary atoms and molecules. Its principal proponent is the Nobel Prize winner Robert Laughlin, who describes his ideas in his book *A Different Universe.*[6] The idea at its core is the old "ether theory" that maintains that the vacuum is some special material. The ether idea was popular in the nineteenth century, when both Faraday and Maxwell tried to think of electromagnetic fields as stresses in the ether. But after Einstein the ether fell into disrepute. Laughlin would like to resurrect the old idea by picturing the universe as material with properties similar to superfluid helium. Superfluid helium is an example of a material with special "emergent" properties, properties that reveal themselves (emerge) only when huge numbers of atoms are assembled in macroscopic amounts. In the case of liquid helium, the fluid has amazing superfluid properties such as flowing without any friction. In a lot of ways, superfluids are similar to the Higgs fluid that fills space and gives particles their properties. Roughly speaking Laughlin's view can be summarized by saying that we live in such a space-filling material. He might say it even more strongly: space *is* such an emergent material! Moreover, he believes that gravity is an emergent phenomenon.

One of the main themes of modern physics is that emergent phenomena have a kind of hierarchical structure. Little collections of molecules or atoms group together to form bigger entities. Once you know the properties of these new entities, you can forget where they came from. The new entities, in turn, combine and cluster into new groups of even larger size. Once again you can forget where they came from and group them into yet bigger groups until the whole macroscopic material is explained. One of the most interesting properties of these systems is that it doesn't matter exactly what you begin with. The original microscopic entities don't make any difference to the emergent behavior — the material always comes out with the same large-scale

6. Robert Laughlin, *A Different Universe: Reinventing Physics from the Bottom Down* (New York: Basic Books, 2005).

behavior — within limits.[7] For this reason Laughlin believes there is no point in looking for the fundamental objects of nature, since a wide variety of basic objects would lead to the same Laws of Physics — gravity, the Standard Model, and so on — in the large-scale world. In fact there are all kinds of "excitations" in materials that do resemble elementary particles but are really collective motions of the underlying atoms. Sound waves, for example, behave as though they were made of quanta called phonons. Moreover, these objects sometimes behave uncannily like photons or other particles.

There are two serious reasons to doubt that the laws of nature are similar to the laws of emergent materials. The first reason involves the special properties of gravity. To illustrate, consider the properties of superfluid helium, although any other material would do as well. All sorts of interesting things take place in superfluids. There are waves that behave similarly to scalar fields and objects called vortices that resemble tornadoes moving through the fluid. But there is no kind of isolated object that moves around in the fluid and resembles a black hole. This is not an accident. Black holes owe their existence to the gravitational force described by Einstein's General Theory of Relativity. But no known material has the characteristics that the General Theory of Relativity ascribes to space-time. There are very good reasons for this. In chapter 10, where I dealt with black holes, we saw that the properties of a world with both quantum mechanics and gravity are radically different from anything that can be produced with ordinary matter alone. In particular, the Holographic Principle — a mainstay of current thinking — seems to require totally new kinds of behavior not seen in any known condensed-matter system. In fact Laughlin himself illustrates the point by arguing that black holes (in his theory) cannot have properties, such as Hawking radiation, that practically everyone else believes them to have.

But suppose one found an emergent system that had some of the features that we want. The properties of emergent systems are not very flexible. There may be an enormous variety of starting points for the

7. Of course too large a change in the microscopic starting point could lead to an entirely different macroscopic result — for example, a crystal instead of superfluid.

microscopic behavior of atoms, but as I said, they tend to lead to a very small number of large-scale endpoints. For example, you can change the details of helium atoms in many ways without changing the macroscopic behavior of superfluid helium. The only important thing is that the helium atoms behave like little billiard balls that just bounce off one another. This insensitivity to the microscopic starting point is the thing that condensed-matter physicists like best about emergent systems. But the probability that out of the small number of possible fixed points (endpoints) there should be one with the incredibly fine-tuned properties of our anthropic world is negligible. In particular, there is no explanation of the most dramatic of these fine-tunings, the small but nonzero cosmological constant. A universe based on conventional condensed-matter emergence seems to me to be a dead-end idea.

Natural Selection and the Universe

Lee Smolin has attempted to explain the very special properties of the world — the Anthropic properties — by a direct analogy with Darwinian evolution — not in the general probabilistic sense that I explained earlier but much more specifically.[8] To his credit Smolin understood early that String Theory is capable of describing a tremendous array of possible universes, and he attempted to use that fact in an imaginative way. Although I feel that Smolin's idea ultimately fails, it is a valiant effort that deserves serious thought. The gist of it follows:

In any universe with gravitational forces, black holes can form. Smolin speculates about what might take place inside a black hole, in particular, at its violent singularity. He believes, in my opinion with no good evidence, that instead of space collapsing to a singularity, a resurrection of the universe takes place. A new baby universe is born inside the black hole. In other words universes are *replicators* that reproduce in the interior of black holes. If this is so, Smolin argues, then by a process of repeated replication — black holes forming inside universes, which are inside black holes, which are inside universes, and so on — an evolution will take place toward maximally fit universes. By fit Smolin means having the ability to produce a large number of black holes and, therefore,

8. See Lee Smolin, *The Life of the Cosmos* (Oxford: Oxford University Press, 1997).

a large number of offspring. Smolin then conjectures that our universe is the most fit of all — the laws of nature in our pocket are such that they produce the maximum possible number of black holes. He claims that the Anthropic Principle is totally unnecessary. The universe is not tuned for life. It is tuned to make black holes.

The idea is ingenious, but I don't think it explains the facts. It suffers from two serious problems. The first is that Smolin's idea of cosmic evolution is too closely patterned on Darwin's and requires changes between generations to be small incremental changes. As I said above, the pattern suggested by the String Theory Landscape is quite the opposite. In Smolin's defense I should point out that almost all of our knowledge of the Landscape was derived after his theory was published. At the time Smolin was formulating his ideas, the working paradigm for string theorists was the flat supersymmetric part of the Landscape, where it is indeed true that changes are incremental.

The other problem is cosmological and has little to do with String Theory. There is no reason whatever to believe that we live in a universe that is maximally efficient at producing black holes. Smolin makes a series of tortured arguments to prove that any changes in our universe would result in fewer black holes, but I find them very unconvincing. We saw in chapter 5 that it is a lucky "miracle" that the universe is *not* catastrophically filled with black holes. A relatively small increase in the early lumpiness of the universe would cause almost all matter to collapse to black holes rather than life-nurturing galaxies and stars. Also, increasing the masses of the elementary particles would cause more black holes to form since they would be more susceptible to gravitational attraction. The real question is why the universe is so lacking in black holes. The answer that seems to me to make the most sense is that many, maybe most, pockets have far more black holes than our pocket, but they are violent places in which life could not have formed.

The whole argument that we live in a world that is maximally fit to reproduce is, in my opinion, also fundamentally flawed. Space does indeed reproduce — one well-understood mechanism is Inflation — but the maximally reproducing universe is nothing like our own. The most fit universe in Smolin's sense, the one that replicates most rapidly, would be the universe with the largest cosmological constant. But there is no overlap between the fitness to reproduce and the fitness to support intelligent life. With its ultrasmall cosmological constant and its paucity of black holes, our universe is particularly unfit to replicate.

Going back to the analogy of the tree of life, in biology there is also no overlap between reproductive fitness and intelligence. The maximally fit creatures are not humans: they are bacteria. Bacteria replicate so rapidly that a single organism can have as many as ten trillion descendants in a twenty-four-hour period! According to some estimates the earth's population of bacteria is more than a thousand billion billion billion. Humans may be special in many ways but not in their ability to reproduce. A world that can support life is also very special — but, again, not in its ability to reproduce.

To put it another way, imagine Gregor Samsa — the hero in Franz Kafka's *The Metamorphosis* — on that fateful day when he awoke as a giant cockroach. Might he have asked himself, while still foggy with sleep, "What kind of creature am I?" Following Smolin's logic, the answer would be, "With overwhelming probability, I must belong to the class of creatures that are most fit to reproduce and are therefore the most populous. In short I must be a bacterium."

But a few seconds of reflection might convince him otherwise. Misquoting Descartes, he would conclude, "I think, therefore I am not a bacterium. I am something very special — a remarkable creature with extraordinary brainpower. I am not average: I am exceedingly far from average." We, also, should take no more than a moment to conclude that we are not average. We do not belong to the branch of the megaverse that is most fit to reproduce. We belong to the branch that can say: "I think, therefore the cosmological constant must be very small."

My reaction to Smolin's idea has been harsh. But the harshness is directed against particular technical points: not against Smolin's overall philosophy. I think Smolin deserves great credit for getting the most important things right. Smolin was the first to recognize that the diversity of String Theory vacuums may play an important positive role in explaining why the world is the way it is. He was also the first to try to use this diversity in a creative way to explain our special environment. And most important — he understood that there was an urgent question to answer: "How can the deepest and most powerful ideas of modern physics provide a truly scientific explanation of the *apparent* 'intelligent design' that we see all around us?" In all of this he was going directly against the strong prejudices of the string theorists, and I think that he was more correct then they were.

As I have repeatedly emphasized, there is no known explanation of the special properties of our pocket other than the populated Landscape — no explanation that does not require supernatural forces. But there are real problems with our current understanding of the populated Landscape, and some are potentially very serious. In my own view the biggest challenges have to do with Eternal Inflation — the mechanism that may populate the Landscape. The cloning of space is not seriously questioned by anyone, and neither is the spinning off of bubbles by the metastable vacuum. Both ideas are based on some of the most trusted principles of general relativity and quantum mechanics. But no one has a clear understanding of how these observations are to be turned into predictions — even statistical guesses — about our universe.

Given a megaverse, endlessly filled with pocket universes, the Anthropic Principle is an effective tool to weed out and eliminate most of them as candidates for our universe. Those that don't support our kind of life can be tossed in the trash. That provides marvelous explanatory power for questions like, why is the cosmological constant small? But much of the controversy over the Anthropic Principle has to do with a more ambitious agenda, the hope that it can substitute for the silver bullet in predicting all of nature.

This is an unreasonable expectation. There is no reason why every feature of nature should be determined by the existence of life. Some features will be determined by mathematical reasoning of the traditional kind, some by anthropic consideration, and some features may just be accidental environmental facts.

As always, the world of the big-brained fish (chapter 5) is a good place to get some perspective. Let's follow the fish as they learn more about their world:

> In time, with the help of the codmologists, the fish came to the realization that they inhabited a planet revolving around a hot glowing nuclear reactor — a star — that provided the heat that warmed their water. The question that had obsessed their best minds would take on an entirely new complexion. Realizing that the temperature depends on how distant the star is, the puzzle

> would be restated: "Why is the orbital distance of our planet from the source of heat so finely tuned?" But the answer of the codmologists is the same. The universe is big. It has many stars and planets, and some small fraction are just the right distance for liquid water and for fish.
>
> But some fyshicists are unhappy with the answer. They correctly claim that the temperature depends on something else besides the orbital distance. The luminosity of the star — the rate that it radiates energy — comes into the equation. "We could be close to a small, dim star or far from a bright giant. There is a whole range of possibilities. The Ickthropic Principle is a failure. There is no way that it can explain the distance to our star."

But it was never the intention of the codmologists to explain every feature of nature. Their claim that the universe is big and contains a wide variety of environments is as valid as ever. The criticism that the Ickthropic Principle can't explain everything is a straw man, set up by the fyshicists just to knock it down.

There are very close parallels between this story and the case of the Anthropic Principle. One example involves both the cosmological constant and the lumpiness in the early universe. In chapter 2 I related how Weinberg explained the fact that the cosmological constant is so incredibly small — if it were much bigger, the very small density contrasts (lumpiness) in the universe could not have grown into galaxies. But suppose the initial density contrasts were a bit stronger. Then a somewhat larger cosmological constant could be tolerated. As in the case of the distance and luminosity of the star, there is a range of possible values for the cosmological constant and lumpiness which permit life, or at least galaxies. The Anthropic Principle by itself is powerless to choose between them. Some physicists take this as evidence against the Anthropic Principle. Once again I regard it as a straw man.

But it is possible that with further input both the fyshicists and we could do better. Let's bring in the astro-fyshicists: the experts on how stars form and evolve. These fishy scientists have studied the formation of stars from giant gas clouds, and as expected, they find that a range of luminosities is possible. There is no way to be certain of the stellar luminosity without getting above the surface and observing the star, but still it seems that some values for the luminosity are more likely than others. Indeed, the astro-fyshicists find that the majority of long-lived stars should have a luminosity of between 10^{26} and 10^{27} watts. Their star is probably in this range.

Now the codmologists are in business. With such luminosity the planet would have to be about 100,000,000 miles from the star in order to have a climate temperate enough for liquid water. That prediction is not as absolute as they might like. Like all probabilistic claims, it could be wrong. But still, it's better than no prediction.

What the two situations — one involving liquid water and the other the formation of galaxies — have in common is that anthropic (or ick-thropic) considerations alone are not enough to determine or predict everything. This is inevitable if there is more than one valley in the Landscape that can support our kind of life. With 10^{500} valleys it seems certain that this will be the case. Let's call such vacuums anthropically acceptable. The usual physics and chemistry may be very similar in many of these — electrons, nuclei, gravity, galaxies, stars, and planets much like we know them in our own world. The differences may be in those things that only a high-energy particle physicist would be interested in. For example, there are many particles in nature — the top-quark, the tau lepton, the bottom-quark, and others — whose detailed properties hardly matter at all to the ordinary world. They are too heavy to make any difference except in high-energy collisions in giant accelerator laboratories. Some of these vacuums (including our own) may have many new types of particles that make little or no difference to ordinary physics. Is there any way to explain in which of these anthropically acceptable vacuums we live? Obviously, the Anthropic Principle cannot help us predict which one we live in — any of these vacuums is acceptable.

This conclusion is frustrating. It leaves the theory open to the serious criticism that it has no predictive power, something that scientists are very sensitive about. To address this deficiency many cosmologists have tried to supplement the Anthropic Principle with additional probabilistic assumptions. For example, instead of asking precisely what the value is of the top-quark mass, we might try to ask what the probability is that the top-quark has a mass in a particular range.

Here is one such proposal: eventually we will know enough about the Landscape to know just how many valleys exist for each range of the top-quark mass. Some values of the mass may correspond to a huge number of valleys — some may correspond to a much smaller number. The proposal is simple enough — the values of the top-quark mass that correspond to a great many valleys are more probable than values that correspond to few valleys. To carry out this kind of program, we would have to know far more about the Landscape than we do now. But let's put ourselves in the future, when the details of the Landscape have been mapped out by String Theory, and we know the number of vacuums with any conceivable set of properties. Then the natural proposal would be that the relative probability for two distinct values of some constant would be the ratio of the number of appropriate vacuums. For example, if there were twice as many vacuums with mass value M_1 as mass value M_2, it would follow that M_1 was twice as likely as M_2. If we are lucky we might find that some value of the top-quark mass corresponds to an exceptionally large number of valleys. Then we might move forward by assuming this value to be true for our world.

No single prediction of this kind, based as it is on probability, can make or break the theory, but many successful statistical predictions would add great weight to our confidence.

The idea I have just outlined is tempting, but there are serious reasons to question the logic. Remember that the Landscape is merely the space of possibilities. If we were fyshicists thinking about the Landscape of possible planets, we might count all sorts of bizarre possibilities as long as they were solutions to the equations of physics: planets with cores of pure gold among others. The equations of physics have just as many solutions corresponding to huge golden balls as iron balls.[9] The logic of counting possibilities would say that an iron-cored planet is no more likely to be the fyshicists' home than a golden planet — obviously a mistake.[10]

9. One might also think that planets shaped in the form of elongated ellipsoids, cubes, or even sea urchins would be solutions of the equations of physics. But this is not so. If the planet were big enough to hold an atmosphere, gravity would quickly pull the material together into the form of a ball. Not everything is possible.
10. The core of the earth is composed mostly of iron.

What we really want to know is not how many *possibilities* of each kind there are: what we want to know is how many *planets* of each kind there are. For this we need much more than the abstract counting of possibilities. We need to know how iron and gold are produced during the slow nuclear burning inside stars.

Iron is the most stable of all the elements. It is more difficult to dislodge a proton or neutron from an iron nucleus than from any other. Consequently, nuclear burning proceeds down the periodic table, hydrogen to helium to lithium, until finally it ends with iron. As a result iron is far more common in the universe than any of the elements with a higher atomic number, including gold. That's why iron is cheap and gold costs almost five hundred dollars an ounce. Iron is ubiquitous in the universe: gold, by contrast, is very rare. Almost all solid planets will have far more iron in their cores than gold. By comparison with iron planets, the number of solid-gold planets in the universe is minute, very possibly zero. We want to count *actualities* rather than possibilities.

The same logic as applies to planets ought to apply to pocket universes. But now we encounter a horrific problem with Eternal Inflation. Because it goes on forever, Eternal Inflation (as it is now conceived) will create an infinite number of pockets — in fact an infinite number of every kind of pocket universe. Thus, we face an age-old mathematical problem of comparing infinite numbers. Which infinity is bigger than which and by how much?

The problem of comparing infinite numbers goes back to Georg Cantor, who in the late nineteenth century asked exactly that question — how do you compare the size of two sets, each of which has an infinite number of elements? First he began by asking how to compare ordinary numbers. Suppose, for example, we have a bunch of apples and a bunch of oranges. The obvious answer is to count both bunches, but if all we want to know is which is bigger, then there is a more primitive thing we can do — something that doesn't even require any knowledge of numbers: line up the apples and next to them line up the oranges, matching each orange to an apple. If some apples are left over, then there are more apples. If oranges are left over, then there are more oranges. If the oranges and apples match, their number is the same.

Cantor said the same thing could be done with infinite (or what he called transfinite) sets. Take, for example, the even integers and the odd

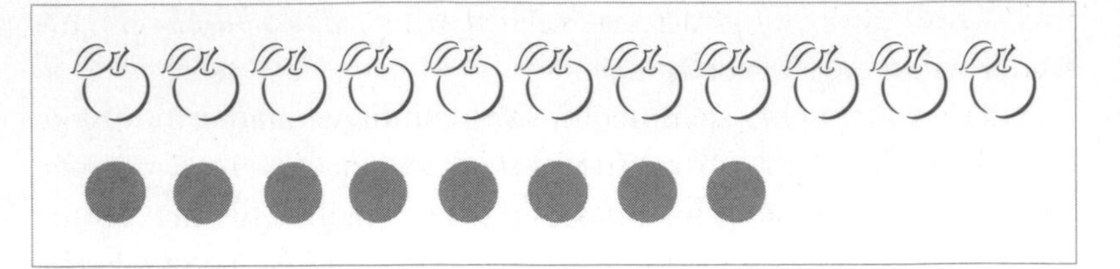

integers. There is an infinite number of each, but is the infinite number the same? Line them up and see if they can be made to match in such a way that there is an even for every odd. Mathematicians call this a *one-to-one correspondence.*

1 3 5 7 9 11 13 . . .
2 4 6 8 10 12 14 . . .

Note that the two lists eventually contain every odd integer and every even integer — none is left out. Moreover, they match exactly, so Cantor concluded that the number of evens and odds are the same even though they are both infinite.

What about the total number of integers, even and odd? That is obviously larger than the number of even integers — twice as large. But Cantor disagreed. The even integers can be matched exactly with the list of all integers.

1 2 3 4 5 6 7 . . .
2 4 6 8 10 12 14 . . .

According to the only mathematical theory of infinite numbers, the theory that Cantor constructed, the number of even integers is the same as the number of all integers! What is more, the set of numbers divisible by 10 — 10, 20, 30, 40, etc. — is exactly the same infinite size. The integers, the even or odd integers, the integers that are divisible by ten —

11. Not all infinites are the same, according to Cantor. The integers — the even integers and the odd integers — are what mathematicians call a countably infinite set. The number of real numbers, all possible decimals, are a much bigger set that cannot be put into

they are all examples of what mathematicians call countably infinite sets, and they are all equally large.[11]

Let's do a thought experiment involving infinite numbers. Imagine an infinite bag filled with all the integers written on scraps of paper. Here is the experiment: first shake the bag to thoroughly mix up the scraps. Now reach in and draw out a single integer. The question is — what is the probability that you've pulled out an even integer?

The naive answer is simple. Since half the integers are even, the probability must be one-half — 50 percent. But we can't really do this experiment because no one can make an infinite bag of integers. So to test the theory, we can cheat a little and use a finite bag, containing, let's say, the first thousand integers. Sure enough if we do the experiment over and over we will, indeed, find the probability to pull an even integer is one-half. Next we can do the same experiment with a bag filled with the first ten thousand integers. Again, since half the scraps are even and half are odd, we will find the probability for an even integer is one-half. Do it again with the first one hundred thousand integers, the first million integers, the first billion, and so on. Each time the probability is one-half. It is reasonable to extrapolate from this that if the bag had an infinite number of scraps, the probability would remain one-half.

But wait. We could modify the contents of the bag in the following way. Start with the first *one* thousand even integers and the first *two* thousand odd integers. Now there are twice as many odds as evens, and the probability to draw an even number is only one-third. Next repeat the experiment with the first ten thousand even integers and the first twenty thousand odd integers. Again the probability is one-third. As before, we can extrapolate to the limit of an infinite bag, but each time the result will be one-third. In fact we can make the answer come out any way we want by varying how we define the limit of an infinite experiment.

The eternally inflating universe is an infinite bag, not of paper scraps with numbers but of pocket universes. In fact it is a bag in which each possible type of universe — each valley of the Landscape — is represented a countably infinite number of times. There is no obvious math-

---

one-to-one correspondence with the integers, but all countable sets are the same size! Pocket universes are like the integers; they are things you can count.

ematical way to compare one kind of pocket universe with another and to declare that one is more probable than the other. The implication is very disquieting: there seems no way to define the relative likelihood of different anthropically acceptable vacuums.

The measure problem (the term *measure* referring to the relative probabilities of different vacuums) has vexed some of the greatest minds in cosmology, Vilenkin and Linde especially. It could prove to be the Achilles heel of Eternal Inflation. On the one hand, it is hard to see how Eternal Inflation can be avoided in a theory with any kind of interesting Landscape. But it is equally hard to see how it can be used to make scientific predictions of the kind that would establish it as science in the traditional sense.

In the past physics has faced many other problems involving infinite numbers: the ultraviolet catastrophe confronting Planck or the strange infinities that were the bane of quantum field theory in the early days. Even the problems of black holes that Hawking, 't Hooft, and I debated are problems of the infinite. According to Hawking's calculations, a black hole horizon is able to store an infinite amount of information without giving it back to the environment. These all were deep problems of transfinite, or infinite, numbers. In each case new physical principles had to be discovered before progress could be made. In Planck's case it was quantum mechanics itself — Einstein's recognition that light was made of quanta. The infinite numbers that plagued quantum field theory were purged only when the new principles of *renormalization theory* were uncovered and eventually understood by Kenneth Wilson. The black hole story is still being understood, but the outlines of a solution in terms of the Holographic Principle are in place. In each case it was found that the classical rules of physics overestimated the number of degrees of freedom that describe the world.

I believe the measure problem will also require a major new idea before we can understand how to make predictions about the Landscape. If I had to make a guess, I would say that it has something to do with the Holographic Principle and the way that information beyond our horizon is contained in the cosmic radiation in our own pocket. But if I were an opponent of the populated Landscape, I would aim my attack at these conceptual problems of Eternal Inflation.

The measure problem aside, the practical difficulty of making testable predictions that can be compared with experiments or observations is a serious problem. But I think the situation is far from hopeless. There are a few types of evidence that could be obtained in the near future.

The Beginning of Inflation

In chapter 4 I explained how the tiny density contrasts in the early universe (observed in the cosmic microwave background) were created during the final Inflation that took place on a ledge overlooking our valley. These were the seeds that evolved into galaxies. There was lumpiness on many different scales, some occupying tiny portions of the sky and some much bigger structures, occupying almost the entire sky. The cosmic lumps and bumps we can observe now are fossil remnants dating to different eras. The important correlation to remember is that the biggest lumps were frozen in at the earliest times.

If we are very, very lucky, the largest lumps in the CMB might date to a time just before the usual Inflation got started — in other words, just as the universe was settling onto the inflationary ledge. If that were the case, the largest lumps would be a bit less lumpy than the slightly smaller lumps, which were produced after the Inflation was going on for a while. Indeed, there is some evidence that the very largest lumps are weaker than the others. It is a long shot, but those large-scale density contrasts could have information about the formation of our bubble from a previous epoch with a larger cosmological constant.

If we are that lucky, then the Inflation did not go on long enough to wipe out evidence for the curvature of space. Here again bubble nucleation has a distinct signature. If our pocket universe was born in a bubble-nucleation event, the universe must be *negatively curved*. The interior angles of cosmic triangles will add up to less than 180 degrees.

At the level of accuracy that the curvature of space has been measured, there is no indication of such curvature. This idea may fail because standard Inflation probably had been going on for a long time when the largest visible lumps were formed. But if we do detect negative curvature, that will be a smoking gun telling us that our universe was born as a tiny bubble in a vacuum with a larger cosmological constant.

Superstrings in the Sky

We have not exhausted all our options for observing the universe yet. Is it possible that we can actually see superstrings? The obvious answer is that they are much too small to be seen. But the same thing could have been said about the tiny quantum fluctuations that occurred during Inflation. In chapter 5 we saw that the expansion of the universe, and the effects of gravity, somehow inflated these fluctuations until they became first the density contrasts in the cosmic microwave background and eventually the eminently visible galaxies in today's sky. That we can see the effects of microscopic quantum phenomena frozen onto the sky like an expanding pointillist abstract painting is an incredible fact. It came as a complete surprise to most physicists, who were used to thinking of the quantum world as strictly microscopic. So perhaps we shouldn't be too quick to assume that small-scale objects like strings can't do something similar: perhaps turn the sky into a giant Jackson Pollock canvas.

Building on the work of their colleagues, Thibault Damour, Alex Vilenkin, Joe Polchinski, and others have begun to explore an enormously exciting new opportunity, once again originating from phenomena connected with Inflation. Inflation is caused by vacuum energy that was present long ago. That vacuum energy disappeared as the universe slid down the Landscape to its present very low altitude, but the vacuum energy didn't depart without leaving something behind. It was converted into more ordinary forms of energy, namely, heat and particles, the stuff of the current universe.

But there is also another form the energy can take. Some of it can get converted into a tangled collection of string resembling an incredibly snarled fishing line or a ball of wool after the cat got a hold of it. The tangle could include not just the ordinary strings of String Theory but also the string-like D1-branes devised by Polchinski.

If such a tangle were created in the early universe, the subsequent expansion would stretch the tangle to enormous proportions: tiny microscopic loops and swirls, growing to hundreds of millions of light-years in size. But some portion of the strings would remain till today, flapping around on a vast scale of space and time. The strings would not be visible by means of light or any other electromagnetic radiation, but

luckily there is another way to detect them. Damour and Vilenkin have demonstrated that such cosmic strings would emit gravity waves (wavelike disturbances of the gravitational field) that may very well be detectable in the next decade. Observing such strings in the sky would be an extraordinary triumph for String Theory.

Studying these cosmic superstrings, if indeed they exist, can tell us a great deal — not about the entire Landscape but at least about our immediate neighborhood. Polchinski and collaborators have studied the detailed conditions under which string tangles occur and the nature of the networks they form. The details are very sensitive to things like the dimensionality of the Landscape, the presence of branes and fluxes in the compact dimensions, and more. The sky, rather than particle accelerators, may well be the place to look for the smoking gun of String Theory.

High-Energy Physics

Astronomical and cosmological observations are probably the wave of the future, but we have not yet reached the limits of laboratory science. Our greatest short-term hope for gaining new groundbreaking information about the Laws of Physics is what it always was — experimental high-energy (elementary-particle) physics done at accelerator laboratories. It may be true that we are reaching the limits of this kind of science, but no doubt we are going to push the frontiers to at least one more level. The largest accelerator in the world and probably the only one big enough to tell us a great deal of new information is presently nearing completion and should be operating by 2007. Geneva, Switzerland, the location of CERN, is the site of the Large Hadron Collider, or LHC, as it is called. Originally conceived for the purpose of studying the Higgs boson, it is also the ideal machine for discovering the supersymmetric twins of the elementary particles.

In chapter 7 I explained why many physicists think that supersymmetry "is just around the corner." The argument, first made twenty-five years ago, is that supersymmetry would ensure that the violent quantum fluctuations of the vacuum do not create an enormous mass for the Higgs boson and, thereby, ruin the Standard Model. Supersymmetry may well be around the next corner. The majority of theoretical physicists expect it to be so, at least if you go by the number of papers published on the subject.

But there is another possibility. Like the vacuum energy (or cosmological constant), too large a Higgs mass would ruin the possibility of life evolving in our pocket universe. So perhaps the answer is not supersymmetry but a more anthropic consideration. If the world is big enough and the Landscape diverse enough, then some tiny fraction of the megaverse will have a small enough Higgs mass for life to flourish — end of story. As in the case of the cosmological constant, supersymmetry would be irrelevant and unnecessary.

The two explanations do not necessarily exclude each other. The most likely possibility for finding a valley with sufficiently small Higgs mass may be to find one with supersymmetry just around the corner. It is even possible that all of the valleys with small Higgs mass are of this type.

Or the opposite may be true: the vast majority of vacuums with small Higgs mass may completely lack any kind of supersymmetry. The exploration of the Landscape is still in its early infancy, and we don't know the answer to this question. My original guess was that supersymmetry was not favored, and I said so in print. But I have since changed my mind — twice — and probably not for the last time.

In trying to predict the relative probability of supersymmetry versus no supersymmetry, we run head on into the measure problem. Perhaps we should stop right there. But there is a strong temptation to dismiss the subtleties and push on. Theoretical physicists such as Michael Douglas, Shamit Kachru, and many others are developing methods to count the number of sites on the Landscape with different properties. Here I mean the number of possibilities, not the number of actual pocket universes. Then, having no other information, we might guess that if there are vastly more anthropic vacuums with approximate supersymmetry than without, approximate supersymmetry is overwhelmingly likely. But the measure problem is another huge elephant in the room that may be quietly laughing at us.

In any case the difficulties in testing the Landscape, Eternal Inflation, and the Anthropic Principle are real, but there are many ways to test a theory. Mathematical consistency may not impress the most hard-nosed experimental physicist, but it should not be underestimated. Consistent theories that combine quantum mechanics and general relativity are far from common. Indeed, this is the reason that String Theory has so little competition. If no alternatives show up and if String Theory proves to have as varied a Landscape as it seems, then the populated Landscape will be the default position — the theory to beat, so to speak.

But giving up on the possibility of more direct tests is certainly premature. It is true that theory and experiment usually proceed "hand in hand," but it's not always the case. It took more than two decades for Alan Guth's inflationary universe to be tested by observation. In the early days almost everyone thought the idea was interesting but could never be tested. I think even Alan himself was skeptical of ever confirming its truth.

Even more extreme was Darwin's theory. It was based on general observations about the world and a very clever hunch. But a direct, controlled, experimental test must have seemed completely impossible — you would need a time machine to take you back millions, if not billions, of years. In fact it took about one hundred years for ingenious biologists and chemists to figure out how to subject the theory to rigorous laboratory tests. Sometimes theory has to forge ahead to light the way.

Epilogue

Just before boarding the giant Hercules aircraft that was to take us to Punta Arenas from the Chilean Antarctic station, I hugged my friend Victor farewell. An emotional and sentimental Russian, Victor was saddened by our leaving. The last thing I said to him before trekking out into the blizzard was, "Victor, don't you think Antarctica is beautiful?" He was lost in melancholy thought for a brief moment and then quietly smiled and said, "Yes, like some women: beautiful — but cruel." Had Victor asked me whether I thought our universe and its Laws of Physics are beautiful, I might have answered, "No, not beautiful. But rather friendly."

Throughout this book I have dismissed beauty, uniqueness, and elegance as false mirages. The Laws of Physics (in the sense that I defined them in chapter 1) are neither unique nor elegant. It seems that the world, or our part of it, is a Rube Goldberg machine. But I confess: I am as vulnerable to the seductive charms of Uniqueness and Elegance as any one of my colleagues. I, too, want to believe that the grand overarching principles that transcend the rules governing any particular pocket of the universe are unique, elegant, and wonderfully simple. But the results of those rules need not be elegant in the least. Quantum mechanics, which rules the microscopic world of atoms, is very elegant — but not everything made of atoms is. The simple laws that give rise to tremendously complicated molecules, liquids, solids, and gases yield stinkweeds as well as roses. I think I might find the universal principles of String Theory most elegant — if I only knew what they were.

I often joke that if the best theories are the ones with the minimum number of defining equations and principles, String Theory is by far the best — no one has ever found even a single defining equation or principle! String Theory gives every indication of being a very elegant mathematical structure with a degree of consistency far beyond any other physical theory. But nobody knows what its defining rules are, nor does anyone know what the basic "building blocks" are.

Remember, building blocks are the simple objects that everything else is made of. For a housing contractor, building blocks may be exactly that — the blocks or bricks that compose the walls and foundation. The relation between building blocks and the composite objects that they compose is very asymmetrical: houses are made of bricks. Only someone with a severe perceptual disorder — perhaps an Oliver Sacks patient, "The Man Who Mistook His House for a Brick" — would get this relationship backward.

The basic building blocks of science depend on context and the state of knowledge at the time. In the nineteenth century the building blocks of matter were the atoms of the periodic table. The same ninety-two elements can be combined into an endless variety of composites called molecules. Later, atoms were discovered to be composite, and they gave way to electrons, protons, and neutrons. The pattern that we have learned to expect is that big things are made up of littler things. For a physicist probing deeper into the laws of nature, this has usually meant uncovering a substructure of smaller building blocks. At the present stage of physics, ordinary matter is believed to be composed of electrons and quarks. The questions that people, both laymen and scientists, often ask is, "Do you think this will go on forever, or do you think there is a smallest building block?" These days the question often takes the form, "Is there anything smaller than the Planck length?" or "Are strings the most fundamental objects or are *they* made of smaller things?"

These may be the wrong questions. The way String Theory seems to work is subtler than this. What we find is that if we focus attention on some particular region of the Landscape, everything is built out of one specific set of building blocks. It may be closed or open strings of some specific type in certain regions. In other regions all matter is composed of D-branes. In yet other parts of the Landscape, objects similar to ordinary field quanta can be assembled into strings, branes, black holes, and more. Whatever is singled out as "most fundamental," the other objects of the theory behave like composites — composites in the same sense that atoms and molecules are composites of electrons, protons, and neutrons.

But as we move through the Landscape, from one location to another, strange things happen. The building blocks change places with composite objects. Some particular composite shrinks, behaving in a simpler and simpler manner, as if it were becoming an elementary building block. At the same time the original building blocks start to grow and show signs of having the structure of composites. The Landscape is a dreamscape in which, as we move about, bricks and houses gradually exchange their role. *Everything is fundamental, and nothing is fundamental.*

What are the basic equations of the theory? Why, they are the equations governing the motion of the basic building blocks, of course. But which building blocks — open strings, closed strings, membranes, D0-branes? The answer depends on the region of the Landscape we are momentarily interested in. What about the regions intermediate between one description and another? In those regions the choice of building blocks and defining equations is ambiguous. We seem to be dealing with a new kind of mathematical theory in which the traditional ideas of fundamental versus derived concepts is maddeningly elusive. Or perhaps 't Hooft is right, and the true building blocks are more deeply hidden. The bottom line is that we have no clear idea how to describe the entire mathematical structure of String Theory or what building blocks, if any, will win the title of "most fundamental."

Still, I hope that the principles of String Theory, or whatever underlies it, will have the elegance, simplicity, and beauty that theorists hunger for. But even if the equations satisfy every esthetic criterion that a physicist could hope for, it does not mean that particular solutions of the equations are simple or elegant. The Standard Model is so complicated — with thirty apparently unrelated parameters, unexplained replication of particle types, and forces whose strength vary all over the map — that the String Theory version of it will almost certainly have Rube Goldberg complexity and redundancy.

For my own tastes, elegance and simplicity can sometimes be found in principles that don't at all lend themselves to equations. I know of no equations that are more elegant than the two principles that underpin Darwin's theory: random mutation and competition. This book is about an organizing principle that is also powerful and simple. I think it deserves to be called elegant, but again, I don't know an equation to describe it, only a slogan: "A Landscape of possibilities populated by a megaverse of actualities."

And what about the biggest questions of all: who or what made the universe and for what reason? Is there a purpose to it all? I don't pretend to know the answers. Those who would look to the Anthropic Principle as a sign of a benevolent creator have found no comfort in these pages. The laws of gravity, quantum mechanics, and a rich Landscape together with the laws of large numbers are all that's needed to explain the friendliness of our patch of the universe.

But on the other hand, neither does anything in this book diminish the likelihood that an intelligent agent created the universe for some purpose. The ultimate existential question, "Why is there Something rather than Nothing?" has no more or less of an answer than before anyone had ever heard of String Theory. If there was a moment of creation, it is obscured from our eyes and our telescopes by the veil of explosive Inflation that took place during the prehistory of the Big Bang. If there is a God, she has taken great pains to make herself irrelevant.

Let me then close this book with the words of Pierre-Simon de Laplace that opened it: "I have no need of this hypothesis."

A Word on the Distinction between Landscape and Megaverse

The two concepts — *Landscape* and *megaverse* — should not be confused. The Landscape is not a real place. Think of it as a list of all the possible designs of hypothetical universes. Each valley represents one such design. Listing the designs one after another, like names in a telephone book, would not capture the fact that space of designs is multidimensional.

The megaverse, by contrast, is quite real. The pocket universes that fill it are actual existing places, not hypothetical possibilities.

Glossary

Absorption lines — Dark lines superimposed on a rainbowlike spectrum of colors. The dark lines are due to absorption of certain colors by gas.

Anthropic Principle — The principle that requires the laws of nature to be consistent with the existence of intelligent life.

Antiparticle — The twin of a particle that is identical except with opposite electric charge.

Boson — A type of particle not constrained by the Pauli exclusion principle. Any number of identical bosons can occupy the same quantum state.

Broken Symmetry — An approximate symmetry of nature that for some reason is not exact.

Calabi Yau manifold — The special six-dimensional geometries that String Theory uses to compactify the extra dimensions of space.

Calabi Yau space — Same as Calabi Yau manifold.

Charge conjugation symmetry — A (broken) symmetry of nature under which every particle is replaced by its antiparticle.

Compactification — The rolling up of extra dimensions of String Theory into microscopic spaces.

Cosmic microwave background (CMB) — The electromagnetic radiation left over from the Big Bang.

Cosmological constant — The term that Einstein introduced into his equations to counter the effect of gravitational attraction.

Coupling constant — The constant of nature that determines the probability for an elementary event.

D-brane — The points or surfaces where the strings of String Theory are allowed to end.

Density contrast — The variations of energy density in the early universe that eventually evolved into galaxies.

De Sitter space — The solution to Einstein's equations with a positive cosmological constant. De Sitter space describes an expanding universe in which space clones itself exponentially.

Domain wall — The boundary separating two phases of a material such as water and ice.

Doppler shift — The shift in the frequency of waves due to the relative motion of the source of the waves and the detector of the waves.

Electric field — The field surrounding charged particles at rest. Along with magnetic fields, electric fields are composed of electromagnetic radiation such as light.

Electron — Elementary charged particle that makes up electric currents and the outer parts of atoms.

Emergent — Refers to properties of matter that manifest themselves only when large numbers of atoms behave in a collective or coordinated manner.

Eternal Inflation — The exponential cloning of space that spawns bubbles, which populate the Landscape.

Exchange diagram — A Feynman diagram in which a particle such as the photon is emitted by one particle and absorbed by another. Such diagrams are used to explain the forces between objects.

Fermion — Any particle that is subject to the Pauli exclusion principle. This includes electrons, protons, neutrons, quarks, and neutrinos.

Feynman diagram — Feynman's pictorial way of explaining the interactions among elementary particles.

Field — An invisible influence in space that affects the motion of objects. Examples include the electric, magnetic, and gravitational fields.

Fine structure constant (0.007297351) — The coupling constant governing the emission of a photon by an electron.

Flux — One of the many components of a string compactification. A flux is similar to a magnetic field except oriented along the compact directions of space.

Glueball — Composite particles made out of collections of gluons and having the structure of closed strings.

Gluon — Particles whose exchange account for the forces between quarks.

Gravitational waves — Disturbances of the gravitational field that propagate through space with the speed of light.

Graviton — The quantum of the gravitational field. Its exchange accounts for the gravitational force.

Harmonics — The patterns of vibration of a string, such as a guitar string.

Heisenberg Uncertainty Principle — The principle that says that it is impossible to determine both the position and momentum of any object.

Higgs boson — The quantum of the Higgs field.

Higgs field — The field in the Standard Model whose value controls the masses of elementary particles such as the electron and quark.

Holographic Principle — The principle that says that a region of space can be completely described by degrees of freedom on its boundary, with no more than one degree of freedom per Planck area — an area equal to one square Planck length.

Homogeneous — Everywhere the same; completely smooth and without variation from point to point.

Horizon — The point of no return at which an observer would be receding with the speed of light. It applies to both black holes and to rapidly inflating cosmic space.

Hubble constant — The constant appearing in Hubble's Law.

Hubble's Law — The law stating that the recessional velocity of galaxies is proportional to their distance. It can be expressed as an equation: V = HD, where V is velocity, D is distance, and H is the Hubble constant.

Inflation — The rapid exponential expansion of space that ironed out all the wrinkles and created a large, smooth universe. Inflation has become the standard theory of the early universe.

Isotropic — The same in every direction.

Joule — An ordinary unit of energy. The quantity of energy needed to heat one gram of water 1 degree centigrade.

Landscape — The space of possible vacuums (environments) allowed by fundamental theory. In practice, the space of vacuums of String Theory.

Long range — Refers to forces that reach out over long distances to pull or push objects. Gravity, electric, and magnetic forces are long range.

Magnetic field — The relative of the electric field which is created by charges in motion (currents).

Matrix Theory — The underlying mathematical framework for M-theory.

Megaverse — The huge vastness of pocket universes.

Mode of oscillation — Same as harmonic.

Moduli — The parameters that determine the size and shape of the compact directions of space, particularly in String Theory.

MRI machine — Medical imaging machine utilizing a space with a large magnetic field in it.

M-theory — The eleven-dimensional theory that unifies many of the diverse String Theories. M-theory has membranes but no strings.

Neutrino — The "ghostly" particle emitted by a neutron, along with an electron, when the neutron decays and becomes a proton.

Neutron — One of two particles composing the nucleus. The neutron is electrically neutral.

Non-abelian gauge theory — A class of quantum field theories that form the basis for the Standard Model of particle physics.

Nucleon — Proton or neutron.

Pauli exclusion principle — The principle that says that no two fermions can occupy the same quantum state.

Photon — Quantum of the electromagnetic field. The basis for Einstein's particle theory of light.

Planck length or Planck distance — The natural unit of length determined by Planck's constant, Newton's gravitational constant, and the speed of light. It is about 10^{-33} centimeters.

Planck mass — The natural unit of mass determined by Planck's constant, Newton's gravitational constant, and the speed of light. It is about 10^{-5} grams.

Planck's constant — Very small numerical constant that determines the limit in the simultaneous determination of position and momentum (Heisenberg Uncertainty Principle).

Planck time — The natural unit of time determined by Planck's constant, Newton's gravitational constant, and the speed of light. It is about 10^{-42} seconds.

Plasma — Gas that has been heated to the point where some or all of the electrons have been torn free of the atoms and are free to move through the material. Plasmas are good electrical conductors and are opaque to light.

Pocket universe — A portion of the universe in which the Laws of Physics take a particular form.

Positron — The electron's antiparticle.

Principle of Black Hole Complementarity — The principle that allows two apparently contradictory descriptions of matter that falls into a black hole.

Propagator — The component of Feynman diagrams that represent the motion of a particle from one space-time point to another; also the mathematical expression that controls the probability for such a process.

Proton — The positively charged nucleon.

Quantum Chromodynamics (QCD) — The theory of quarks and gluons that explains the existence and properties of nucleons and nuclei. The modern nuclear physics.

Quantum Electrodynamics (QED) — The theory of electrons and photons. The basis for all atomic physics and chemistry.

Quantum field theory — The mathematical theory of elementary particles that originated by combining quantum mechanics with the Special Theory of Relativity.

Quantum jitters — The unpredictable fluctuating motion of particles or fields that derive from the principles of quantum mechanics.

Quark — The elementary particles that combine together, three at a time, to make up nucleons.

Reductionism — The philosophy that says that nature can be understood by reducing all phenomena to ultimately simple microscopic events.

Rube Goldberg machine — An overly complicated, inelegant solution to an engineering problem. Named after the cartoonist Rube Goldberg, whose cartoons depicted fantastic and silly Rube Goldberg machines.

Scalar field — A field that has a magnitude (strength) but no sense of direction. The Higgs field is a scalar; the electric and magnetic fields are not.

Short range — Refers to forces that do not reach out over long distances, i.e., forces between objects that act only when the objects are in contact or almost in contact.

Space-time — The four-dimensional world including time in which all phenomena take place.

Spectral lines — The discrete sharp lines in the spectrum of light that arise from atomic transitions in which an electron makes a quantum jump from one energy level to another and in the process emits a photon.

Standard Model — The currently accepted quantum field theory that describes elementary particles. It includes QED, QCD, and the weak interactions as well as phenomena involving the Higgs boson.

Supercooled water — Water that has been cooled below the freezing temperature but that has remained liquid.

Supernova — The final event in the life of certain stars that results in a collapse to a neutron star. At the same time an explosion sprays chemical elements into the surrounding space.

Supersymmetry — A mathematical symmetry relating fermions and bosons.

Symmetry — An operation that leaves the laws of nature unchanged.

Vacuum — A background or environment in which the Laws of Physics take on a certain form.

Vacuum energy — Energy stored in the quantum fluctuations of empty space.

Vacuum fluctuation — The jittery fluctuation of quantum fields in empty space.

Vacuum selection principle — A mathematical principle that would select a single unique String Theory vacuum out of all the diverse vacuums that the theory describes. Thus far, no such principle has ever been found.

Vector field — A field that has, in addition to a strength, a direction in space. The electric and magnetic fields are vectors.

Vertex diagram — The Feynman diagram depicting the elementary event in which a particle is emitted by another particle.

Virtual particle — A particle in the interior of a Feynman diagram. Not one of the particles that enters or leaves at the beginning or end of the process.

W-boson — One of the particles whose exchange gives rise to the weak interactions.

Weak interactions — Phenomena that are similar to the decay of the neutron.

Weinberg's bound — The bound on the size of the cosmological constant that derives from the condition that galaxies could form in the early universe.

Yang Mills theory — Same as non-abelian gauge theory.

Z-boson — A close relative of the W-boson, also involved in weak interactions.

Note on Terminology

When I first began to write this book, I encountered a problem of terminology that I'm still wrestling with. I didn't know what to call the new vastness that is replacing the old concept of universe. The term that was (and is) most common is *multiverse*. I have no objection to *multiverse* except that I just don't like the sound of it. It reminds me of multiplex cinemas, which I try to avoid. I experimented with a number of other possibilities, including *polyverse*, *googolplexus*, *polyplexus*, and *googolverse*, without success. I eventually settled on *megaverse*, knowing full well that I was committing the linguistic crime of combining the Greek prefix *mega* with the Latin *verse*.

After deciding to use the term *megaverse*, I looked it up on Google and found that I was far from the first to use it. I got 8,700 results for *megaverse*. On the other hand, the same technique applied to *multiverse* got 265,000 results.

Finally, I should add that some of my best friends are users of the term *multiverse*, and so far, we haven't come to blows over it.

Index

Abbott, Edwin, 220
absorption lines, 132
accelerators, 28, 59, 75, 87, 95, 203, 207, 218, 229, 230n1, 248, 249, 261, 365, 373, 374
Aharonov-Bohm effect, 65
Aharonov, Yakir, 65
air, 119–20
Allen, Woody, 131
alpha (α), 48, 49, 54, 175
alpha rays, 56
Alpher, Ralph, 156
American Society of Engineers, 125
Andromeda Galaxy, 145, 154
Antarctica, 3–5, 377
Anthropic Cosmological Principle, The (Barrow and Tipler), 81n
"Anthropic Landscape of String Theory, The" (Susskind), ix
Anthropic Principle, 7, 10–11, 14, 21, 22, 109, 111, 154, 161, 169–97, 199, 250, 300, 304, 323, 324, 346, 350–51, 352, 353, 355, 356, 357, 363, 365, 374–75, 380; fish story and, 170–72, 187–88, 364; Guth and, 353–54, 354n4; many-worlds interpretation and, 316, 322; Smolin and, 360, 361; Weinberg and, 79, 80–81, 83–84, 349
"Anti De Sitter Space and Holography" (Witten), 339
antimatter, 43–48, 184, 245, 267
antiparticles, 53, 162, 184, 200n, 245
Aristotle, 119
astronomers, 14
astronomy, 13, 21, 27, 60, 119
astrophysics archive, 251
atomic nuclei, 9, 20, 20n, 34, 50
atomic physics, 116, 121
atoms, 8, 37, 50, 72, 76, 85, 88, 93, 98, 100–101, 107, 117–18, 133, 173, 174, 183, 189, 195, 201, 207, 229, 235, 240, 241, 250, 264–65, 267n, 275, 358, 359, 360, 377, 378, 379; cosmological constant and, 81–82, 88; electrons and, 27, 30, 34, 36, 55, 79, 132, 177, 264, 266, 267; Higgs field and, 104–5, 106; Laws of Physics and, 9, 59, 60, 91, 129; nuclei and, 38, 268; photons and, 37, 47, 53, 175, 221; Rutherford and, 264, 264n2; vacuum energy and, 76, 77

bacteria, 362
Banks, Tom, 188, 188n, 256, 310n, 351
Barrow, John D., 81, 81n
baryons, 207, 207n, 208, 211, 219, 229

baryosynthesis, 162
beauty, 12, 112, 114, 116–18, 121, 122, 377, 379
Becquerel, Antoine-Henri, 56, 58, 59, 263
behaviorism, 193, 202–3
Bekenstein, Jacob, 329
Belfer Graduate School of Science, 63–65
Bell Labs, 156, 157
beryllium 8, 182
beta rays, 56, 58
Big Bang, 9, 13, 18–19, 108, 123, 131, 143, 146, 155, 156, 163, 177, 185, 293, 304, 312, 322, 346, 347, 380
billiard table metaphor, 24–27, 28–29
binary star system, 5, 5n
biology, 6, 34, 61, 343–44, 347, 362
bits, 332, 332n
blackbody radiation, 262–63
black holes, 65, 75, 87, 88, 103, 116, 122, 129, 157, 177, 180, 183, 185, 247, 257, 269, 310, 359, 360–62, 370, 379. *See also* Black Hole War
Black Hole War, 14–15, 326–41. *See also* black holes
Blind Watchmaker: Why the Evidence of Evolution Reveals a Universe without Design, The (Dawkins), 18n
Bohr, Niels, 22, 30, 49n, 121, 132, 186, 235, 264, 265, 266, 320, 321, 324, 326, 334, 336
bola, 208
Boltzmann, Ludwig, 24, 195
Bondi, Herman, 19
Boomerang!, 158
Bose, Satyendra Nath, 240
bosonic String Theory, 246
bosons, 76, 77, 83, 88, 124, 177, 239, 240, 241, 248, 250, 251
bottom-quarks, 50, 51, 53, 57, 128, 176, 365
Bousso, Raphael, xii, 200, 289, 291, 350, 350n
branes, 310, 355, 373. *See also* D-branes
Brer Rabbit, 54
broken symmetries, 243, 244, 245, 248
Brout, Robert, 94n
Brownian Motion, 195, 239n
bubble nucleation, 298, 306, 309, 371

Calabi Yau manifolds, 199, 237, 238, 253, 274, 276, 283, 286, 287, 289, 349, 354
Callan, Curt, 3, 4
Candelas, Philip, 354
Cantor, Georg, 367–68, 368n
carbon, 10, 133, 177, 182, 183, 183n, 250, 268, 344
Carroll, Lewis, 325
Carter, Brandon, 316, 322
Casimir, Gerhard, 326
cat experiment, 318–21
Cat's Cradle (Vonnegut), 294–95, 296–97
Cerebrothropic Principle, 196, 197
CERN (European Organization for Nuclear Research), 95, 96n, 374
chaos, 25, 49
charge conjugation symmetry, 77, 245
charmed quarks, 50, 51, 53, 57, 128, 176, 231
chemistry, 8, 9, 61, 79, 117, 121, 147, 172–73, 176, 177, 250, 266, 313
Christian Internet sites, 7
Churchill, Winston, 355
circles, 68, 119, 120
cloning space, 298–301, 305–7, 363
closed-and-bounded space, 67, 144, 231
closed universe, 144, 148–49, 150, 151
Coleman, Sidney, 284, 304, 305
collective/emergent phenomena, 117–18

compactification, 231–38, 253, 256, 257–58, 274, 276, 278, 281–85, 289, 310, 349, 354
competition, 18, 345, 380
complementarity, black hole, 334–36, 339
condensed-matter physics, 126
conical singularities, 287
conifold singularities, 286–89
consciousness, 34n
constants of nature, 21, 48, 59, 59n, 60, 92, 109, 128, 351
continuum, 291
conventional-particle physics, 246
Copenhagen rule, 321
Cornell University, 18
"Cosmic Blueprint, The" (Davies), 7
cosmic friction, 197n8
cosmic microwave background (CMB), 148–49, 156–58, 160, 162, 163, 166, 341, 371, 372
"Cosmological Considerations on the General Theory of Relativity" (Einstein), 66
cosmological constant, 11–12, 13, 20, 22, 66, 70–71, 72, 74, 76, 78, 83, 88, 89, 125, 136, 143, 143n, 150–51, 152, 154, 161, 163, 164, 165, 166, 185, 199, 201, 239, 252, 289, 290, 291, 292, 298, 299, 301, 303, 313, 315, 351, 352, 357, 360, 371, 374; Anthropic Principle and, 81, 172; Eternal Inflation and, 307, 308; expansion of universe and, 15, 197n8; KKLT construction and, 288, 289; Smolin and, 361–62; supersymmetry and, 247, 250; Weinberg and, 79, 81, 82, 84, 187, 196, 349, 356, 364
cosmological principle, 138, 139
cosmologists, 14, 129, 352–56
cosmology, 13, 17–18, 19, 20, 60, 128, 162, 347
Coulomb force, 47n
coupling constants, 48, 50–51, 59, 60, 61, 88, 118, 128, 175, 238
creation, 17, 18, 18n, 21, 194, 348
creationism/creationist, x, 6–7, 84, 196–97
Crick, Francis, 18

Dalton, John, 147, 264, 268
Damour, Thibault, 372, 373
dark energy, x, 72n, 147n
dark matter, x, 128, 147, 147n, 151, 160–61, 164, 283
Darwin, Charles, x–xi, 6, 6n, 17–18, 17n, 194, 197, 343–44, 345, 375–76, 380. *See also* evolution; natural selection
"Darwin Triumphant" (Dawkins), 344n
Davies, Paul, 7, 8, 8n, 343n
Dawkins, Richard, 18, 18n, 344n
D-branes, 276–85, 288, 289, 298, 339, 372, 378
Death Valley, 5
de Broglie, Louis, 240
De Luccia, Frank, 305
Deser, Stan, 4
de Sitter, Willem, 299, 301, 326
destructive interference, 31–32, 33
determinism, 49
deuterium, 178
Devil's Chaplain: Reflections on Hope, Lies, Science, and Love, A, 344n
Dicke, Robert, 156, 157
Different Universe: Reinventing Physics from the Bottom Down, A (Laughlin), 358, 358n
Dimopoulos, Sava, 162
Dine, Mike, 188, 188n
Dirac, Paul A. M., 65, 75, 111, 265, 266–67
Dirichlet, Peter, 277
discretuum, 291
DNA, 8, 18, 34, 172, 173

domain wall, 298
Doppler effect, 133–34, 153
Doppler shift, 155
double helix, 18
Douglas, Michael, 352, 375
down-quarks, 50, 51, 52, 53, 57, 128, 176, 176n
Dreams of a Final Theory (Weinberg), 349
Duff, Mike, 257

earth, 119–20
"*Edge* Annual Question – 2005," 353
edge.org, 193, 353
Einstein, Albert, 8n, 23, 26, 27, 32, 34, 49, 61, 65, 67, 69, 84, 86, 94, 95, 103, 111, 114, 115–16, 122, 125, 137, 146, 151, 220, 231–32, 240, 263, 264, 321, 326, 329, 346–47, 348, 358, 370; Brownian Motion and, 195, 239n; cosmological constant and, 11, 22, 66, 70–71, 74, 299; static universe and, 69, 71. *See also* General Theory of Relativity; Relativity, Theory of
electrical attraction, 173
electric fields, 29, 49, 94, 95, 97, 100, 101, 106–7, 107n7, 285–86
electric forces, 9, 46–47, 50, 69, 175, 184, 201, 202, 221–22, 223
electricity, 10, 36–37
electrodynamics, 114–15, 269
electromagnetic field, 94, 95, 100, 115, 266, 358
electromagnetic forces, 201, 222, 229, 234
electromagnetic radiation, 26, 26n, 27, 30, 36, 154, 262, 372
electromagnetic theory, 99
electromagnetic waves, 103
electron(s), 9n, 10, 20, 24, 26, 28, 38, 48, 49, 50, 52, 56, 57, 80, 81, 91, 92, 93, 100, 101, 121, 124, 132, 133, 175, 176–77, 180, 189, 200, 201, 202, 216, 221, 223, 224, 230, 231, 235, 240, 263–64, 266–67, 268–69, 333, 365, 378, 379; antimatter and, 43, 44, 45, 48; atoms and, 20n, 30, 34, 53, 55, 79, 132, 264, 266, 267; fermions and, 77, 176–77, 240, 246; Feynman diagrams and, 39, 41, 42, 43; fields and, 95, 96, 107; Laws of Physics and, 9, 91; mass and, 103, 104, 127; nuclei and, 38, 79, 268; Pauli's exclusion principle and, 133, 177, 180; photons and, 27, 46, 132, 240, 266, 267; Quantum Electrodynamics (QED) and, 35–37, 273; Rutherford and, 264, 264n2; spin and, 224, 224n, 266; supersymmetry and, 241, 248, 250–51; vacuum energy and, 72–73, 74, 75, 76; valence, 8, 173, 174, 175; wave function and, 317, 318, 322; W-bosons and, 57, 59
electrostatic force, 47n
elegance, 12–13, 50, 59, 112–13, 114, 117, 118–19, 120, 121, 124, 125, 127, 129, 130, 202, 250, 251, 377, 379–80
Elegant Universe: Superstrings, Hidden Dimensions, and the Quest for the Ultimate Theory, The (Greene), 28n
elementary particle physicists, x, 118, 162
elementary particle physics, 122–23, 124, 202, 246, 374
elementary particles, 10, 11, 14, 18, 20, 33–35, 50, 53, 90–91, 97, 118, 121, 147, 160, 200–201, 205, 231, 236, 245, 250, 281, 283n, 354, 359; gravity and, 9n, 61, 74, 88; Higgs particle, 94, 95, 96, 230, 249; laws of, 11, 12–13; Laws of Physics and, 9, 12, 59, 173; mass and, 102, 361; neutrons and, 28, 57, 176; protons and, 28, 57, 176; S-matrix theory and,

elementary particles (*continued*) 195, 203; Standard Model and, 50, 59–60, 88, 128, 230; supersymmetry and, 246, 247, 248, 374; vacuum energy and, 73, 74–75. *See also names of individual particles*
elements, 119–20, 121, 378
ellipsoid, 139–40
$E=mc^2$, 73, 208, 249
emergent, laws of nature as, 358–60
empirical numbers, 48
energy-time uncertainty principle, 29
Englert, François, 94n
Epitaph for Isaac Newton (Pope), 117
equivalence principle, 114, 265
escape velocity, 142, 143
Escher, M. C., 311
Eternal Inflation, 14, 21, 194, 195, 302–10, 302n, 317, 323–24, 325, 341, 345, 348, 351, 363, 367, 369, 370, 375
ether, 263, 263n, 358
Euclid, 118, 148, 149, 160
Euclidean geometry, 119, 148
events, 38–39, 41
Everett III, Hugh, 316–17, 320, 324
evolution, 17n, 18, 18n, 46, 79, 129, 308–10, 345, 347, 356, 360, 361. *See also* natural selection
exchange diagrams, 46
exist, meaning of, 177
experimental physics, 246
experiments, discovery in physics and, 261–71, 375–76

falsifiability, in science, 192–96
Faraday, Michael, 94, 99–100, 285–86, 358
Fermi, Enrico, 240
Fermilab conference, 218
fermions, 76–77, 83, 88, 124, 177, 230n1, 239, 240, 241, 245–46, 248, 250, 251
Feynman diagrams, 35, 38–47, 53, 58, 60, 72, 92–93, 104, 175, 178, 212, 215, 216, 221, 252, 267
Feynman, Richard, 24–25, 34–35, 37, 45, 46, 49, 50, 59, 61, 174, 175, 192, 219, 316; on hadrons and partons, 216, 217; Laws of Physics and, 11, 61; positrons and, 43, 267; Quantum Electrodynamics (QED) and, 35, 50. *See also* Feynman diagrams
Feynman rules, 94
fields, 12, 93–98, 102. *See also names of individual fields*
final theory, 113
fine structure constant, 48–51, 48n, 54, 59, 60, 183, 350
Finkelstein, Dave, 65
fire, 119–20
"Fire and Ice" (Frost), 144
Fischler, Willy, 256
Fishnet diagram, 216n
fish story, 170–72, 187–88, 363–65
Flatland (Abbott), 220
flat universe, 144, 148–49, 150, 151–52
fluxes, 285–86, 287, 289, 290, 310, 355, 373
forces, 14, 45–46, 118, 122, 202, 221, 222. *See also names of individual forces*
Franklin, Benjamin, 36, 36n
Friedmann, Alexander, 136, 231–32
Frost, Robert, 144
fundamental constant, 80

galaxies, 60, 66, 68, 69, 70–71, 82, 116, 131, 134–36, 137–38, 143, 146–47, 150, 152, 153, 154, 157, 166, 183, 264, 364, 365, 371, 372. *See also* Andromeda Galaxy
gamma rays, 27, 56
Gamow, George, 19, 156
gauge hierarchy problem, 9n, 184n

Gell-Mann, Murray, 189, 200, 216, 217–18, 219, 316
general relativists, 326, 326n
General Relativity and Quantum Cosmology archive, 251
General Theory of Relativity, 9, 12, 23, 61, 66, 68, 69, 86, 113, 114, 116, 123, 143, 151, 265, 269, 294, 302, 304, 310, 340, 347, 359, 363; black holes and, 332, 333; compactification and, 231, 234; general relativists and, 326, 326n; quantum mechanics and, 204, 265, 375. *See also* Relativity, Theory of
genus one surface, 284
genus two surface, 284
geometry, 119, 139–42, 143–44, 148–49
glueballs, 55, 207, 211, 215, 223, 229
gluons, 50, 51, 53–55, 57, 77, 95, 124, 127, 200, 215, 216, 217, 222, 230, 231, 240, 241, 247, 270, 273
God, 5, 8, 8n, 15, 49, 49n, 187, 196–97, 321, 341, 355, 380
God and the New Physics (Davies), 343n
God Particle, The (Lederman), 95n
gold, 367
Goldberg, Rube, 112, 128, 283, 285, 352
Gold, Thomas, 19
googol, 313n
googolplex, 313, 313n
Grand Unified Theories (GUTs), 191
gravitons, 74, 75, 77, 103, 122, 222, 223, 229–30, 249, 252, 256, 269, 278, 279, 280, 281–82. *See also* gravity
gravity, 9, 10, 11, 20, 22, 45–46, 61, 69–70, 73, 74, 82, 84, 86, 106, 114, 116, 123–24, 126, 128, 129, 143, 145, 146, 147, 154, 166, 174, 180, 183, 186, 201, 221–23, 226, 229, 234, 243, 247, 249, 252, 269, 346, 365, 366n, 372, 380; as attractive, 69, 70; black holes and, 14, 359; D-branes and, 280, 281–82, 283; elementary particles and, 11, 74, 88; emergent phenomena and, 358, 359; gravitational waves, 9, 103, 115, 116, 269, 373; quantum mechanics and, 86, 204, 207, 330, 331, 335, 339; Standard Model and, 61, 122, 191, 252; supersymmetry and, 247, 249; vacuum energy and, 76, 77; as weak, 9n, 184–85, 201. *See also* General Theory of Relativity; gravitons; Newton, Sir Isaac
Greeks, 119, 120, 121
Greene, Brian, 28, 28n
Greenstein, George, 8, 8n
Gross, David, 354–56, 354n5
Gruber, Cary, 64
Guth, Alan, xii, 13, 14, 20, 161–62, 163, 164, 165–66, 193, 294n, 304, 353–54, 354n4, 375

hadrons, 200–202, 203–8, 212, 215–19, 221, 223, 224, 224n, 226–27, 229, 246, 265, 354
harmonic oscillators, 205, 210, 264, 265
Harvey, Jeff, 354
Hawking radiation, 329–30, 331, 332, 333, 335, 335n, 340, 341, 359
Hawking, Stephen, ix, 3, 4, 15, 302, 316, 326–27, 329–30, 331–32, 333, 336, 353, 370
heat, 29
Heisenberg Uncertainty Principle. *See* Uncertainty Principle
Heisenberg, Werner, 25–26, 84, 121, 264, 265, 336
helium, 9n, 10, 82, 129, 133, 150, 154, 178, 179, 182, 266, 358, 359, 360, 367

Herman, Robert, 156
Heterotic String Theory, 252, 253, 349, 354–55
hierarchies, 33
Higgs bosons, 50, 57, 77, 239, 240, 248, 249, 270, 374
Higgs field, 94, 94n, 95–97, 102–5, 107–8, 166, 273
Higgs mass, 250, 252, 374–75
Higgs particle, 94, 95, 96, 230, 249
Higgs, Peter, 94n
high-energy physics, 126–27, 202, 218, 246, 374–76
Holmes, Sherlock, 333
Holographic Principle, 126, 226, 336–40, 351, 359, 370
homogeneous, 19, 137–41, 146, 157, 161, 162, 166, 310
horizons, 15, 194–95, 300, 301, 314–15, 316, 323, 324, 325–26, 328, 329, 330, 332, 333, 335, 337, 340–41, 348, 370
Horowitz, Gary, 354
Hoyle, Fred, 7, 8, 19, 182, 183, 184
Hubble constant, 136, 136n, 145, 148, 153
Hubble, Edwin, 66, 71, 83, 131, 134–36, 145, 264
Hubble expansion, 185
Hubble's Law, 136, 145, 149–50, 155, 299–300, 341
Huxley, Aldous, 111
Huxley, Thomas Henry, 6n, 7, 126
Huygens, Christiaan, 326
hydrodynamics, 270
hydrogen, 9n, 10, 37, 48n, 82, 129, 133, 150, 154, 178, 178n5, 189, 264, 345, 367
hyperbolic geometry, 140, 141, 144
hyperons, 121

infinite flat plane, 140
infinite numbers, 368–69, 370
Inflation, 13, 21, 161–67, 187, 193, 194, 199, 288, 299–300, 301–2, 315, 353, 361, 371, 372, 375, 380
"Inflationary Cosmology: Exploring the Universe from the Smallest to the Largest Scales" (Guth and Kaiser), 354n4
infrared radiation, 41
Institute for Advanced Study (Princeton), 125, 254, 352
intelligence, 197
intelligent design, xi, 6, 8, 21, 81, 304, 343, 362, 380
intelligent life, 11
interference, 30, 40
iron, 177, 178, 366, 366n, 367
isotropic, universe as, 138–40, 157

joule, 75
Joyce, James, 200

Kachru, Shamit, xii, 200, 287n, 352, 375
Kafka, Franz, 362
Kaiser, David I., 354n4
Kallosh, Renata, xii, 200, 287n
Kaluza Klein theories, 234
Kaluza, Theodor Franz Eduard, 234, 235, 237, 269
Kepler, Johannes, 120
KKLT construction, 287–89, 287n, 357
Klein, Oscar, 234, 235
Krueger National Park, 137

Lamarck, Chevalier de, 194
Lamarckian theory, 194
lambda (λ), 70, 79, 81, 83–84
Landscape(s), ix–x, 21, 90, 91, 96, 97–102, 105–9, 125, 165, 165n, 166,

187, 191, 194, 197, 238, 239, 250, 253, 254, 256, 273–76, 278, 285–86, 287–89, 291–92, 293–94, 304, 315, 322, 324, 345–46, 347, 350, 351–53, 355–57, 361, 365–66, 370, 374–75, 378–79, 380; Anthropic, 172–85; Darwin and, 344, 344n; D-branes and, 283, 285; defined, 12, 20; Eternal Inflation and, 302, 304–5, 307n, 308, 309, 310n, 348; fields and, 99, 101; Higgs field and, 94, 105–6; Linde/Vilenkin and, 274, 353; megaverse and, 21, 22, 381; pocket universes and, 98–99, 369; superstrings and, 373; supersymmetry and, 239, 251, 254. *See also* populated Landscape
Laplace, Pierre-Simon de, 23, 380
Large Hadron Collider (LHC), 374
"Large N Limit of Superconformal Field Theories, The" (Maldacena), 339
laser beam, 76
Laughlin, Robert, 358, 359
laws of motion, 17
laws of nature, 5, 7, 8n, 12, 17, 20, 21, 23, 60, 80, 84, 91, 121, 130, 261, 330, 358–60, 361
Laws of Physics, 6, 7, 11, 12, 19, 22–24, 34n, 35, 49, 50, 59–61, 91, 97, 99–100, 118, 124, 129, 130, 172, 173, 177, 189, 197, 203, 212, 237, 248, 250, 254, 261, 285, 294, 310, 322, 331, 343, 344, 351, 355, 359, 374, 377; Anthropic Principle and, 21, 349; elementary particles and, 9, 34; the Landscape and, 20, 108; as variable, 90–91, 92, 94, 96, 101, 106, 107, 109, 237, 238
Lebowitz, Joel, 65
Lederman, Leon, 95n
leptons, 128, 230, 230n1, 365
Lieb, Elliot, 65
Life of the Cosmos, The (Smolin), 360n
light, 26, 27, 30, 32–33, 60, 131–32, 154–61, 240, 263, 263n, 334, 370
light-years, 19–20, 19n
Limit Circle IV (Escher), 311–12
Linde, Andrei, xii, 14, 81n, 200, 274, 287n, 294n, 302, 302n, 304, 353, 370
local minimum, 108, 109
long-range forces, 174, 177, 221, 222, 223
Loop Gravity, 357
Lorentz, Hendrik Antoon, 263, 326
luminosity, 153, 153n

Mach, Ernst, 195
magnetic field lines. *See* magnet lines of force
magnetic fields, 12, 29, 49, 91–93, 95, 97, 99–100, 101, 105, 106–7, 107n7, 285–86
magnetic forces, 47, 174, 222
magnet lines of force, 93
Maldacena, Juan, 339, 351, 353
Man and Beasts Revisited (Robinson and Tiger, eds.), 344n
many worlds interpretation, 316–24
Martinec, Emil, 354
mass, 59, 60, 73, 85, 86, 87, 88, 91n, 96, 102–3, 118, 127–28, 143–45, 146, 148, 149, 163–64, 175, 183, 222, 238, 249, 329; elementary particles and, 127–28, 173, 249; energy and, 74, 151; Higgs, 249, 252, 374–75; Higgs field and, 104, 106; Laws of Physics and, 9, 61; neutrons and, 189, 268–69; quarks and, 176, 176n
mathematics, 24, 34, 124, 355
matrix mechanics, 264

Matrix theory. *See* M-Theory
Maxwell, James Clerk, 24, 26, 34, 94, 114–15, 234, 263, 285, 358
megaparsec, 145
megaverse, 20, 20n, 21, 22, 60, 88, 202, 251, 294, 300, 306, 315, 316, 317, 323, 324, 340, 341, 345–46, 347, 348, 353, 363, 374, 381, 389
membranes, 257, 258, 298, 379. *See also* branes; D-branes
Mendeleyev, Dmitry Ivanovich, 266
mesons, 121, 193, 207, 207n, 209, 210, 211, 214, 219, 223, 229
Metamorphosis, The (Kafka), 362
metastability, 294–97, 301
MeV, 91, 91n
microwaves, 27
Milky Way, 137–38
Millikan, Robert, 263
Mills, Robert, 270n
Minkowski, Hermann, 106, 212, 220
moduli, 238, 255
molecules, 50, 59, 88, 98, 105, 107, 129, 175, 239n, 275, 358, 379
moon, 158–59, 188, 189
Morley, Michelson, 263
Mount Palomar, 71n
Mount Wilson, 71, 71n
MRI (Magnetic Resonance Imaging) machines, 91–93, 99–100
M-Theory, 251–60, 276, 278, 279, 353
Mukhanov, Slava, 166
multiverse, 20n, 381, 389
muons, 50, 77, 121, 128, 230, 230n1, 269, 270, 283n
music, 118–19
mutations, 18

Nagaoka, Hantaro, 264–65
Nambu, Yoichiro, 206, 207, 215, 245
natural selection, 6, 17n, 197, 347, 356, 360–71
neurons, 197
neutrinos, x, 10, 56–57, 74, 121, 127, 178, 180–82, 201–2, 230n1, 231, 235, 269; discovery of, 50, 121, 147; fermions and, 77, 240, 246; Standard Model and, 230, 270; wave function and, 317, 318; W–bosons and, 57, 59
neutrons, 9, 10, 14, 27, 51, 53, 56, 59, 88, 96, 121, 124, 175, 176, 177–78, 180, 189, 200, 202, 208–9, 222, 241, 268–69, 367, 378, 379; elementary particles and, 28, 57–58; nuclei and, 34, 37, 223; quarks and, 52, 53, 176, 193; vacuum energy and, 76, 77; wave function and, 317–18, 320, 322
New Scientist, The, 170n
Newton, Sir Isaac, 17, 23, 24, 26, 34, 49, 61, 69, 73, 85, 131–32, 143, 146, 186, 321–22, 346
Niels Bohr's Times: In Physics, Philosophy, and Polity (Pais), 264n3
Nielsen, Holger Bech, 206, 216, 216n
NMR (Nuclear Magnetic Resonance), 91
No Cloning Theorem. *See* No Quantum Xerox Principle
non-abelian gauge theory, 269–70, 270n
No Quantum Xerox Principle, 333
nuclear forces, 176, 176n, 202, 221–23
nuclear fusion, 178
nuclear physics, 10, 48, 51, 116–17, 223, 226
nuclei, 27, 37–38, 53, 59, 76, 88, 91, 96, 100–101, 116, 121, 129, 175, 177–78, 178n4, 181, 201, 202, 223, 240, 241, 268, 270n, 365; atoms and, 9, 20, 20n, 34, 50, 268; electrons and, 79, 132, 174, 268; Rutherford and, 264, 264n2
nucleons, 37, 51, 52n, 53, 175, 176n, 177, 200, 207, 207n

1-sphere, 68
Onnes, Heike Kamerlingh, 326
On the Origin of Species by Means of Natural Selection (Darwin), 6n
open universe, 144, 148, 149, 150, 151
Oppenheimer, J. Robert, 125
oxygen, 10, 177, 344–45

Pais, Abraham, 264n3
parity symmetry, 244
partons, 216, 219, 256
Pauli's exclusion principle, 76, 133, 177, 180, 240, 251, 266
Pauli, Wolfgang, x, 75, 265, 269, 270n
Peebles, Jim, 352
Penrose, Roger, 65
Penzias, Arno, 156, 157
periodic table, 121, 266, 367, 378
phenomenology, 247
phonons, 359
photon(s), 10, 30, 32, 38, 39, 40, 41, 42, 43, 48, 49, 53, 55, 57, 58, 91, 93, 95, 121, 124, 133, 156, 173, 174, 176, 184, 190, 215, 221, 222, 223, 224, 224n, 226, 230, 231, 240, 249, 262, 263, 270, 283, 322, 334, 336–37, 359; antimatter and, 44, 45, 46; bosons and, 76, 77, 240, 241; electrons and, 132, 241, 266, 267; Hawking radiation and, 329, 331, 341; Laws of Physics and, 9, 91; light and, 26n, 27, 32–33; mass and, 103, 127, 175; Quantum Electrodynamics (QED) and, 35, 37, 50, 273; supersymmetry and, 241, 248, 250; vacuum energy and, 73, 74, 75, 76
pi (π), 48, 128
pions, 176n, 190, 222
Planck length/distance, 84–88, 122, 207, 229–30, 233, 261, 315, 339, 378
Planck mass, 86–87, 249
Planck, Max, 54–86, 95, 229, 264, 370
Planck scale, 230n1, 261
Planck's constant, 25, 25n7, 30, 84–88, 224n, 262
Planck time, 86–87
planets, 120, 183, 365, 366, 366n, 367
plasma, 155, 155n, 158
Plato, 119
Platonic solids, 120, 120n
pocket universes, 14, 15, 20, 21, 83, 98–99, 138, 187, 195–96, 197, 294, 298, 300, 303, 304, 308, 313, 314, 315, 316, 323–24, 340, 341, 345, 348, 353, 363, 367, 368n, 369, 370, 374–75, 381
Polchinski, Joseph, xii, 200, 276–78, 289, 291, 350, 372, 373
Pollock, Jackson, 372
Poor Richard's Almanack (Franklin), 36n
Pope, Alexander, 117
Popper, Karl, 192, 195
populated Landscape, 294, 294n, 297, 314, 345, 350, 353, 355, 356, 363, 370, 375
positive current, 36
positrons, 43, 44, 45, 48, 50, 72–73, 74, 91, 93, 178, 180, 189, 190, 216, 235, 267
Princeton University, 156, 157, 352. *See also* Institute for Advanced Study (Princeton)
probability, 18, 24, 49, 321–22, 324, 355, 365, 366
prolate ellipsoid, 139–40
propagators, 39–41, 43, 45, 49, 50, 92, 212, 213, 214, 215, 216
protons, 9, 10, 14, 27, 43, 46, 51, 53, 56, 58, 59, 88, 103, 121, 124, 163–64, 175, 177–78, 180, 189–91, 200, 201, 202, 207, 208–9, 222, 223, 224, 226, 230, 240, 241, 246, 249, 268–69, 314–15, 367, 378, 379; elementary particles and, 28, 57,

protons (*continued*)
176; lifetime of, 190–91, 190n; nuclei and, 34, 37, 268; quarks and, 52, 53, 96, 176, 193; vacuum energy and, 76, 77; wave function and, 317, 318
psychology, 193, 202–3
Pythagoras, 118–19, 120

quanta, 26, 26n, 28, 30, 263, 359, 370, 379
Quantum Chromodynamics (QCD), 51–55, 56, 121, 200, 215–17, 273, 354, 354n5
Quantum Electrodynamics (QED), 35–37, 41, 48, 50, 53–54, 55, 56, 123, 252, 266, 268, 268n, 273
quantum fields, 166–67, 212
quantum field theory, 11, 34, 35, 48, 50, 53, 65, 66, 72, 74, 75, 104, 123, 247, 267, 268, 269, 273, 274, 290, 305, 370
quantum hologram, 339, 340
quantum jitters, 24, 28, 29, 30, 72, 74, 86, 103, 166–67, 189, 221, 224–26, 249–50, 296, 297–98, 305, 313, 331, 372, 374
quantum mechanics, 12, 23–33, 37, 39–40, 49, 72, 84, 86, 103, 121, 123, 126, 132, 156, 166–67, 189, 205, 208, 209, 210, 235, 240, 247, 256, 262, 265–66, 294, 300–301, 302, 304, 313, 322, 324, 331–32, 334, 337, 343, 347, 363, 370, 375, 377, 380; billiard table metaphor and, 24–27, 28–29; black holes and, 15, 332, 333, 359; gravity and, 61, 204, 207, 330, 331, 335, 339; Heisenberg and, 84, 264; many–worlds interpretation and, 316, 321; wave function and, 317, 318, 320
quantum numbers, 230n1
quantum tunneling, 296, 302
quarks, 9, 20, 28, 53, 54, 57, 95, 96, 104, 124, 128, 200, 200n, 207, 229, 231, 240, 269, 365, 366, 378; falsifiability and, 193, 194–95; fermions and, 77, 240, 241, 246; hadrons and, 200, 202, 215; origins of string theory and, 208, 209–10, 214, 216–17, 224, 227; protons/neutrons and, 34, 57–58, 121, 176, 176n; Quantum Chromodynamics (QCD) and, 51–53, 273; Standard Model and, 230, 270; supersymmetry and, 241, 247, 248; types of, 50, 51–53, 176; vacuum energy and, 74, 76
quasars, 116

Rabi, I. I., 283, 283n
radioactivity, x, 56, 263
radio astronomy, 27
radio waves, 27, 41, 106
random mutation, 343–44
randomness, 24, 29, 32, 49, 59, 331–32
reductionism, 33–34, 34n, 38
Rees, Sir Martin, xii, 81n, 352–53
reflection symmetry, 77, 244, 245
reheating, 303
Relativity, Theory of, 220, 262, 336; Special Theory of Relativity, 38, 106, 240, 265–66, 267, 346. *See also* General Theory of Relativity
renormalization theory, 268, 268n, 370
rest mass, 102
RNA, 8, 34, 172
Robinson, M. H., 344n
Roentgen, Conrad, 263
Rohm, Ryan, 354
rotation symmetry, 243
rubber band theory, 206–7, 216, 217, 219
Rube Goldberg machines, 12, 12n, 13, 14, 112, 125, 199, 288, 289, 310, 377, 379
Rubinstein, Hector, 204–5

Rutgers University, 352
Rutherford, Ernest, 264, 264n2

Sakharov, Andrei, 162
scalar fields, 102, 237, 238, 285, 359
Scattering Matrix. *See* S–matrix theory
Schrödinger, Erwin, 34, 84, 121, 267, 317, 318–21
Schwarz, John, 230
Schwarzschild black hole, 269
Schwarzschild, Karl, 269
Schwinger, Julian, 50
"Scientific Alternatives to the Anthropic Principle" (Smolin), 192
"see," as used by physicists, 335, 335n
Self-Reproducing Universe, 302n
sets, 367–69, 368–69n
Shenker, Steve, 256, 346
Sherk, Joël, 230
short-range forces, 177, 201, 202, 221, 223
Silverstein, Eva, 352
simplicity, 113, 113n, 118, 119, 121, 128, 130, 377, 379–80
singularities, 144, 286–89, 302, 355, 360
Skinner, B. F., 193, 202–3
slogans, 346–47, 380
S-matrix theory, 195, 203, 204, 206, 217, 220
Smolin, Lee, 192, 192n, 193, 195, 357, 360–62
Smoot, George, 341
solar systems, 79, 82, 182, 186
sound waves, 115n
South Africa, 137–38
South Pole, 157–58
space, as flat/curved, 159–60
space-time, 38–39, 41, 49, 106, 221, 359
spectral lines, 60, 60n, 132, 135, 264
spectroscopy, 264–65, 266
Spergel, David, 353
spheres, 67–68, 284
spin, 224n, 266
spontaneously broken dilatation symmetry, 218
Standard Model, 11, 12, 50, 52, 56, 59, 59n, 61, 91, 95, 97, 118, 121, 124, 125, 163, 170, 178, 191, 215, 239, 240, 267, 269, 270, 273, 285, 355, 359, 374, 379; elementary particles and, 88, 230; gravity and, 61, 122, 252; Higgs field and, 94, 95, 104, 106; KKLT construction and, 288, 289; Weinberg and, 80, 349
Stanford Linear Accelerator Center (SLAC), 28, 161–62
Stanford University, 351
Starobinsky, Alexei, 161, 163, 164, 165, 166
stars, 10, 133, 150, 153, 178–83, 185, 365. *See also* supernova
statistics, 352
Steady State theory, 19
Steinhardt, Paul, 304, 353
strange particles, 52, 121
strange-quarks, 50, 51, 52, 52n, 53, 57, 128, 176
String Theory, 12, 14, 21, 75, 78, 81, 88, 106, 109, 122–30, 185, 186–87, 191, 199, 200, 202–27, 229–31, 239, 246–48, 251–60, 268, 273, 274, 276, 278–81, 289, 291–92, 293, 295, 302, 304, 309–10, 313, 324, 326, 339, 349–50, 351, 352, 354, 355–56, 357, 361, 366, 372–73, 375, 377–78, 380; compactification and, 235–38, 284; experiments and, 261, 270; hadrons and, 212, 265; Heterotic, 252, 253, 349, 354–55; Kaluza's theory and, 235–36, 237; Laws of Physics and, 97, 247; meeting of string theorists, 253–54, 256; Smolin and, 360, 362, 363. *See also* superString Theory
Strominger, Andy, 354

strong interactions, theory of, 54
sun, 5, 155, 179, 180, 182, 188, 189
supercooled water, 297–98, 301
supermegaverse, 187
supermoduli space. *See* supersymmetric moduli space
supernova, 10, 130, 152–53, 153n, 161, 180
superstrings, 372–73
superString Theory, 251, 251n
supersymmetric moduli space, 251
supersymmetry, 77, 89, 89n, 128, 185, 191, 239–51, 251n, 253, 254, 289, 308, 313, 315, 352, 361, 374–75
surface of last scattering, 193
symmetry, 77, 241–45

tau particles, 50, 230, 230n1
Tegmark, Max, 353
Teitelboim, Claudio, 4
television, 49
theoretical physicists, 12, 14, 24, 34, 127, 217–18
theory, 246
Thomson, J. J., 35, 56, 263
't Hooft, Gerard, xi–xii, 170, 326, 327, 330, 331–32, 333, 339, 350–51, 370, 379
Thorne, Kip, 3, 4
3-sphere, 67, 68, 136, 141, 144, 146
Tiger, L., 344n
Tipler, Frank J., 81, 81n
Tomanaga, Sin-Itero, 50
topology, 284, 287
top-quarks, 50, 51, 53, 57, 128, 176, 365, 366
torus, 233, 238, 284
translation, 242, 243
Trivedi, Sandip, 200, 287n
two-slit experiment, 30, 322
2-sphere, 67, 68, 139, 231–32, 233, 269

Uhlenbeck, George, 326
ultraviolet catastrophe, 263
Uncertainty Principle, 25, 26, 28, 29, 322, 333, 334, 347
unified theory, 119
uniqueness, 12, 113–14, 118–19, 120, 121, 124–25, 127, 129, 253, 351, 377
universe, 6, 9–10, 82, 83, 91, 131–36, 142–49, 154–61, 172, 184, 262, 293–324, 340, 343, 346, 349, 360–71, 372, 380; age of, 149–51, 161, 182; Anthropic Principle and, 79, 80; closed/open/flat, 67, 144, 148–49, 150, 151, 152, 160–61, 310–11; Eternal Inflation and, 302, 304–10; expansion of, 15, 66, 116, 123, 129, 131, 143, 144, 145–46, 152, 154, 156, 158, 161, 164, 197n8, 288, 310, 313, 372; gravity and, 9, 123; homogeneous, 19, 138–40, 146, 157, 161, 162, 166, 310; Inflation and, 161–67, 371; isotropic, 138–40, 157; parallel, 316, 323; static, 69, 71. *See also* Big Bang; galaxies; pocket universes
Unruh, Bill, 328
up-quarks, 50, 51, 52, 53, 57, 128, 176, 176n, 231
uranium, 58

vacuum energy, 29, 65, 72–78, 84, 87, 88, 108, 147n, 151, 152, 154, 163, 164, 165, 170, 185, 199, 249, 250, 251, 268, 288, 290, 291, 303, 306, 313, 314, 315, 374
vacuum(s), 90, 92, 97, 108, 200, 254, 273, 274, 276, 289, 293–94, 294–97, 298, 305, 308, 350, 351, 353, 355, 358, 362, 363, 365, 366, 370, 371, 374, 375
van der Waals, Johannes Diderik, 326
vector fields, 102
Veltman, Martinus (Tini), 169–70, 326

Veneziano amplitude, 204
Veneziano formula, 211, 217
Veneziano, Gabriele, 204, 206
vertex diagrams, 41, 43–46, 48, 50, 53, 56, 59, 93, 175, 214, 216
vertices, 49, 212, 216
Vilenkin, Alexander, xii, 14, 81n, 274, 294n, 302, 304, 353, 370, 372, 373
virtual particles, 73, 74–75, 87, 104
Vonnegut, Kurt, 294–95, 296–97
vortices, 359
voxels, 338, 339

Wagoner, Bob, 162
Wallace, Alfred Wallace, 17–18, 17n
Washington Heights, 63
water, 119–20, 297–98, 298n
Watson, James, 18
wave equations, 115n
wave functions, 317–21, 322, 323, 324
W-bosons, 50, 57–59, 77, 104, 128, 176, 178, 230, 240
weak interactions, 56–59
Weinberg, Steven, xi, 79, 80–82, 83–84, 109, 113, 154, 161, 170, 172, 183n, 187, 196, 250, 291, 316, 349, 356, 364
Weizmann Institute, 204
Wheeler, John, 316
When I Heard the Learn'd Astronomer (Whitman), 117
white dwarf stars, 153, 180
White, T. H., 189
Whitman, Walt, 117
Wilberforce, Sam, 6, 6n, 7
Wilkinson Microwave Anisotropy Probe (WMAP), 158, 160
Wilson, Kenneth, 268n, 370
Wilson, Robert, 156, 157
Witten, Edward, 125, 254–56, 257, 276, 278, 339, 349–50, 354
Wittgenstein, Ludwig, 195
world line, 212
world sheets, 213–17, 222, 223
W-particles, 270

X-rays, 27, 41, 263

Yang, Chen Ning, 270n
Yang Mills theory, 269
Yeshiva University, 63

Z-bosons, 50, 57, 77, 104, 128, 176, 178, 230, 240
zero genus surface, 284
zero-point energy, 29
Z-particles, 270

About the Author

Leonard Susskind grew up in the South Bronx, where he worked as a plumber and steam fitter during his early adult years. As an engineering student in CCNY, he discovered that physics was more to his liking than either plumbing or engineering. He later earned a PhD in theoretical physics at Cornell University.

Susskind has been a professor of physics at the Belfer Graduate School in New York City, Tel Aviv University in Israel, and since 1978 at Stanford University, where he is the Felix Bloch Professor of Physics. During the past forty years he has made contributions to every area of theoretical physics, including quantum optics, elementary-particle physics, condensed-matter physics, cosmology, and gravitation. In 1969 Susskind and Yoichiro Nambu independently discovered string theory. Later he developed the theory of quark confinement (why quarks are stuck inside the nucleus and can never escape), the theory of baryogenesis (why the universe is full of matter but no antimatter), the Principle of Black Hole Complementarity, the Holographic Principle, and numerous other concepts of modern physics.